FUNDAMENTALS OF ENGINEERING THERMODYNAMICS

V. Babu
Professor
Department of Mechanical Engineering
Indian Institute of Technology, Madras, INDIA

**Ane Books
Pvt. Ltd.**

Ane Books Pvt. Ltd, New Delhi

CRC Press
Taylor & Francis Group
Boca Raton London New York

CRC Press is an imprint of the
Taylor & Francis Group, an **informa** business

MATLAB® is a trademark of The MathWorks, Inc. and is used with permission. The MathWorks does not warrant the accuracy of the text or exercises in this book. This book's use or discussion of MATLAB® software or related products does not constitute endorsement or sponsorship by The MathWorks of a particular pedagogical approach or particular use of the MATLAB® software.

CRC Press
Taylor & Francis Group
6000 Broken Sound Parkway NW, Suite 300
Boca Raton, FL 33487-2742

First issued in paperback 2023

© 2020 by V. Babu and Ane Books Pvt. Ltd.
CRC Press is an imprint of Taylor & Francis Group, an Informa business

No claim to original U.S. Government works

ISBN 13: 978-1-03-265418-8 (pbk)
ISBN 13: 978-0-367-36321-5 (hbk)
ISBN 13: 978-0-367-81608-7 (ebk)

DOI: 10.1201/9780367816087

Library of Congress Cataloging in Publication Data
A catalog record has been requested

Visit the Taylor & Francis Web site at
http://www.taylorandfrancis.com

and the CRC Press Web site at
http://www.crcpress.com

**Ane Books
Pvt. Ltd.**

*Dedicated to my wife
Chitra without whose
patience, understanding
and support, this book too
would never have been
written*

PREFACE

I am happy to come out with the second edition of the book on the fundamentals of Engineering Thermodynamics. Five new chapters have been added and the chapter on Thermodynamic cycles (Chapter 10 in the previous edition) has been completely rewritten. The new chapter on Exergy discusses the notion of exergy and how it can be used as a performance metric for any process - cyclic or otherwise, in a much more general sense than the isentropic efficiency. The revised chapter on Thermodynamic cycles now includes, the Otto, Diesel and Dual cycles. Second law analysis of all the cycles as well as the individual components are also presented now, which is a unique feature. In the chapter on psychrometry, rather than giving illustrations of the psychrometric chart alongside each example, cropped images of the relevant part of the chart itself has been given. The image also shows how the values for the various properties are retrieved. Hopefully, students would find this helpful when they are using the chart to solve the examples on their own. The chapter on Combustion, contains worked examples on solid, liquid (including blends) and gaseous fuels and combustors in practical devices such as gas turbines and IC engine. The chapter on Gas phase chemical equilibrium starts with a broad introduction to the notion of equilibrium – mechanical, thermal and chemical. Some of the examples in this chapter build on the examples from the previous chapter with the objective of demonstrating the difference between a stoichiometric analysis and an equilibrium analysis. Several examples involving two and three reactions simultaneously in

Once again, I would like to thank my teachers from whom I have benefited greatly. In particular, I would like to thank Sri. S. Sundaresan, my high school teacher, Prof. R. Bodonyi, Prof. M. Foster and Prof. T. Scheick of the Ohio State University, who, as teachers, inspired me to a great extent. I would also like to place on record my appreciation of my advisor, friend and mentor, Prof. S. Korpela of the Ohio State University. He has always been and will always be a source of strength and support.

I wish to express my heartfelt gratitude to my parents who endeavored so much to give me a good education. They have given a lot to me but received very little in return. Thanks are also due to my son Aravindh for many engaging discussions on the topics discussed in this book and also for his help with some of the illustrations in Chapter 12 as well as in solving some of the worked examples using Cantera and MATLAB.

I wish to thank the readers who pointed out errors in the previous edition. These have been corrected in this edition. If you find any more errors or if you have any suggestions for improving the exposition of any topic, please feel free to communicate them to me via e-mail (vbabu@iitm.ac.in). I also wish to place on record my sincere appreciation and gratitude to Linrick.com for allowing me to use their psychrometric chart in the book.

V. Babu

PREFACE TO THE FIRST EDITION

I am happy to come out with this book on the fundamentals of Engineering Thermodynamics. The material in this book has grown out of my lecture notes for the freshman undergraduate course on Engineering Thermodynamics that I have been teaching at IIT Madras for nearly twenty years now. The decision to finally write this book was in large part due to the positive feedback that I have received during these years from students who took the course and their expressed desire to see the material taught in class being published as a book.

Since the course was a freshman course and almost all students in India study thermodynamics as a part of their Physics and Chemistry courses at the Higher Secondary level, perhaps the biggest challenge that I faced while teaching this course was trying to make them "unlearn" many concepts that they had learnt incorrectly. Consequently, readers from other countries may find t he excessive emphasis on concepts that are taken for granted elsewhere, puzzling or surprising or even unwarranted! I have not changed this emphasis while writing the book because of the belief that it will still benefit m any s tudents w ho, f or one reason or another, do not ask such questions in class and are unable to find t he a nswers o utside the classroom either.

One of the unique features of the book is the primary importance given to the system reservoirs in Chapter 8. I had two objectives in mind for doing this - first, in real life, all reservoirs are finite and so students should be able to carry out analysis involving them quite comfortably and second, this allows the notion of exergy to be introduced quite naturally, thereby preparing the students for the next level course where they will learn this in detail. I have also not given any end of chapter problems, choosing, instead to work out numerous examples in each chapter. Many of the examples are continued across chapters, in order to illustrate increasing levels of complexity with each new chapter. It is my sincere wish that the readers will benefit a lot from attempting the examples on their own and then comparing their work with what is given in the book.

I have given a list of books at the end (in alphabetical order) that I like. In this context, I would like to single out the book by Prof. Moran and co-authors for special mention. I had the fortune to attend Prof. Moran's lectures in 1987 while I was preparing for my PhD qualifying exam as a graduate student at the Ohio State University in the Department of Mechanical engineering. The clarity of the lectures - what was said and what was written on the board, left me awe-struck. Not surprisingly, the book that Prof. Moran has written on the subject also has the same clarity and is a particular favorite of mine. This, combined with the fact that there are quite a few excellent books on the subject (a few of which are given in the Suggested Reading) had made me reluctant for long to undertake the task of writing a book on the same topic. However, as I mentioned earlier, the encouragement from students who have gone through my lectures has finally made me overcome this reluctance. It is my hope that students will find this book a stepping stone to read, learn from and appreciate other wonderful text books available on the same topic.

Once again, I would like to thank my teachers from whom I have benefited greatly. In particular, I would like to thank Sri. S. Sundaresan, my high school teacher, Prof. R. Bodonyi, Prof. M. Foster and Prof. T. Scheick of the Ohio State University, who, as teachers, inspired me to a great extent. I would also like to place on record my appreciation of my advisor, friend and mentor, Prof. S. Korpela of the Ohio State University. He has always been and will always be a source of strength and support.

I wish to express my heartfelt gratitude to my parents who endeavored so much to give me a good education. They have given a lot to me but received very little in return. Thanks are also due to my son Aravindh for many engaging discussions on the topics discussed in this book while he was an undergraduate student in Mechanical engineering. He was able to provide a unique perspective on the difficulties and challenges faced by students trying to learn Engineering Thermodynamics. Accordingly, I have made subtle changes in the manner in which the material is developed. Hopefully, students trying to learn the subject for the first

for numerous valuable suggestions. If any errors remain, I am entirely responsible. If you find any errors or if you have any suggestions for improving the exposition of any topic, please feel free to communicate them to me via e-mail (vbabu@iitm.ac.in).

V. Babu

CONTENTS

INTRODUCTION

Thermodynamics originated as a subject in mechanical engineering at the start of the industrial revolution when, steam generated from burning coal was utilized to run machinery. The quest to improve the efficiency of devices that convert heat into useful work led to the evolution of engineering thermodynamics. Today, thermodynamics is almost synonymous with mechanical engineering! The laws of thermodynamics, based on experiments, have now come to be regarded as being among the fundamental laws of the Universe. They are now applied in diverse areas such as financial markets, cosmology, computing, bio-technology and agriculture, to name a few. The second law has a particularly special place in our minds because it seems to explain so many occurrences in our daily life. For instance, everyone can relate to statements like *For something to become cleaner something else must become dirtier* and *Left to themselves things tend to go from bad to worse* from the popular Murphy's laws. These may quite rightly be viewed as statements of the Principle of Increase of Entropy, which we will discuss later. For readers who are interested in the philosophical aspects of the laws of thermodynamics, I strongly recommend the two books by Peter Atkins that are given in the Suggested Reading list. He summarizes the three laws of thermodynamics in the following manner:

1.1 Macroscopic approach

In engineering thermodynamics, we are concerned with the conversion of heat into useful work. In this context, we will try to answer the following questions:

- How much of the input heat is converted into work by the engine under consideration?

- Everything else remaining the same, what is the maximum possible work output?

- What are the factors that affect the performance of the engine and by how much?

The framework that we develop will not only be useful for analyzing engines that convert heat into useful work but also engines that utilize work to produce a useful effect (like, for instance, a refrigerator or a heat pump). In addition, it will also allow us to evaluate the performance of individual devices that make up the engine.

The analysis that we are interested in, utilizes a macroscopic or black box approach that ignores internal details. For instance, when we say that a certain amount of heat is transferred to or from a device, details regarding how exactly the heat transfer takes place is immaterial to a thermodynamic analysis. Similarly, when work (or power) is supplied to a compressor, details of how this is utilized in the compression process is immaterial. Details of how a turbine converts the enthalpy of a fluid into power are similarly immaterial to the analysis. Such details are the subject matter of courses on Heat Transfer and Fluid/Turbo Machines.

Furthermore, molecular level details are also ignored in the macroscopic approach. The working substance is assumed to be a single entity with a unique value for the properties - pressure, density, temperature and so on. Mixing and stirring processes are assumed to be macroscopic in nature and molecular effects in such processes are neglected.

1.2 Continuum hypothesis

An important requirement of the macroscopic approach that we have adopted here is that continuum must prevail. Only then, properties such as pressure, density, temperature and so on of the thermodynamic systems under consideration will be known without any ambiguity. In other words, a statement that the pressure of air in

port to allow observations of the contents within a fixed observation volume. We now propose to measure the density of the gas at an instant as follows - count the number of molecules within the observation volume; multiply this by the mass of each molecule and then divide by the observation volume.

To begin with, let there be 100 molecules inside the vessel. We would notice that the density values measured in the aforementioned manner fluctuate wildly going down even to zero at some instants. If we increase the number of molecules progressively to 10^3, 10^4, 10^5 and so on, we would notice that the fluctuations begin to diminish and eventually die out altogether. Increasing the number of molecules beyond this limit would not change the measured value for the density.

We can carry out another experiment in which we attempt to measure the pressure using a pressure sensor mounted on one of the walls. Since the pressure exerted by the gas is the result of the collisions of the molecules on the walls, we would notice the same trend as we did with the density measurement. That is, the pressure measurements too exhibit fluctuations when there are few molecules and the fluctuations die out with increasing number of molecules. The measured value, once again, does not change when the number of molecules is increased beyond a certain limit.

We can intuitively understand that, in both these experiments, when the number of molecules is less, the molecules travel freely for a considerable distance before encountering another molecule or a wall. As the number of molecules is increased, the distance that a molecule on an average can travel between collisions (which is termed as the mean free path, denoted usually by λ) decreases as the collision frequency increases. Once the mean free path decreases below a limiting value, measured property values do not change any more. The gas is then said to behave as a *continuum*. The determination of whether the actual value for the mean free path is small or not has to be made relative to the physical dimensions of the vessel. For instance, if the vessel is itself only about 1 μm in dimension in each side, then a mean free path of 1 μm is not at all small! Accordingly, a parameter known as the Knudsen number (Kn) which is defined as the ratio of the mean free path (λ) to the characteristic dimension (L) is customarily used. Continuum is said to prevail when $Kn \ll 1$. In reality, once the Knudsen number exceeds 10^{-2} or so, the molecules of the gas cease to behave as a continuum.

BASIC CONCEPTS

In this chapter, certain basic concepts which are essential for carrying out a thermodynamic analysis of devices, are discussed. Although these concepts are relatively simple and easily understood by and large, they are discussed at length here in order to clearly bring out certain subtle aspects that are generally overlooked. A clear understanding of these basic concepts will form a strong foundation for the material developed later.

2.1 System

Following Spalding and Cole, we define a thermodynamic system as *a quantity of matter of fixed mass and identity on which attention is focussed for study*. Everything external to the system is referred to as the *surroundings*. Defining a system is a crucial first step in any thermodynamic analysis. In some cases, it may be relatively straightforward to define a system, while in some other cases, it may not be. Furthermore, a valid thermodynamic system need not necessarily be useful for analysis. It is possible to define more than one valid system for a given problem. The choice of which one to use is guided by the information given, the information sought

Heat

Initial   Final

Figure 2.1: A simple thermodynamic system

and the ease of analysis. Let us explore these aspects through several examples.

We start with a very simple problem. A gas is contained within a piston cylinder assembly as shown in Fig. 2.1. Heat is added to the gas until it expands to a certain volume. We wish to define a system suitable for a thermodynamic analysis of this problem.

It is quite easy (perhaps trivial) to define a system for this problem. This is indicated as the shaded region in Fig. 2.1 and the system boundary is shown using a dashed line. The simplicity of the problem is deceptive and it is necessary to make the following observations:

- This system contains the same amount of matter from the beginning to the end of the process and thus satisfies the definition given above.

- During the process, parts of the system boundary adjacent to the walls of the cylinder remain fixed while that part adjacent to the piston moves along with it. In other words, the system boundary deforms in such a manner as to always contain the same mass throughout. Hence, it is essential to know the system boundary *throughout* the process - not just at the beginning and the end of the process. This is an important requirement since it implicitly demands that the process take place slowly. This also ensures that the pressure, temperature and volume are measurable at every instant and will be uniform throughout the system.

- Intuitively, it is easy to appreciate that wherever there is deformation of the system boundary, there is a work interaction between the system and the surroundings - either the system is doing work, as in this example, where the piston and the mass are being lifted and the atmosphere is being pushed upwards, or the surroundings do work on the system. The system boundary expands in the case of the former and contracts in the case of the latter. Such a work interaction is termed *displacement work* and an expression for evaluating the same is developed in the next chapter.

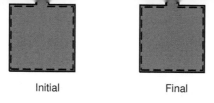

Initial   Final

Figure 2.2: Thermodynamic system for analyzing the inflation process of a balloon

- It is possible to define other, equally valid systems for this problem. For instance, a system that contains the gas and the piston, or one that contains the gas, piston and the mass or one that contains just the atmosphere are all valid.

Our next example looks at inflating an initially empty balloon from a rigid vessel that contains air at a pressure higher than the atmospheric pressure. A suitable thermodynamic system for analyzing this problem is shown in Fig. 2.2. It is clear from this figure that this system contains the same amount of matter throughout and hence is a valid thermodynamic system. The part of the system boundary which is outside the vessel expands during the process, from which it may be inferred that the air in the vessel is doing work to expand the balloon. If the balloon material is thin and inextensible, then the pressure inside the balloon is the same as atmospheric pressure and the work is done entirely on the atmosphere. On the other hand, if the balloon material is elastic in nature (such as a rubber sheet), then the pressure inside the balloon will be higher than the atmospheric pressure and the work done by the air is partly utilized to stretch the balloon material and partly to push the atmosphere aside. In both cases, the process is guaranteed to take place slowly by the presence of the valve, which provides the required resistance. Unlike the previous example, here, different parts of the system, namely, the air inside the cylinder and the air inside the balloon, are at different pressures. At first sight, this may appear to violate the framework of the macroscopic approach that we have adopted here. This is not so, since the valve is a mechanical device that can support the pressure difference across these two parts of the system. Note that, in this example, the atmosphere may also be identified as a thermodynamic system.

The next example involves the filling of a rigid vessel from the atmosphere (Fig. 2.3). The process may be described as follows: the vessel is initially evacuated. The valve is opened *slightly* to allow atmospheric air to flow into the vessel. The valve is closed after, say, 1000 cc of air has flowed in.

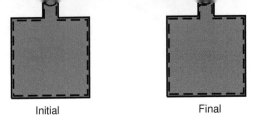

Initial   Final

Figure 2.3: Thermodynamic system for analyzing the filling of a rigid vessel from the atmosphere

In contrast to the previous examples, in this case, the thermodynamic system is defined as the air that is *finally* in the vessel. The initial state of this system is shown in Fig. 2.3. Note that the actual shape of the part of the system boundary that is in the atmosphere at any instant during the process is immaterial [†] - the only requirement is that it must initially enclose 1000 cc of air. Defined in this manner, the system shown in Fig.2.3 is a valid thermodynamic system. It should be noted that by opening the valve only slightly, sufficient resistance is provided to ensure that the process takes place slowly (irrespective of whether the vessel is initially evacuated or not) and the system boundary outside the vessel is known at all instants during the process. Since this part of the system boundary contracts during the process, it may be inferred that work is done *by* the atmosphere to push the air into the vessel against the resistance provided by the valve.

The next example (Fig. 2.4) provides a small modification in that the vessel is filled from a line in which air flows rather than from the atmosphere. Once again, the air that is finally in the vessel is identified as the thermodynamic system. The initial configuration of this system is as shown in this figure. It is easy to see that the part of the system boundary inside the line shrinks during the process as a result of the work done to push it inside the vessel against the resistance provided by the valve.

The next example involves the emptying of a rigid vessel containing air at a high pressure slowly into the atmosphere (Fig. 2.5). In this example also, the air that finally remains in the vessel is identified as the system. The initial configuration of this system is also shown in this figure. It is depicted as a circle for illustrative purpose only; the actual shape is immaterial. The system expands from this initial configuration during the process thereby doing work as the rest of the air in the vessel

---

[†] This is justified in the next chapter

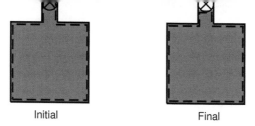

Initial   Final

Figure 2.4: Thermodynamic system for analyzing the filling of a rigid vessel from a line

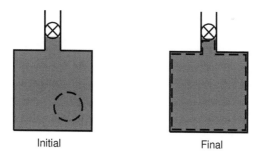

Initial   Final

Figure 2.5: Thermodynamic system for analyzing the emptying of a rigid vessel containing air into the atmosphere

is pushed out. Once again, in this case also, the presence of the valve ensures that the process takes place slowly.

The next example illustrates deformation of more than one part of the system boundary. The intake stroke of an air compressor is studied here (Fig. 2.6). Initially, the intake valve is opened and as the piston moves to the right, a certain quantity of air, say, 100 cc, is drawn in from the atmosphere. The intake valve is closed at the end of the intake stroke when the piston reaches its extreme position. The exhaust valve remains closed during the entire intake stroke.

The thermodynamic system in this case, initially comprises the air already present inside the cylinder and 100 cc of the air from the atmosphere that is going to be drawn in. The final configuration of this system is shown in Fig. 2.6. Note that the part of the system boundary in the atmosphere shrinks in volume indicating that work is being done by the atmosphere to push the air inside the cylinder against the resistance provided by the valve. The part of the system boundary adjacent to the

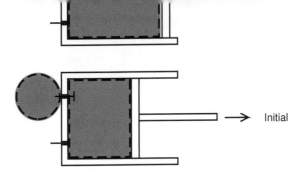

Initial

Figure 2.6: Thermodynamic system for analyzing the intake stroke of a reciprocating compressor

piston expands and does work against the resistance provided by the atmosphere as well as the external agent that powers the compressor. The system deforms while always containing the same mass. Note once again that the exact shape of the system boundary in the atmosphere is immaterial.

In all the examples above, the process was guaranteed to take place slowly (in other words, *a fully resisted process*) owing to the resistance provided by the atmosphere, mass, valve and/or an external agent. In contrast, our next and last example involves a partially resisted or an unrestrained expansion process.

Air is contained in the left half of a rigid container by means of a partition (Fig. 2.7). The right half is either fully or partially evacuated. The partition is removed and the air expands rapidly to fill the entire container and eventually attains an equilibrium state.

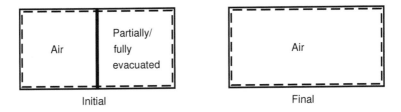

Figure 2.7: Thermodynamic system for analyzing the unresisted/partially resisted expansion inside a rigid container

partially evacuated. It is, of course, tempting to define the air initially on the left hand side as the system and then allow this system to expand and finally fill the container. This is not practicable since the expansion process is rapid. Consequently, precisely locating the expanding part of the system boundary at every intermediate instant is extremely difficult if not impossible. In addition, the rapid expansion causes the air near the expanding front to move rapidly in contrast to the rest of the air. This non-uniform inertia effect results in pressure and possibly some other properties being non-uniform across the system. This clearly falls outside the macroscopic framework that we have adopted here.

In summary, we can state that heat and work may cross a system boundary but not mass.

2.2 Control volume

It must be clear from the examples discussed in the previous section that, defining an appropriate thermodynamic system requires careful consideration of the process at hand from beginning to end. This provides useful insights on the nature of the interaction between the system and the surroundings that simplify the analysis. In contrast, a control volume suitable for a thermodynamic analysis may be defined quite simply in most cases as the device under consideration itself. Mass, energy, heat and work may cross the boundary of the control volume, usually referred to as the control surface.

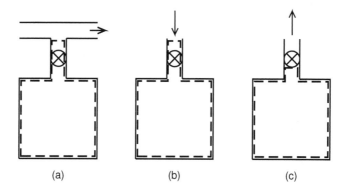

Figure 2.8: Control volume for analyzing filling and emptying of a tank. (a) From a line, (b) from the atmosphere and (c) into the atmosphere.

the analysis. Note that the control surface is rigid in all the cases shown in Fig. 2.8. Although this is not mandatory, it is advantageous to do so. The next example illustrates this point.

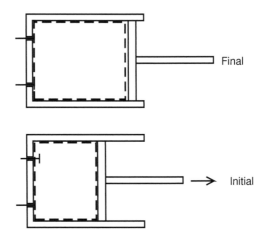

Figure 2.9: Control volume for analyzing the intake stroke of a reciprocating compressor

A suitable control volume for analyzing the intake stroke of a reciprocating compressor is shown in Fig. 2.9. Note that, in this case, the control surface adjacent to the piston must deform along with it. A comparison of this control volume with the corresponding thermodynamic system defined in Fig. 2.6 shows that the control volume with a deformable control surface is quite similar to the system and hence the former provides very little benefit over the latter.

In summary, we can state that mass and the associated energy, heat and work can cross a control surface.

2.2.1 System or control volume?

The choice of adopting a system or control volume approach is, in general, determined by whether the device under consideration is a flow or non-flow device. In other words, if the device works with a fixed amount of matter during the process as in the reciprocating compressor mentioned earlier, a system approach is better. On the other hand, if there is a continuous flow of matter in and out of the device during its operation, then a control volume approach is preferable. This is discussed next in the context of two practical devices.

processes, namely, intake, compression, combustion, expansion and exhaust. In the case of the piston engine, a certain amount of air is drawn into each cylinder and it undergoes each of these processes in succession during each stroke of the piston - one process per stroke (combustion and expansion occur in succession within one stroke). One stroke of the piston refers to the travel of the piston from top to bottom or *vice versa*. Hence, the system approach is best suited for analyzing each of the four strokes. Of course, complications such as injection of fuel and combustion do arise but a thermodynamic analysis idealizes these by invoking the so-called air standard assumption.

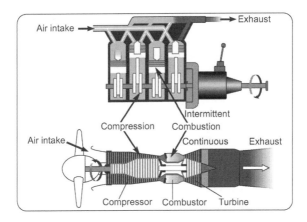

Figure 2.10: Illustration of piston engine and a gas turbine engine. *Adaptation Courtesy: Rolls-Royce plc*

Since all the processes are executed in sequence in each cylinder, the work produced by the piston engine is intermittent - during each expansion (power) stroke, which is once in every four strokes. In practical applications, this is usually overcome by having a flywheel of large mass. In addition, reducing the number of strokes per power stroke by executing more than one process during a stroke (as is done in a two stroke engine) or increasing the number of cylinders so that at any given instant, there is always a power stroke being executed (as shown in Fig. 2.10), are two strategies that are widely employed. The shortcoming of the former is that the efficiency, measured as the work produced per unit weight of fuel consumed, decreases, while that of the latter is that the power produced per unit engine weight decreases.

The gas turbine engine shown in Fig. 2.10 overcomes these shortcomings by segmenting the engine into different sections, namely, intake, compressor, combustor, turbine and exhaust and devoting each section to carry out a single process. The

2.3 Property, state of a system and process

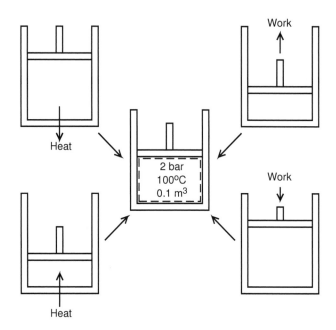

Figure 2.11: Properties and path dependent quantities

A particular quantity may be termed a property if its value depends only on the state of the system and not on the path by which the system attained that state. At a given instant, let the pressure, temperature and volume of the system in Fig. 2.11 be, say, 2 bar, 100°C and 0.1 m³. These values are unique to this state and would have been the same irrespective of whether the system reached this state through a compression or an expansion process from a previous state. In contrast, the work interaction to reach this state depends on the process since a compression process requires work to be done on the system while an expansion process will cause the system to do work. The same is true of heat as well - the current state could have been attained as a result of heat addition or heat removal[†].

[†]provided, of course, that it does not undergo a phase change - evaporation or condensation

specific total energy, specific internal energy and specific enthalpy, may be obtained after dividing by the mass of the system. In such cases, upper case letters are used to denote extensive properties (V, E, U, H) while lower case letters are used for the corresponding intensive properties (v, e, u, h).

The thermodynamic state of a system is fixed by a *certain number* of measurable and *independent* properties such as pressure, volume, temperature and so on. The actual number is equal to the number of possible ways by which the energy of the system may be changed. For the system shown in Fig. 2.1, this would be 2, since its energy may be changed through heat and work interactions. This is the simplest possible system and and hence it is termed a *simple system*. If, in addition, it is possible to change the elevation of the system or impart kinetic energy to it, then additional properties, namely, elevation and speed, would have to be specified. The minimum number required to fix the state is very important since any other property must then be expressed in terms of these properties through appropriate relations[‡].

The state of the system in Fig. 2.1 may be fixed at any instant by knowing pressure and volume or pressure and temperature or temperature and volume. It is well known from high school physics that, for an ideal gas, any two out of these three properties are independent and the third one may be determined using the other two by means of the equation of state. Density and specific volume are properties but the combination cannot be a part of a set of independent properties since one is the reciprocal of the other. Properties that are normally independent may become dependent under certain conditions. For instance, if the working substance in the aforementioned system is a mixture of water in liquid and vapor phase, then, pressure and temperature are not independent. In other words, if the pressure in such a case is measured to be 100 kPa, then the temperature has to be 100°C. This follows from the fact that water can exist as a mixture of liquid and vapor at 100 kPa only, so long as the temperature is 100°C.

Air at room temperature is contained in a piston cylinder arrangement with a mass placed on the piston as shown in Fig. 2.12. A force balance on the piston will show that the pressure of the air is higher than atmospheric pressure by an amount equal to the combined weight of the mass and the piston divided by the area of the piston. The initial state of the system, denoted *1*, is shown in in Fig. 2.13 using P, V coordinates. Starting from this initial state, the system is now made to undergo two different processes as follows:

[‡]For instance, if pressure and temperature are used to fix the state of the system, then other properties such as internal energy, enthalpy, entropy and so on have to be evaluated using the known values of pressure and temperature

Figure 2.12: Illustration of an unresisted (left) and a fully resisted (right) process

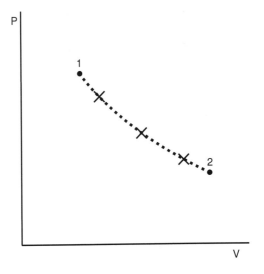

Figure 2.13: A fully resisted process in P-V coordinates

- The mass is suddenly removed

- The mass is divided into a large number of small pieces and the pieces are removed one after another with a sufficiently long interval in-between

In the first case, it is easy to see that the air undergoes a rapid expansion (not fully resisted) and after a sufficiently long period of time, settles down at a final state. This state is marked 2 in Fig. 2.13. The final temperature of the air is the same as the ambient temperature and the final pressure is higher than atmospheric pressure by an amount equal to the weight of the piston divided by the area of the piston. In this case, only the initial and final states are known, since properties such as pressure, temperature and volume are measurable with certainty only in these states and not at any of the intermediate instants. This is because the initial and final states are

This causes the pressure and temperature to be different in different parts of the system and so the state of the system is indeterminate.

In contrast, in the second case, as the weights are removed, the system goes from one equilibrium state to another. These intermediate states are denoted by dots in Fig. 2.13. The pressure, volume and temperature at each intermediate state is measurable. The system goes through these intermediate states to reach the same final state as in the previous case. As the number of the known intermediate states increases, the path by which the system goes from the initial to the final state (termed the *process*) also becomes known. For instance, if the mass were divided into just four pieces, then, only 3 intermediate states will be known (denoted using Xs in Fig. 2.13) and the path connecting them will not be known since they are too far apart. As the mass of each piece becomes infinitesimally small, the number of known intermediate states becomes infinitely large and they are close enough to be connected together to obtain the process. Such a process is referred to as a *fully resisted process* or a *quasi-equilibrium process*, since the system is in mechanical equilibrium throughout. It is important to note that this is also a *reversible process*, for, if we stop removing the weights at any instant and start putting them back, the system will follow the *same* path back to the initial state.

In view of these observations, the definition given earlier for the state of a thermodynamic system may be modified as follows: the state of a thermodynamic system is defined by a set of independent, measurable properties and for a property to be measurable, the system must be in equilibrium. A path or process is the locus of a set of thermodynamic states that are located only infinitesimally apart. The system is out of equilibrium in between two successive states, however close they may be. This departure from equilibrium is essential for a change of state, and hence any process, to occur. The departure is quite small when the successive states are infinitesimally apart. The origin of the departure from equilibrium may be mechanical as in this example or thermal as we will see later.

2.4 Temperature and the Zeroth law

In our discussion so far, it has been tacitly assumed that readers are familiar with properties such as density, pressure, volume and temperature. While this is reasonable since these are encountered in everyday life, the property temperature alone warrants closer attention[†]. It is perhaps surprising, upon reflection, to

---

[†]Peter Atkins quite appropriately states temperature to be perhaps the most familiar but most enigmatic of these properties

Measurement of temperature also poses some fundamental issues, in addition. Two methods are usually employed to measure temperature. These are the direct contact method and the non-contact method. In the former, a thermometric substance in the measuring device (thermometer) is brought in contact with the system whose temperature is to be measured. Change in a particular property of the thermometric substance, referred to as the thermometric property, is monitored until it attains a steady state value indicating that the system and the thermometric substance are in thermal equilibrium. The actual value for the temperature will have to be evaluated from the measured steady state value of the thermometric property. This is made possible through a calibration procedure. A few direct contact thermometers are listed in Table 2.1.

Table 2.1: Examples of direct contact thermometers

	Thermometric substance	Thermometric property	Calibration relation
Liquid-in-glass	Mercury Alcohol	Volumetric thermal expansion	Linear
Resistance	Platinum	Electrical resistance	Polynomial
Thermocouple	Platinum, Platinum-Rhodium Copper-Constantan Iron-Constantan	Seebeck effect	Polynomial

Not surprisingly, it is not possible to measure high temperatures using direct contact thermometers. In such cases, pyrometers, which measure the radiation emitted by the system and then determine the temperature, are used. Constant volume gas thermometers rely on the well known ideal gas equation of state to measure temperature and do so very accurately.

However, irrespective of the method used, it is known that the measured value of temperature, depends to some irreducible extent on the thermometric substance used or the assumed calibration relation. As we will see in section 8.5, the only device that does not suffer from this fundamental limitation is the Carnot engine.

WORK AND HEAT

In this chapter, fundamental ideas relating to work and heat in engineering thermo-dynamics are developed. Different types of work and expressions for calculating the same are discussed. While the notion of work is dealt with in detail, this is not the case for heat. This is because, in the macroscopic approach, heat interaction is either known or is an outcome of the analysis. Most importantly, an appropriate system must first be defined since the determination of whether an interaction is a heat or work interaction as well as its sign and magnitude depend on where the system boundary lies. This is the reason why the familiar definition of work in mechanics, namely, that it is the product of the applied force and the distance moved along the direction of the force, is inadequate in the context of engineering thermodynamics.

3.1 Definition of thermodynamic work

In engineering thermodynamics, work is said to be done *by* a system if the *combined* effect of its interactions with the surroundings is the raising of a mass. A system may interact with the surroundings across several parts of the system boundary (Fig. 2.6), which is why the combined effect of the interactions must be considered.

system has a negative sign if work is done *on* the system by the surroundings. It is good practice not to mix the signs and the words on/by. For instance, the following statements are acceptable:

- Work done *by* the system is 10 J, or, alternatively, work interaction for the system is 10 J

- Work done *on* the system is 10 J, or, alternatively, work interaction for the system is -10 J

On the other hand, the following statements are confusing and must be avoided:

- Work done *by* the system is -10 J

- Work done *on* the system is -10 J

Let us explore the definition in detail now.

Consider the arrangement shown in Fig. 3.1. A battery is connected to a motor and a pulley is mounted on the output shaft of the motor. A mass is suspended by means of a rope from the pulley.

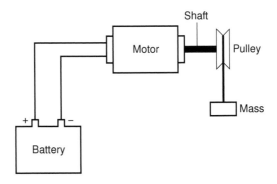

Figure 3.1: Illustration of thermodynamic work

Let us suppose that electric current flows from the battery, turning the output shaft of the motor and hence the pulley, thereby raising the mass *slowly*. If the battery alone is taken as the system, then its interaction with the surrounding is seen to be the raising of a mass and so the work interaction for the battery is positive. If the motor alone is taken as the system, then it is easy to see that it interacts with the surroundings mechanically through the output shaft and electrically though the wires that connect

On the other hand, if the motor is not ideal, then the outgoing work will be less than the incoming work and the work interaction for the motor is thus negative[†]. The difference is converted into heat through friction and internal dissipation, resulting in an increase in the temperature of the motor. The work interaction for the pulley is also zero (assuming it to be frictionless), as the incoming shaft work is used entirely to lift the mass.

The work interaction for the mass may easily seen to be negative by the above definition of work. The same inference may be drawn in an alternative manner as follows. It is clear that, in this case, the potential energy of the mass has been increased during the process, since it is being raised slowly. This increase has occurred because of an external agent raising the mass, thereby doing work. Hence, work is done on the mass. If the mass is raised rapidly so that it acquires kinetic energy as well, then the increase in the potential and kinetic energy of the mass is accomplished by the external agent. Hence, work is done on the mass in this case also.

Let us now look at the situation when the mass is lowered *slowly*. In this case, the work interaction for the mass is positive since this interaction may be used to raise a mass. Or, its potential energy decreases and this may be utilized to do work. If the pulley is still assumed to be frictionless, then its work interaction is zero in this situation as well. However, the motor must now act as a generator and charge the battery since the shaft rotates in the opposite direction. As before, the work interaction will be zero for an ideal generator or negative otherwise. The work interaction for the battery in this situation is negative[‡].

On the other hand, if the mass is lowered rapidly, then a part of the potential energy of the mass is converted into its own kinetic energy and only the rest is available for doing work. Hence, the work interaction for the mass is positive but less in magnitude than before. In the limiting case when the rope snaps, the potential energy of the mass is *entirely* converted into its own kinetic energy and the work interaction for the mass is thus zero in this case.

For a system that includes everything, *i.e.,* the battery, motor, pulley and the mass, the work interaction is zero since such a system does not interact with the surroundings at

---

[†] It must be remembered that incoming work is work done on the system and is thus negative; outgoing work is work done by the system and is thus positive

[‡] In real life, in this situation, the work done by the mass is entirely dissipated by the friction in the pulley and the bearings in the motor. However, the strategy of using a motor/generator combination for motion in opposite directions is utilized in the so-called regenerative braking system used in electric cars and trains.

It is important to note that the definition given above for thermodynamic work is applicable only in cases when the work interaction for the system is positive. Negative work is deliberately not defined. The fact that the algebraic sum of the work interaction for the system, denoted W_{sys}, and the surroundings, denoted W_{surr}, is zero

$$W_{sys} + W_{surr} = 0$$

can be used to evaluate negative work.

3.2 Forms of work

In the previous section, the general definition of thermodynamic work was given. In this section, expressions are developed for evaluating the magnitude of the work, for different types of work interaction. These are not the only forms of work interaction, rather they are the most commonly encountered ones in engineering thermodynamics.

3.2.1 Displacement work

Displacement work is that form of work that occurs by virtue of the deformation of the system boundary. This has already been alluded to in section 2.1 and is now discussed in greater detail.

Consider the piston cylinder arrangement in Fig. 3.2 which contains a working substance initially at pressure P_1 and occupying a volume V_1. An external agent initially provides a resisting force F_1 which is gradually diminished causing the system shown in the figure to undergo an expansion process to a final state labelled 2. At an intermediate instant, let the pressure and volume of the system be P and V respectively. If the piston is now displaced by an incremental distance dx, the work done by the system to accomplish this is given as

$$\delta W = PA\,dx$$

where A is the cross-sectional area of the cylinder. Since the product $A\,dx$ is the incremental change in the volume of the system, dV, the above expression may be written as

$$\delta W = P\,dV$$

The displacement work done by the system during the entire process may be written as

$$W = \int_1^2 P\,dV \tag{3.1}$$

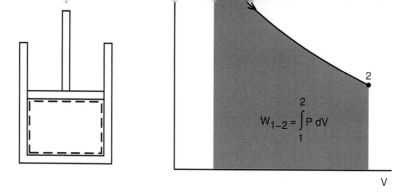

Figure 3.2: Illustration of displacement work

Graphically, this integral is the area under the process curve in $P - V$ coordinates, as shown in Fig. 3.2 on the right. Consequently, displacement work cannot be calculated unless the process curve is known. The process curve is known only for a fully resisted or quasi-equilibrium process. This is the reason why displacement work cannot be calculated for an unrestrained or a partially restrained expansion. It is important to note that displacement work is zero only for the system shown in Fig. 2.7, not for any system.

One advantage with the expression given in Eqn. 3.1 is that it not only gives the magnitude of the displacement work but does so with the correct sign. Since the process is a quasi-equilibrium process, at any instant, the pressure inside the cylinder may be written as

$$P = \frac{F}{A} + P_{atm} + \frac{M_p g}{A}$$

where M_p is the mass of the piston. Substituting this into Eqn. 3.1, we get

$$W = \int_1^2 \frac{F}{A} dV + \int_1^2 P_{atm} dV + \int_1^2 \frac{M_p g}{A} dV$$

This may be simplified to read

$$W = \int_1^2 F dx + P_{atm}(V_2 - V_1) + M_p g(z_2 - z_1)$$

where z is the height of the piston above the bottom surface of the cylinder. The individual terms in the right hand side may be identified respectively as the negative

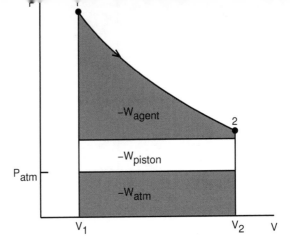

Figure 3.3: Components of the total displacement work done by the system

It is important to realize that the pressure P in the integrand in Eqn. 3.1 is the pressure in that part of the system boundary that is deforming during the process and not necessarily the system pressure. These two are the same for the situation illustrated in Fig. 3.2 but not so for the examples illustrated in Figs. 2.2 - 2.6. The dependence of the displacement work on where the system boundary lies is illustrated in Fig. 3.4 for the example at hand.

Note that, when a part of the system boundary deforms at constant pressure as in Fig. 3.4 and in Figs. 2.2 - 2.6, the magnitude of the corresponding displacement work is simply the product of the pressure and the change in volume. This is the reason why we asserted in Section 2.1 that the exact shape of such a boundary is immaterial.

If multiple parts of the system boundary undergo deformation simultaneously, then the total displacement work is simply

$$W = \sum_b \int_1^2 P_b \, dV_b$$

where the subscript b denotes the boundary.

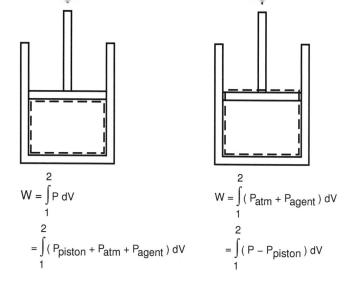

$$W = \int_1^2 P \, dV$$

$$= \int_1^2 (P_{piston} + P_{atm} + P_{agent}) \, dV$$

$$W = \int_1^2 (P_{atm} + P_{agent}) \, dV$$

$$= \int_1^2 (P - P_{piston}) \, dV$$

Figure 3.4: Illustration of the dependence of displacement work on the system boundary. Here, P_{atm}, P_{agent} and P_{piston} are the pressures exerted by the atmosphere, external agent and the piston respectively.

■ EXAMPLE 3.1

A steam main carries steam at a pressure of 1.5 MPa and a temperature of 250°C. A rigid vessel that initially contains steam at a pressure of 0.4 MPa is connected to the main through a valve. The valve is now opened and 0.1 kg of steam from the main slowly flows into the vessel. Determine the work interaction for this process considering the steam finally in the vessel (Fig. 2.4). The specific volume of the steam in the line may be taken to be 0.142 m³/kg.

Solution: The part of the system boundary inside the line alone deforms during the process. Since 0.1 kg of steam from the line enters the vessel, the initial volume of this part of the system boundary is 0.1 kg × 0.142 m³/kg = 0.0142 m³. The final volume of this part of the system boundary is zero. Hence, the displacement work is given as

$$
\begin{aligned}
W &= \int_1^2 P_{line} \, dV_{line} \\
&= P_{line} \, (V_2 - V_1)_{line} \\
&= (1500)(0.0 - 0.0142)
\end{aligned}
$$

EXAMPLE 3.2

A certain amount of air is enclosed inside a vertical cylinder by a frictionless, leak-proof piston as shown in the figure. The cross-sectional area of the piston is 20 cm². Initially, the pressure of the air in the cylinder is 60 kPa. The cylinder is connected through a valve to a rigid vessel which initially contains air at a pressure of 30 kPa. The lower face of the piston is exposed to the atmosphere at 100 kPa. The valve is now opened, and the piston slowly moves a distance of 12 cm from its original position. Calculate the work done by each of the following systems: (a) the piston, (b) the atmosphere and (c) the air.

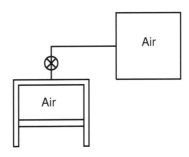

Solution: Force balance on the piston at any instant gives $P_{atm} A = P A + M_p g$, where P is the pressure of the air in the cylinder, A is the cross-sectional area of the piston and M_p is the mass of the piston. This may be rewritten to give $P = P_{atm} - M_p g \div A$. Since none of the quantities on the right hand side of this expression change during the process, it may be inferred that the pressure of the air in the cylinder (not the vessel) remains constant.

Given that the initial pressure P_1 inside the cylinder is 60 kPa, the mass of the piston may be evaluated as

$$M_p = \frac{A}{g} (P_{atm} - P_1) = 8.155 \, \text{kg}$$

(a) It is clear that the piston moves *up* during the process. Hence

$$W_{piston} = -M_p g (z_2 - z_1) = (8.155)(9.81)(0.12) = -9.6 \, \text{J}$$

(b) $W_{atm} = P_{atm} (V_2 - V_1)_{atm} = (100 \times 10^3)(20 \times 10^{-4} \times 0.12) = 24 \, \text{J}.$

■ EXAMPLE 3.3

Consider the frictionless piston cylinder arrangement shown in the figure, in which two pistons are connected by a thin rod of negligible mass and volume and open to the atmosphere at the top and bottom. A certain amount of air is enclosed in the space between the pistons such that the piston assembly is initially in equilibrium. The cross-sectional area of the upper piston is 10 cm² greater than the lower one. The combined mass of the pistons is 5 kg. A very slow heating process now takes place as a result of which the piston assembly moves upward by 25 cm. Determine (a) the initial pressure of the air, (b) the work interaction for the piston assembly, (c) the work interaction for the atmosphere and (d) the work interaction for air. Take the atmospheric pressure to be 10^5 Pa.

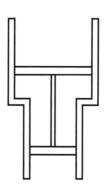

Solution: Force balance on the piston assembly at any instant gives

$$(P - P_{atm}) A_u = (P - P_{atm}) A_l + M_p g$$

where the subscripts u and l stand for upper and lower respectively and the other terms have the same meaning as before. Rearranging, we get

$$P = P_{atm} + \frac{M_p g}{A_u - A_l}$$

Since the right hand of this expression remains the same during the process, the pressure of the enclosed air is constant.

(a) Initial pressure, $P_1 = 10^5 + (5)(9.81) \div (10 \times 10^{-4}) = 149.05$ kPa.

(b) $W_{piston} = -M_p g (z_2 - z_1) = -(5)(9.81)(0.25) = -12.2625$ J.

(c)

$$= \quad -25 \text{ J}$$

(d) Since $W_{air} + W_{atm} + W_{piston} = 0$, W_{air} = 37.2625 kJ. Alternatively,

$$
\begin{aligned}
W_{air} &= P_{air} (V_2 - V_1)_{air} \\
&= P_{air} (A_u - A_l)(z_2 - z_1) \\
&= (149.05 \times 10^3)(10 \times 10^{-4})(0.25) \\
&= 37.2625 \text{ J}
\end{aligned}
$$

3.2.2 Spring work

When a linear spring of stiffness k N/m is compressed by an amount x m, its potential energy increases by $1/2 \, k \, x^2$. Hence

$$W_{spring} = \frac{1}{2} k \, x^2$$

EXAMPLE 3.4

A gas is contained in a frictionless piston cylinder mechanism at an initial pressure of 100 kPa. The piston has a mass of 10 kg and it rests on stops initially as shown in the figure. The cylinder is open to the atmosphere at

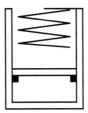

the top and a pressure of 115 kPa is required for the piston to be lifted from the stops. The gas is now heated causing the piston to rise slowly. After a rise of 12 cm, the piston touches a linear spring with a stiffness value of 100 N/mm. The heating of the gas is stopped when the spring is compressed by 2 cm. Determine the work interaction for the atmosphere, piston, spring and the gas. Also sketch the process undergone by the gas on a P-V diagram and show the four work interactions mentioned above by hatching the appropriate areas. Take the atmospheric pressure to be 100 kPa.

from which we can evaluate the area of the piston A_p to be 0.00654 m^2.

The piston rises through a total distance of 12+2 = 14 cm. Thus

$$
\begin{aligned}
W_{atm} &= P_{atm}\left(V_2 - V_1\right)_{atm} \\
&= -P_{atm}\, A_p\left(z_2 - z_1\right) \\
&= -(10^5)(0.00654)(0.14) \\
&= -91.56 \text{ J}
\end{aligned}
$$

$W_{piston} = -M_p\, g\left(z_2 - z_1\right) = -(10)(9.81)(0.14) = -13.734$ J.

Since the spring is compressed by 2 cm,

$W_{spring} = -\frac{1}{2}k\, x^2 = -\frac{1}{2} \times 100 \times 10^3 \times 0.02^2 = -20$ J.

Since $W_{gas} + W_{atm} + W_{piston} + W_{spring} = 0$, $W_{gas} = 125.294$ J.

The gas enclosed in the cylinder undergoes a three step process: a constant volume process *1-2* during which the pressure increases from 100 kPa to 115 kPa followed by a constant pressure process *2-3* during which the piston moves up by 12 cm which is followed by a process *3-4* where the pressure and volume both increase while the spring is compressed by 2 cm. The displacement work during the first step is obviously zero. The displacement work during the second step is easily evaluated to be equal to $(115 \times 10^3)(0.00654)(0.12) = 90.252$ J. At any instant during the third step, the pressure P of the gas is given as

$$
P = P_{atm} + \frac{M_p\, g}{A_p} + \frac{k\, x}{A_p} = 115 \times 10^3 + \frac{k}{A_p^2}\left(V - V_3\right)
$$

where V_3 is the volume occupied by the gas just when the piston touches the spring. The work interaction for this step may then be evaluated as

$$
W_{3-4} = \int_{V_3}^{V_4} P\, dV = \int_0^{0.02\, A_p}\left(115 \times 10^3 + \frac{k}{A_p^2}\, V'\right) dV'
$$

where $V' = V_4 - V_3$. It is left as an exercise to the reader to plot the process undergone by the gas on a $P - V$ diagram and identify areas associated with work interactions for the various components.

If, under the action of this applied torque, the shaft begins to rotate N revolutions per minute ($2\pi N$ radians per minute), then the shaft power is given as $2\pi NT/60$ W. When shaft power is supplied for t seconds, shaft work supplied is equal to $2\pi NTt/60$ J.

3.2.4 Electrical work

We have already seen electrical work supplied by the battery to the motor in the discussion relating to Fig. 3.1. When current I (amps) is supplied from a source across a voltage difference V (volts) for a duration of t seconds, the electrical work is given as VIt J.

3.2.5 Paddle work

In many applications in engineering, stirring of the contents of a vessel is required to ensure homogeneous mixing. The arrangement (Fig. 3.5) consists of a paddle or stirrer which is driven by an external source such as an electric motor or an engine. It is convenient to take the system as the one shown in this figure. The paddle work is usually given or is an outcome of the analysis.

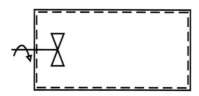

Figure 3.5: Illustration of paddle work

3.3 Heat

Heat transfer between two bodies takes place because of a temperature difference between them and a means to communicate. Heat can be supplied to the system from a source which is at a higher temperature or heat may be rejected by the system to a sink which is at a lower temperature. In many engineering applications, the ambient itself is usually the sink to which heat is rejected. Depending on how the two bodies communicate, heat transfer can occur due to conduction, convection or radiation. In the first, the two bodies are in physical contact with each other; in the second, a fluid such as water or air flowing between the two bodies is the medium

heat *given* to a system is positive and heat *removed* from a system is negative.

In many engineering applications, systems (or devices) are covered with insulating material such asbestos, glass wool and so on to minimize heat loss to the surroundings. In such a case, the system is termed adiabatic and the heat interaction is taken to be zero. At the other end, in some applications, it may be desirable to maximize the heat transfer to the surroundings so that the system remains at the same temperature as the ambient temperature.

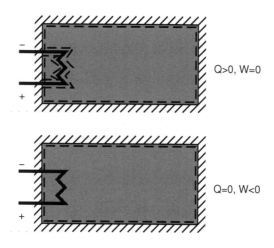

Figure 3.6: Dependance of the nature of the interaction - whether work or heat, on the system boundary

The magnitude of heat interaction depends on where the system boundary is, in the same manner as for work. In addition, the nature of the interaction, whether heat or work, also depends on the system boundary. This is illustrated in Fig. 3.6. Here, an electric resistor is placed inside an insulated vessel. When an electric current flows through the resistor, heat is generated due to ohmic heating which is then absorbed by the contents of the vessel. If the system includes the resistor as in the bottom in Fig. 3.6, then, electric work crosses the system boundary but not heat. If the system excludes the resistor as in the top in Fig. 3.6, then, heat is supplied to the system but not work.

We close this chapter by reiterating that both heat and work are non-zero only when there is an interaction of the system with the surroundings. Any change that happens

FIRST LAW OF
THERMODYNAMICS

In the previous chapters, concepts such as system, control volume, work and heat were discussed. We are now in a position to connect them together in the form of the First law of thermodynamics. To begin with, the first law is written for a thermodynamic system. Subsequently, this is modified suitably for application to a control volume.

4.1   First law of thermodynamics for a system

The first law of thermodynamics for a system undergoing a cyclic process states that the net heat and net work interactions are equal. Mathematically,

$$\oint \delta Q = \oint \delta W \qquad (4.1)$$

This may also be written as

$$\oint (\delta Q - \delta W) = 0$$

where E is a property [†]. Note that an incremental change in a property is denoted using lowercase Roman letter d, whereas the same for a path dependent quantity is denoted using its Greek counterpart, δ. The interesting point about the above expression is that, although Q and W are not properties, their difference is. If we integrate this equation along a process between states 1 and 2, we get

$$\int_1^2 dE = \int_1^2 \delta Q - \int_1^2 \delta W$$

$$\Delta E = E_2 - E_1 = Q_{1-2} - W_{1-2} \tag{4.3}$$

Here, we have used uppercase Greek letter D, Δ, to denote a finite change in a property. Equation 4.3 is the first law applied to a non-cyclic process. It is important to realize that, although the change in total energy between two states (the left hand side in Eqn. 4.3) is independent of the process, information about the process is required in order to calculate it (the right hand side of Eqn. 4.3).

The property E has been identified to be the total energy of the system, and is equal to the sum of the energy stored by the system in different modes such as internal energy, potential energy, kinetic energy and so on. Internal energy (denoted U) is the energy possessed by the molecules that comprise the system in the form of translational and rotational kinetic energies, and other modes such as latent heat. When the system *as a whole*, experiences a change in elevation, then, potential energy also becomes a relevant mode. Similarly, when the system *as a whole*, starts moving with a velocity, then, the system can store energy in the form of kinetic energy. This should not be confused with the kinetic energy of the molecules inside the system since this already has been accounted for in the internal energy. It emerges that internal energy may be categorized as a "disordered" mode since it is associated with the energy possessed by the molecules which are always in random motion. In contrast, both potential and kinetic energies may be deemed to be "ordered" modes since they are associated with the energy possessed by the system *as a whole*. In addition, among these modes, internal energy is unique since it is the only mode that heat can access. In other words, when heat is supplied to a system, it causes a change in the internal energy of the system, which is a disordered mode and hence only a part of the heat supplied may then be converted into work. In contrast, all the modes including internal energy can be accessed by work transfer. This is explored in detail next using an example inspired by one given by Peter Atkins in his book.

Consider a closed, rigid vessel containing, say, 10 kg of a working substance (Fig.

[†] It would have been equally correct to have written $\oint (\delta W - \delta Q) = 0$ and hence $dE = \delta W - \delta Q$. However, this is inconsistent with the sign convention that we have adopted for heat and work.

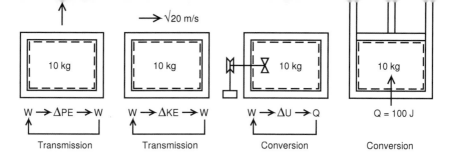

Figure 4.1: Energy transmission and conversion using a thermodynamic system

4.1). Let the vessel along with its contents be the thermodynamic system for the present discussion. A certain amount of energy, say, 100 J is now transferred to the system in one of the following three ways:

- Raising the vessel through a height of 1 m (taking $g = 10\,m/s^2$)

- Causing the vessel to move with a speed equal to $\sqrt{20}\,m/s$

- Work transfer

- Heat transfer

In the first two cases, since the energy is transferred to an ordered mode of energy of the system, it can be converted *entirely* into work, which is also an ordered form of energy. The system returns to its initial state after the work transfer and can repeat the sequence indefinitely. In essence, these two cases may be thought of as simply *transmitting* work. In the third case, the energy is transferred in the form of paddle work by lifting the mass attached to the pulley to a certain height. As the mass is allowed to descend slowly, the paddle wheel is turned causing the temperature of the system to increase. If we force the system to remain at the same temperature by exchanging heat with a reservoir, then *all* of the input work (ordered) may be converted into heat (disordered). The mass can once again be raised as a result of work transfer and the process can be repeated. In the last case, the energy is transferred to a disordered mode of energy of the system. It is possible to convert the heat entirely into work even in this case by using an ideal gas as the working substance and causing it to execute an isothermal process. However, irrespective of the process used to convert the heat to work, the resultant change in the thermodynamic state of the system demands a certain amount of work to be transferred to the system to return it to its initial state. Consequently, energy supplied in the form of heat *cannot be entirely* converted into work. It is important to note

work depends on the system boundary (Fig. 3.4). For the system shown on the left in this figure, the total energy of the system consists only of the internal energy of the working substance and the change in the potential energy of the piston is accounted for in the displacement work. For the system on the right, the total energy of the system now includes the potential energy of the piston also and the displacement work is modified suitably to account for this. An effect (in this case, raising of the piston) may be accounted for either as a work interaction or as a change in the total energy of the system, depending upon where the system boundary is.

4.2 First law of thermodynamics for a control volume

Consider the device shown in Fig. 4.2. Fluid enters at the rate of $\dot{m}_i$ kg/s and leaves at the rate of $\dot{m}_e$ kg/s, where $\dot{m}_i$ need not necessarily be equal to $\dot{m}_e$. Subscripts i and e here denote inlet and exit respectively. For instance, $\dot{m}_i \neq 0$ and $\dot{m}_e = 0$ for the tank in Figs. 2.8a,b while $\dot{m}_i = 0$ and $\dot{m}_e \neq 0$ for the situation depicted in Fig. 2.8c. Heat and work interactions for the device amount to $\dot{Q}$ J/s and $\dot{W}_x$ J/s respectively. We wish to relate the changes in the properties of the fluid between the inlet and the outlet to the heat and the work interactions.

We freeze the device at a time instant t and identify a system as shown in Fig. 4.2. At this instant, there is an incremental amount of mass δm_i that is about to enter the device. Incremental amounts of heat, δQ and work, δW_x are exchanged between the device and the surroundings. As a result, this system evolves and assumes the form shown in the figure at time $t + \delta t$. It can be seen that an amount of mass δm_e is about to leave the device. Upon applying first law to this system (Eqn. 4.2), we get

$$dE = E_{t+\delta t} - E_t = \delta Q - \delta W$$

It can be seen from Fig. 4.2 that the total energy of the system at the instants shown is the sum of the total energy of the fluid within the control volume and the total energy of the fluid that is about to enter or leave. Accordingly, at time t, the total energy of the system at this instant may be written as $E_t = E_{CV,t} + e_i \, \delta m_i$. Here, e is the specific total energy of the *fluid* defined as $e = u + \frac{V^2}{2} + gz$, where u is the specific internal energy, V is the velocity and z is the elevation with respect to an arbitrary datum. Similarly, for the system at time $t + \delta t$, we can write $E_{t+\delta t} = E_{CV,t+\delta t} + e_e \delta m_e$. Of course, E_{CV} the total energy contained in the control volume and is equal to $(U + KE + PE)_{CV}$.

Considering the system at the two instants shown in the figure, the net work interaction may be written as

$$\delta W = \delta W_x + \text{Disp. work at the inlet} + \text{Disp. work at the exit}$$

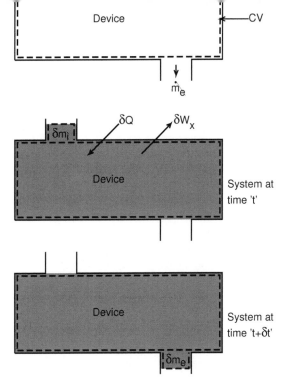

Figure 4.2: Derivation of the first law for a control volume

$$
\begin{aligned}
&= \delta W_x + P_i\left(0 - v_i \delta m_i\right) + P_e\left(v_e \delta m_e - 0\right) \\
&= \delta W_x - P_i v_i \delta m_i + P_e v_e \delta m_e
\end{aligned}
$$

Here, v is the specific volume. Note that there is no change in the property of the fluid while it remains in the inlet or exit. The displacement work at the inlet and the exit is usually called the flow work. This is the amount of work that is required to push the fluid in and out of the device. Upon combining the above expressions, we get

$$
\left(E_{CV,t+\delta t} + \mathbf{e}_e \delta m_e\right) - \left(E_{CV,t} + \mathbf{e}_i \delta m_i\right) = \delta Q - \left(\delta W_x - P_i v_i \delta m_i + P_e v_e \delta m_e\right)
$$

This may be rearranged to give

$$
E_{CV,t+\delta t} - E_{CV,t} = \delta Q - \delta W_x + \delta m_i(\mathbf{e}_i + P_i v_i) - \delta m_e(\mathbf{e}_e + P_e v_e)
$$

$$E_{CV,t+\delta t} - E_{CV,t} = \delta Q - \delta W_x + \delta m_i \left(h_i + \frac{V_i^2}{2} + g z_i \right) - \delta m_e \left(h_e + \frac{V_e^2}{2} + g z_e \right)$$

where, we have introduced a new property, namely, the specific enthalpy, h, which is equal to $u + Pv$. Upon dividing both sides by δt, and taking the limit as $\delta t \to 0$, we finally get

$$\frac{dE_{CV}}{dt} = \dot{Q} - \dot{W}_x + \dot{m}_i \left(h_i + \frac{V_i^2}{2} + g z_i \right) - \dot{m}_e \left(h_e + \frac{V_e^2}{2} + g z_e \right) \qquad (4.4)$$

This is the form in which the first law can be conveniently used for a control volume analysis. This is also known as the Unsteady Flow Energy Equation (UFEE).

Since the mass contained in a system must remain the same throughout by definition, we may write for the system shown in the figure,

$$m_{CV,t+\delta t} + \delta m_e = m_{CV,t} + \delta m_i$$

If we rearrange this equation and divide both sides by δt and take the limit, we get

$$\frac{dm_{CV}}{dt} = \dot{m}_i - \dot{m}_e \qquad (4.5)$$

4.2.1 Steady flows

In most engineering applications, devices attain a steady state after the initial startup transient period. In such a situation, there are no changes in the property values with time. Consequently, for devices that operate at steady state, the time derivative in Eqns. 4.4 and 4.5 may be dropped. This results in

$$\dot{Q} - \dot{W}_x + \dot{m}_i \left(h_i + \frac{V_i^2}{2} + g z_i \right) - \dot{m}_e \left(h_e + \frac{V_e^2}{2} + g z_e \right) = 0$$

and

$$\dot{m}_i - \dot{m}_e = 0 . \qquad (4.6)$$

It follows that we may write $\dot{m}_i = \dot{m}_e = \dot{m}$ and so

$$\dot{Q} - \dot{W}_x + \dot{m} \left[\left(h_i + \frac{V_i^2}{2} + g z_i \right) - \left(h_e + \frac{V_e^2}{2} + g z_e \right) \right] = 0 . \qquad (4.7)$$

This equation is called the Steady Flow Energy Equation (SFEE).

The development above assumed that the device had a single inlet and a single exit.

and

$$\sum_i \dot{m}_i - \sum_e \dot{m}_e = 0. \tag{4.9}$$

4.3 Pure substance

The forms of first law derived above, relate the work interaction and heat interaction to the changes in the internal energy or enthalpy of the working substance. Determination of work and heat interaction was discussed in detail in Chapter 3. The next chapter is devoted to the evaluation of properties of a system at a given state. However, before this can be done, a discussion relating to the kind of working substances that can be dealt with within our present framework, is necessary.

A pure substance is a thermodynamic system that is

A. uniform in composition

B. uniform in chemical aggregation

The first requirement refers to the relative proportion of the individual elements that make up the substance and the second requirement refers to the manner in which the elements are chemically combined. In a given system, when samples drawn from different spatial locations have the same composition and aggregation, then, the system may be treated as a pure substance. This is best understood through the following examples:

Table 4.1: Determination of whether a substance may be treated as a pure substance or not

Substance	A	B	Comments
Water (liq)	√	√	Single component in the liquid phase
Water (vapor)	√	√	Single component in the gaseous phase
N_2	√	√	Gas
Air (1 kmol O_2 + 3.76 kmol N_2)	√	√	Mixture of gases
Water (liq) + Water (vapor)	√	√	Two phase mixture
2 kmol H_2 + 1 kmol O_2 + Water (vapor)	√	√	Mixture of gases
2 kmol H_2 + 1 kmol O_2 + Water (liq)	√	×	A sample containing only the liquid has a different aggregation of H_2 and O_2 since the atoms are chemically bonded. In a sample containing only the gases, they are not combined
1 kmol H_2 + 1 kmol O_2 + Water (liq)	×	×	In addition to the above, the proportion of the hydrogen and oxygen atoms are different in samples containing only liquid or only gas.

above may be interpreted as additional requirements for the macroscopic approach to be valid.

Once it is established that the working substance is a pure substance, we need to develop methods to evaluate properties such as internal energy, enthalpy and entropy. As mentioned in section 2.3, two properties are used to fix the state of the system and so we must be able to calculate the internal energy or the enthalpy, given the values for these two properties, whatever they may be. This is taken up in the next chapter.

---

†The internal energy will not be uniform in the case of a two phase mixture also. However, the difference in this case is due to the latent heat that accompanies a change of phase and not due to different chemical bonds between the constituent elements.

PROPERTIES OF PURE SUBSTANCES

As mentioned in the previous chapter, analysis of a device using first law requires the evaluation of properties such as internal energy and enthalpy from given values of measurable properties such as P, v or T of the working substance. It is clear from Table 4.1 that for this to be a pure substance, it must either be a two phase mixture (liquid+vapor) or a gas. Examples of interest in the former category are water and refrigerant such as R134a. Examples of the latter are oxygen, nitrogen, argon and so on. Evaluation of properties of two phase mixtures as well as gases are discussed in this chapter. In the case of the former, although the development is for water, there is no loss of generality since the concepts and ideas carry over without any change to two phase mixtures of other substances.

5.1   P-v and T-v diagram of liquid water-water vapor mixture

Consider a piston cylinder apparatus shown in Fig. 5.1. The cylinder contains liquid water initially at a high pressure (say, 10 MPa) and a temperature of 100°C. The entire apparatus is maintained at a constant temperature in an isothermal bath as shown in the figure. We now carry out an experiment by reducing the pressure

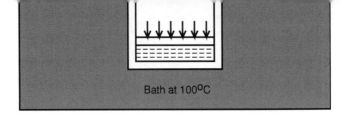

Figure 5.1: Isothermal expansion of water

gradually and allowing the piston to move up slowly. In the beginning, we notice that even a small upward movement of the piston results in a large reduction in pressure. This trend continues until the pressure reaches 100 kPa. If we now attempt to reduce the pressure by moving the piston upward, we notice that the liquid inside the cylinder begins to evaporate and the vapor occupies the extra volume that becomes available. Consequently, the pressure does not change but remains constant at 100 kPa. It must be recalled that the temperature is being held constant at 100°C during the process. As the piston is moved further upward, more liquid evaporates and occupies the extra volume that is available thereby keeping the pressure constant. This continues until all the liquid has evaporated. Once this happens, any further upward movement of the piston results in an appreciable reduction in pressure and hence a commensurate change in the volume of the vapor that is enclosed inside.

The process that we have described above is depicted as the process curve $LMNO$ in the $P - v$ diagram shown in Fig. 5.2. Points M and N respectively denote the onset and completion of vaporization. It should also be noted that since the temperature is kept constant during the entire process, $LMNO$ is an isotherm and in addition, along the segment MN of the process curve, pressure remains constant as well. In other words, the isotherm corresponding to 100°C and the isobar corresponding to 100 kPa coincide along the segment MN. Pressure and temperature remain constant when phase change (liquid to vapor or vapor to liquid) takes place. States M and N are referred to as saturated liquid and saturated vapor states respectively. The term saturated is used to highlight the fact that these states indicate the beginning or termination of a change of phase. For a given temperature T, the pressure at which phase change takes place is called the saturation pressure corresponding to that temperature, denoted $P_{sat}(T)$. Alternatively, for a given pressure P, the temperature at which phase change takes place is called the saturation temperature corresponding to that pressure and is denoted $T_{sat}(P)$. When water undergoes a change of phase (be it evaporation or condensation) at 100 kPa, the temperature must be 100°C and *vice versa*. In other words, pressure and temperature are not independent during

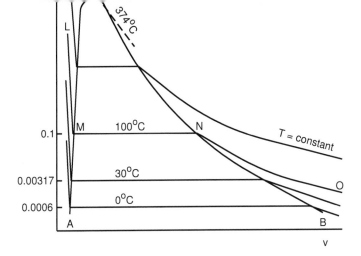

Figure 5.2: $P - v$ diagram for liquid water-water vapor mixture. Not to scale.

phase change.

If the experiment described above were to be conducted for different bath temperatures, the corresponding process curves (isotherms) would appear as shown in Fig. 5.2. Note that as the temperature increases, the segment corresponding to the change of phase becomes shorter and it disappears altogether when the temperature attains a "critical" value equal to 374°C for water. State point C corresponding to 22 MPa and 374°C is called the critical state.

Curve CA connects all the saturated liquid states and curve CB connects all the saturated vapor states. These curves are respectively called the saturated liquid line and saturated vapor line. The "dome" shaped region bounded by the curve ACB delineates the region occupied by a two phase mixture (liquid+vapor) and that occupied by a single phase (liquid or vapor). States that lie to the left of CA and below the critical isotherm are referred to as compressed liquid states or sub-cooled states. This is because, for such states, the given (P, T) is such that $P > P_{sat}(T)$ or, alternatively, $T < T_{sat}(P)$. States that lie to the right of CB and above the critical isotherm represent superheated vapor (steam).

The information presented in Fig. 5.2 using $P - v$ coordinates is given in Fig. 5.3 using $T - v$ coordinates. It is easy to see that the salient features mentioned in

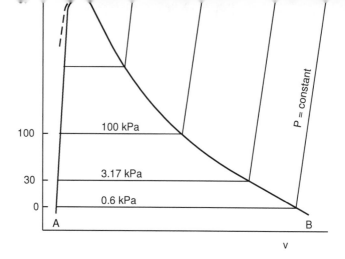

Figure 5.3: $T - v$ diagram for liquid water-water vapor mixture. Not to scale.

connection with the $P - v$ diagram carry over to the $T - v$ diagram without change. Students must thoroughly familiarize themselves with both these diagrams and learn to draw the saturation lines, isobars and isotherms. In addition to these two diagrams, it is also possible to plot a $P - T$ diagram to depict the same information[†]. If we imagine a three dimensional coordinate space with P, v and T as the coordinates, then, the curves shown in Figs. 5.2 and 5.3 may be interpreted as the projections of a three dimensional surface.

5.1.1 Location of state

As mentioned at the beginning of this chapter, calculating properties such as u, h (and later on the specific entropy s) for a given thermodynamic state is essential for thermodynamic analyses of devices. The first step towards this goal is locating a given thermodynamic state in a $P - v$ or $T - v$ diagram such as Fig. 5.2 or 5.3. In the context of these two diagrams, locating a state simply means the determination whether the given state lies to the left of the saturated liquid line CA (compressed/sub-cooled liquid) or to the right of the saturated vapor line CB (superheated vapor) or in between the two (two phase mixture). This quantitative

---

[†]However, as this is not of any use for the applications that we will be looking at, this is not given here. Interested readers are urged to consult the text books suggested at the end for more information.

the corresponding saturated liquid state (subscript f) and the saturated vapor state (subscript g). The same information is provided in Table B for values of pressure from the saturation pressure at room temperature to the critical pressure.

For our present purpose, a given thermodynamic state is uniquely determined by specifying the values for two *independent* properties (usually measurable). In practice, one of three combinations, namely, $P - v$, $T - v$ or $P - T$ is specified. We now consider each one of these in turn using specific examples.

To locate a state given P and T Let the given state A be specified as P=250 kPa and T=200°C. From Table B, we note that, corresponding to 250 kPa, the saturation temperature is 127.4°C. The isobar corresponding to P=250 kPa and the isotherm corresponding to T=127.4°C are both shown in Fig. 5.4 (left). It is now apparent that the isotherm corresponding to the given state, namely, T=200°C will lie above as shown in this figure. Hence, state A lies at the point of intersection of the isobar corresponding to P=250 kPa and the isotherm corresponding to T=200°C, in the superheated region, as shown in the figure.

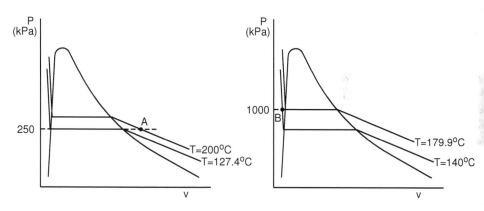

Figure 5.4: Locating a state specified by a given P and T. Diagram not to scale.

In the same manner, it is easy to show that a state B corresponding to P=1000 kPa and T=140°C lies in the compressed liquid region, as illustrated in Fig. 5.4 (right). It should be noted that, as mentioned before, the isobar and the isotherm coincide in the mixture region.

To locate a state given P and v Let the given state Q be specified as P=250 kPa and v=0.5 m³/kg. From Table B, we note that, corresponding to 250 kPa, the

circle in this figure.

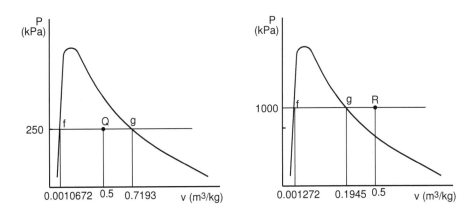

Figure 5.5: Locating a state specified by a given P and v. Diagram not to scale.

Consider another state R for which P=1 MPa and v=0.5 m³/kg. From Table B, we note that, for 1 MPa, the specific volume of the corresponding saturated liquid state is $v_f = 0.001272$ m³/kg and that of the saturated vapor state is $v_g = 0.1945$ m³/kg. Since the given value for the specific volume, 0.5 m³/kg is greater than v_g, the given state lies in the superheated region (Fig. 5.5, right).

To locate a state given T and v Let the given state be specified as T=200°C and v=0.1 m³/kg. From Table A, we note that, for T=200°C, the specific volume of the saturated liquid state is $v_f = 0.0011564$ m³/kg and that of the saturated vapor state is $v_g = 0.1273$ m³/kg. Since the given value for the specific volume, 0.1 m³/kg lies in between these two values, the state falls in the two phase mixture region.

Let us consider another example, where T=250°C and v=0.2 m³/kg. From Table A, we note that, for T=250°C, the specific volume of the saturated liquid state is $v_f = 0.0012515$ m³/kg and that of the saturated vapor state is $v_g = 0.0501$ m³/kg. Since the given value for the specific volume, 0.2 m³/kg is greater than v_g, the state is superheated.

It is left as an exercise to the reader to draw the $T-v$ diagrams for these two examples and locate the states.

property called the dryness fraction or quality[†] (x) is introduced. This is defined as

$$x = \frac{m_g}{m} = \frac{m_g}{m_f + m_g} \tag{5.1}$$

where the subscripts f and g refer to the liquid and vapor phase respectively and m denotes the mass. It is easy to see that $x = 0$ corresponds to a saturated liquid and $x = 1$ corresponds to a saturated vapor state. Thus, the saturated liquid line CA is the locus of all states for which $x = 0$ and the saturated vapor line CB is the locus of states for which $x = 1$. The dryness fraction is indeterminate at the critical state C. Physically, this means that, when a substance exits at the critical state, it is impossible to distinguish whether it is in the liquid or vapor state. It is important to realize that the concept of dryness fraction is meaningless outside the dome region.

Any specific property of a two phase mixture may be evaluated as the weighted sum of the respective values at the saturated liquid and vapor states, where the weights are expressed in terms x. For instance, let a two phase mixture of mass m contain m_f and m_g of saturated liquid and vapor respectively. The specific volume of the mixture (v) is given as

$$
\begin{aligned}
v &= \frac{V}{m} \\
&= \frac{m_f v_f + m_g v_g}{m} \\
&= (1 - x)v_f + x v_g \\
&= v_f + x(v_g - v_f)
\end{aligned}
$$

Any specific property of the mixture (ϕ) in the two phase region may be written in the same manner as

$$\phi = \phi_f + x(\phi_g - \phi_f) \tag{5.2}$$

where, ϕ can be the specific volume v, specific internal energy u or specific enthalpy h.

Superheated and compressed liquid states Property values of superheated states may be retrieved directly from Tables C and F. Tables are also available for compressed/sub-cooled liquid states. However, for the present purpose, it suffices to use an engineering approximation to compute these property values rather than resorting to tables. To this end, we use the fact that for a liquid (a) specific volume

---

[†]This has no relation to the purity of the water

$$v(T, P) \approx v_f(T)$$
$$u(T, P) \approx u_f(T) \tag{5.3}$$
$$h(T, P) \approx u_f(T) + P v_f(T)$$

It should be noted that the approximation given above for the specific enthalpy is not equal to $h_f(T)$, since the latter is equal to $u_f(T) + P_{sat}(T) v_f(T)$ by definition. However, in most situations, $h(T, P)$ is reasonably well approximated by $h_f(T)$. The properties on the right hand side of the above equation are retrieved from the tabulated data using the given temperature alone and neglecting the given pressure[†].

5.2 Ideal gases

It is known from high school physics that an ideal gas obeys the relation $P v \div (R T) = 1$. Before we proceed to use this equation, let us take a closer look at what is meant by "ideal" and under what conditions does a gas behave as an ideal gas.

We start by evaluating $|1 - P v \div (R T)|$ for water vapor using the data in Table C for different values of pressure and temperature and then superimposing the calculated values on the $T - v$ diagram for water (Fig. 5.3). The outcome is shown in Fig. 5.6. The region within which the deviation from the ideal gas relation is 10 percent or less for water vapor is shown in this figure qualitatively. Indeed, this figure will not be different for any other gas except for the numerical values of the properties at the critical state. This suggests that a better alternative may be to plot this data by somehow incorporating the critical temperature and pressure.

Figure 5.7 shows the same data plotted against the so-called reduced pressure, $P_r = P \div P_{critical}$ for different values of the reduced temperature, $T_r = T \div T_{critical}$ for gases such as O_2, N_2, H_2O, CH_4, CO_2, C_2H_4, C_2H_6, C_3H_8, to name a few. It

---

[†]If we consider a length of streamtube in a steady, isothermal flow of an incompressible fluid to be the control volume and apply Eqn. 4.7, we get

$$\left(h_i + \frac{V_i^2}{2} + g z_i \right) - \left(h_e + \frac{V_e^2}{2} + g z_e \right) = 0$$

Since the fluid is incompressible, the density (or equivalently, the specific volume) remains constant and since the flow is isothermal, the specific internal energy also remains constant. Hence, $h_i - h_e = u_i + P_i v_i - u_e - P_e v_e$ may be simplified to be equal to $P_i/\rho - P_e/\rho$, and the above expression becomes

$$\left(\frac{P_i}{\rho} + \frac{V_i^2}{2} + g z_i \right) - \left(\frac{P_e}{\rho} + \frac{V_e^2}{2} + g z_e \right) = 0$$

This is easily recognizable as the famous Bernoulli's equation.

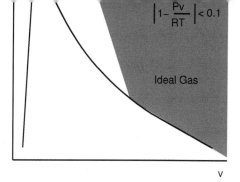

$$\left|1 - \frac{Pv}{RT}\right| < 0.1$$

Ideal Gas

Figure 5.6: Region of ideal gas behaviour. Adapted from *Thermodynamics: An Engineering Approach* by Cengel and Boles.

is quite startling to see the $T_r = constant$ curve for so many gases collapsing into a single curve. The shaded region from Fig. 5.6 is shown in this figure as well and it can be inferred that ideal gas behaviour is exhibited for all values of pressure but only for high temperatures ($T_r > 2$). Within the shaded region shown in Fig. 5.7, it is safe to assume that the volume occupied by the molecules is negligible and also that the inter-molecular forces are negligible. For other regions, the simple ideal gas relation no longer holds and much more complicated relations have to be used. Interested readers are referred to the books suggested at the end for details on equations of state for non-ideal *i.e.,* real gases.

5.2.1 Perfect gas equation of state

The equation of state for an ideal gas may be written as

$$Pv = RT \tag{5.4}$$

where T is the temperature. R is the particular gas constant and is equal to $\mathcal{R} \div M$ where $\mathcal{R} = 8314$ J/kmol.K is the Universal Gas Constant and M is the molecular weight of the gas in units of kg/kmol. Equation 5.4 may be written in many different forms depending upon the application under consideration. A few of these forms are presented here for the sake of completeness. Since the specific volume $v = 1/\rho$, we may write

$$P = \rho RT$$

or, alternatively as

$$PV = mRT$$

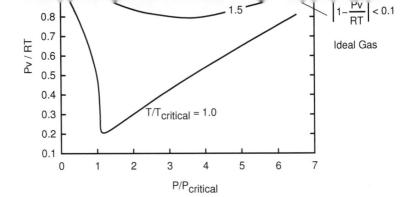

Figure 5.7: Compressibility chart for gases

where m is the mass and V is the volume. If we define the concentration c as $(m \div M)(1/V)$, then,

$$P = c\mathcal{R}T$$

Here, c has units of kmol/m^3. The mass density ρ is be related to the particle density n (particles/m^3) through the relationship $\rho = n M \div N_A$. Here, we have used the fact that 1 kmol of any substance contains Avogadro number of molecules ($N_A = 6.023 \times 10^{26}$). Thus

$$P = n\frac{\mathcal{R}}{N_A}T \; = \; nk_BT$$

where k_B is the Boltzmann constant.

5.2.2 Calorically perfect gas

In thermodynamics, we need, in addition to the equation of state, an equation relating the internal energy to other measurable properties. The internal energy, strictly speaking, is a function of two thermodynamic properties, namely, temperature and specific volume. In reality, the dependence on specific volume is very weak for gases and hence is usually neglected. Such gases are called *thermally perfect* and for them $u = u(T)$. The exact nature of this function is examined next.

From a molecular perspective, it can be seen intuitively that the internal energy will depend on the number of modes in which energy can be stored (also known as degrees of freedom) by the molecules (or atoms) and the amount of energy that can be stored in each mode. For monatomic gases, the atoms have the freedom to move

additional degrees of freedom are possible. These molecules, in addition to translational motion along the three axes, can also rotate about these axes. Hence, energy storage in the form of rotational kinetic energy is also possible. In reality, since the moment of inertia about the "dumb bell" axis is very small, the amount of kinetic energy that can be stored through rotation about this axis is negligible. Thus, rotation adds essentially two degrees of freedom only. In the "dumb bell" model, the bonds connecting the two atoms are idealized as springs. When the temperature increases beyond 600 K or so, these springs begin to vibrate and so energy can now be stored in the form of vibrational kinetic energy of these springs. When the temperature becomes high (> 2000 K), transition to other electronic levels and dissociation take place and at even higher temperatures, the atoms begin to ionize. These effects do not represent degrees of freedom.

Having identified the number of modes of energy storage, we now turn to the amount of energy that can be stored in each mode. The classical equipartition energy principle states that each degree of freedom, when "fully excited", contributes $1/2\ RT$ to the internal energy per unit mass of the gas. The term "fully excited" refers to a threshold beyond which a new storage mode begins to get excited *i.e.,* energy can be stored in an additional mode. For example, the translational mode becomes fully excited at temperatures as low as 3K itself. The rotational modes begin to get excited beyond this temperature. For diatomic gases, beyond 600 K, the rotational mode is fully excited and the vibrational mode begins to get excited. Vibrational modes become fully excited beyond 2000 K or so. Strictly speaking, all the modes are quantized and so the energy stored in each mode has to be calculated using quantum mechanics. However, the spacing between the energy levels for the translational and rotational modes are small enough, that we can assume equipartition principle to hold for these modes.

We can thus write

$$u = \frac{3}{2}RT$$

for monatomic gases and

$$u = \frac{3}{2}RT + RT + \frac{h\nu/k_B T}{e^{h\nu/k_B T} - 1}RT$$

for diatomic gases. In the above expression, ν is the fundamental vibrational frequency of the molecule. Note that, for large values of T, the last term approaches RT. We have not derived this term formally as it is well outside the scope of this book. Interested readers may consult textbooks on Gas Dynamics for full details.

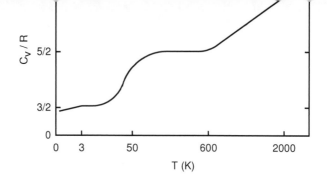

Figure 5.8: Variation of C_v/R with temperature for diatomic gases

The enthalpy per unit mass may now be calculated by using the fact that

$$h = u + Pv = u + RT$$

We can calculate C_v and C_p from these equations by using the fact that $C_v = \partial u/\partial T$ and $C_p = \partial h/\partial T$. Thus

$$C_v = \frac{3}{2}R$$

for monatomic gases and

$$C_v = \frac{5}{2}R + \frac{(h\nu/k_BT)^2\,e^{h\nu/k_BT}}{\left(e^{h\nu/k_BT} - 1\right)^2}R$$

for diatomic gases. The variation of C_v/R is illustrated schematically in Fig. 5.8. It is clear from this figure that $C_v = 5/2\,R$ in the temperature range 50 K $\leq T \leq$ 600 K. In this range, $C_p = 7/2\,R$, and thus the ratio of specific heats $\gamma = 7/5$ for diatomic gases. For monatomic gases, it is easy to show that $\gamma = 5/3$. In this temperature range, where C_v and C_p are constants, the gases are said to be *calorically perfect.* We will assume calorically perfect behavior in all the subsequent chapters. Also, for a calorically perfect gas, since $h = C_pT$ and $u = C_vT$, it follows from the definition of enthalpy that

$$C_p - C_v = R \tag{5.5}$$

This is called Meyer's relationship. In addition, it is easy to see that

$$C_v = \frac{R}{\gamma - 1}, \qquad C_p = \frac{\gamma R}{\gamma - 1} \tag{5.6}$$

Consider a gas mixture composed of m_i kg each of a number of components each of which is an ideal gas. The number of moles of species i is $n_i = m_i \div M_i$, where M_i is its molecular weight. The total number of moles in the mixture is then $n = \sum_i n_i$. Mole fraction of species i is given as $y_i = n_i \div n$ and mass fraction as $x_i = m_i \div m$. Here $m = \sum_i m_i$ is the total mass of the mixture.

The mixture molecular weight M is given as $M = m \div n$. From the definitions of the mole fraction and mass fraction given above, it is easy to show that

$$M = \frac{m}{n} = \frac{\sum_i m_i}{n} = \frac{\sum_i n_i M_i}{n} = \sum_i y_i M_i$$

$$= \frac{m}{n} = \frac{m}{\sum_i n_i} = \frac{m}{\sum_i m_i/M_i} = \frac{1}{\sum_i (x_i/M_i)} \tag{5.7}$$

Equation of state for a mixture is defined in the same manner as that of a single component gas, with the particular gas constant being evaluated using the mixture molecular weight. Thus

$$PV = mRT = n\mathcal{R}T$$

where $R = \mathcal{R} \div M$ and $\mathcal{R}$ is the universal gas constant.

Two new concepts, namely, partial pressure and partial volume are commonly used when dealing with a mixture of gases. The Dalton's model of a mixture of gases assumes each component i to be at its own partial pressure, P_i but occupying the same volume as that of the mixture and at the same temperature (Fig. 5.9). Thus, $P_i V = m_i R_i T$, where $R_i = \mathcal{R} \div M_i$. Since $PV = mRT$ for the mixture, it follows that

$$P_i = \frac{m_i}{m} \frac{R_i}{R} = x_i \frac{M}{M_i} \tag{5.8}$$

Alternatively, $P_i V = n_i \mathcal{R}T$ and since $PV = n\mathcal{R}T$, it follows that

$$P_i = \frac{n_i}{n} = y_i \tag{5.9}$$

In the Amagat's (Fig. 5.9) model, each component occupies a partial volume V_i, but has the same pressure and temperature as that of the mixture. Hence $PV_i = m_i R_i T$ and it is easy to show that

$$V_i = x_i \frac{M}{M_i} = y_i \tag{5.10}$$

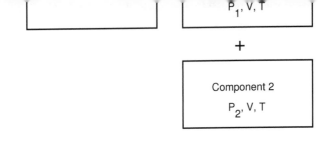

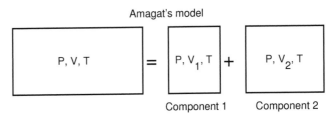

Figure 5.9: Illustration of the Dalton and Amagat models for a mixture consisting of two ideal gases

Obviously, $\sum_i P_i = P$ and $\sum_i V_i = V$.

Properties of mixtures can be calculated in a straightforward manner as follows. Any specific property ϕ of the mixture (on a mass basis) is given as

$$\phi = \frac{\sum_i m_i \phi_i}{\sum_i m_i} = \sum_i x_i \phi_i \tag{5.11}$$

and on a molar basis as

$$\bar{\phi} = \frac{\sum_i n_i \bar{\phi}_i}{\sum_i n_i} = \sum_i y_i \bar{\phi}_i \tag{5.12}$$

Mixture specific heats on a mass basis may easily be shown to be

$$C_p = \sum_i x_i C_{p,i}$$

and

$$C_v = \sum_i x_i C_{v,i}$$

where we have used Eqn. 5.7. Consequently, Eqn. 5.6 is true for a mixture as well. Note that, the ratio of specific heats for the mixture, $\gamma = \left(\sum_i x_i C_{p,i}\right)/\left(\sum_i x_i C_{v,i}\right)$.

FIRST LAW ANALYSIS OF SYSTEMS

In this chapter, first law analysis of systems containing different pure substances, namely, ideal gases, ideal gas mixtures and two phase mixtures of water and refrigerant R134a is presented.

■ **EXAMPLE 6.1**

Two kg of ice at $0°C$ is dropped into 5 kg of liquid water initially at $20°C$ in a large closed container. Simultaneously, 3500 kJ of heat is slowly added to the container. Assuming that the pressure remains constant and there is no heat loss from the container to the surroundings, determine (a) the final temperature and (b) the mass of vapor formed (if any). The latent heat of melting of ice may be taken to be 334 kJ/kg, latent heat of vaporization of water to be 2260 kJ/kg and specific heat of water to be 4.2 kJ/kg.K. The boiling point of water may be taken as $100°C$.

Solution : The system for this problem is comprised of the ice and the liquid water. Since it is not known whether any of the water evaporates, let us determine the heat required for the final state to be water at $100°C$. To this end, we apply first law to the

Here, $W = 0$ as there is no displacement or any other form of work. The ice at $0°C$ melts to form water at $0°C$ and is then heated to $100°C$. The corresponding change in internal energy is

$$m_{ice} L + m_{ice} c_{liq} (100 - 0)$$

where L denotes the latent heat of melting and c denotes the specific heat capacity. The change in internal energy for the water to go from $20°C$ to $100°C$ is $m_{liq} c_{liq} (100 - 20)$. Thus,

$$Q = m_{ice} L + m_{ice} c_{liq} (100 - 0) + m_{liq} c_{liq} (100 - 20) = 3188 \, kJ$$

Since the heat added is higher than this value by 312 kJ (= 3500-3188), some amount of water will definitely vaporize. Heat required to vaporize *all* of the water at $100°C$ is given as $(m_{ice} + m_{liq}) L' = 15820$ kJ, where L' denotes the latent heat of vaporization. It is thus clear that the heat supplied is not sufficient to evaporate all the water. Hence, the final temperature is $100°C$ and the mass of water evaporated is equal to $312 \, kJ \div L' = 0.1381$ kg.

▮ EXAMPLE 6.2

Two kg of air contained in a piston cylinder assembly undergoes a compression that obeys $PV^n = constant$. Such a process is called a polytropic process and the exponent, the polytropic index. The air is initially at 100 kPa, 300 K and is compressed to a final pressure of 800 kPa. Determine, for $n = 1.27$, (a) work done and (b) heat interaction.

Solution : Polytropic processes corresponding to different indices are shown in Fig. 6.1.

(a) Taking the air as the system, the work interaction for such a process may be evaluated as

$$W = \int_1^2 P \, dV = \frac{P_2 V_2 - P_1 V_1}{1 - n} = \frac{m R (T_2 - T_1)}{1 - n}$$

Here, we have used the fact that $PV^n = P_1 V_1^n = P_1 V_2^n$ and also used the equation of state.

For air, $M = 28.8$ kg/kmol and $\gamma = 7/5$. Therefore, $R = 8314 \div 28.8 = 288.68$ J/kg.K and $C_v = R \div (\gamma - 1) = 721.7$ J/kg.K.

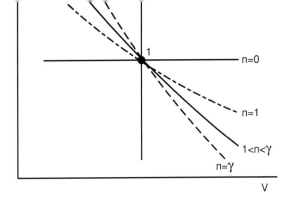

Figure 6.1: Polytropic processes for an ideal gas corresponding to various polytropic indices. Starting from the initial state denoted 1, final states in the second quadrant are attained as a result of compression and those in the fourth quadrant are attained after expansion.

We may write $PV^n = constant$ as, $P^{(1-n)/n} T = constant$, after using the equation of state. The final temperature may thus be evaluated as

$$T_2 = T_1 \left(\frac{P_2}{P_1}\right)^{(n-1)/n} = 467 \, \text{K}$$

Therefore, the work interaction,

$$W = \frac{m R (T_2 - T_1)}{1 - n} = \frac{2 \times 288.68 \times (467 - 300)}{1 - 1.27} = -357.11 \, \text{kJ}$$

(b) Application of first law to this system gives

$$\Delta E = \Delta U = Q - W \Rightarrow Q = m C_v (T_2 - T_1) + W$$

Upon substituting the numerical values, we get Q = -116.06 kJ. The sign indicates that heat is lost by the system to the surroundings.

▣ EXAMPLE 6.3

An insulated vessel is divided into two compartments A and B by a thin membrane which is designed to rupture at 1 MPa. Compartment A occupies 0.5 m³ and is initially evacuated. Compartment B occupies 0.25 m³ and contains 2 kg of air initially at a pressure of 700 kPa. The air is stirred until the membrane

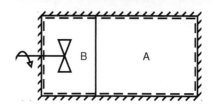

Solution : The initial temperature of the air in compartment B may be calculated using the ideal gas equation of state as

$$T_1 = \frac{P_1 V_1}{mR} = \frac{700 \times 10^3 \times 0.25}{2 \times 288.68} = 303\,\text{K}$$

(a) The pressure in compartment B when the membrane ruptures is 1000 kPa. Hence, the temperature at this instant is given by

$$T_2 = \frac{P_2 V_2}{mR} = \frac{1000 \times 10^3 \times 0.25}{2 \times 288.68} = 433\,\text{K}$$

(b) We take the entire vessel to be the system. Applying first law for process *1-2*, we get

$$\Delta E = \Delta U = \overset{0}{\cancel{Q}} - W_{1-2} = -\overset{0}{\cancel{W}}_{disp} - W_{stirrer}$$

Here, we have taken Q to be zero since the vessel is insulated. Displacement work, W_{disp} is also zero since the system boundary does not deform during the process. Therefore,

$$\begin{aligned} W_{stirrer} = -\Delta U &= -m\,C_v\,(T_2 - T_1) \\ &= -(2)(721.7)(433 - 303) = -187.64\,\text{kJ} \end{aligned}$$

Hence, work done by the stirrer is 187.64 kJ.

(c) Let the final state be denoted *3*. Applying first law for process *2-3*, we get

$$\Delta E = \Delta U = \overset{0}{\cancel{Q}} - \overset{0}{\cancel{W}}_{2-3} = 0$$
$$\Rightarrow \quad m\,C_v\,(T_3 - T_2) = 0$$
$$\Rightarrow T_3 = T_2 = 433\,\text{K}$$

For process *2-3*, there is no work interaction, since the stirrer is stopped once the membrane ruptures. The final temperature is 433 K. The final pressure may be

EXAMPLE 6.4

Consider the frictionless piston-cylinder arrangement shown. Initially, the air contained in cylinder A occupies 0.0864 m^3 and is at 200 kPa and 300 K, while the air in cylinder B occupies 0.0398 m^3, and is at 300 K. The cross-sectional areas of cylinders A and B are, respectively, 0.00786 m^2 and 0.00393 m^2. Heat is now supplied to cylinder A causing the piston to move slowly downwards by 2 m. The total mass of the pistons and connecting rod is 10 kg. Cylinder B is completely insulated. Determine: (a) Mass in cylinder A (b) Initial pressure in cylinder B (c) Mass in cylinder B (d) W_{atm} (e) W_{piston} (f) Final pressure in B (g) Final temperature in B (h) W_B (i) Final pressure in A (j) Final temperature in A and (k) Heat supplied, Q. Take $P_{atm} = 100$ kPa and $g = 10 \text{ m/s}^2$.

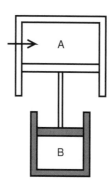

Solution : (a) Mass of air in cylinder A is given as

$$m_A = \frac{P_{1,A}V_{1,A}}{RT_{1,A}} = \frac{200 \times 10^3 \times 0.0864}{288.68 \times 300} = 0.2 \text{ kg}$$

(b) At any instant during the process, a force balance on the piston assembly gives

$$(P - P_{atm})\,A_p|_A + M_p\,g = (P - P_{atm})\,A_p|_B$$

The initial pressure in B is thus

$$P_{1,B} = P_{atm}\left(1 - \frac{A_{p,A}}{A_{p,B}}\right) + P_{1,A}\frac{A_{p,A}}{A_{p,B}} + \frac{M_p\,g}{A_{p,B}}$$

Upon substituting the known values into the right hand side, we get $P_{1,B} = 325.45$ kPa.

(d) The work interaction for the atmosphere may be evaluated as

$$W_{atm} = P_{atm} \left(\Delta V_{atm,A} + \Delta V_{atm,B} \right)$$

Given that the piston assembly moves *down* by 2 m, $\Delta V_{atm,A} = -2 \times A_{p,A}$ and $\Delta V_{atm,B} = 2 \times A_{p,B}$. Therefore, W_{atm} = -786 J.

(e) $W_{piston} = -M_p\, g \left(z_2 - z_1 \right) = $ -200 J.

(f) At any instant, considering the air in cylinder B as the system, we may write (Eqn. 4.2),

$$dE = dU = \overset{0}{\cancel{\delta Q}} - \delta W = -P\, dV$$

where we have set $\delta Q = 0$, since cylinder B is insulated. This expression may be written as

$$
\begin{aligned}
m\, C_v\, dT &= -m\, P\, dv \\
C_v\, dT &= -\frac{RT}{v}\, dv \\
\frac{dT}{T} &= -(\gamma - 1)\frac{dv}{v}
\end{aligned}
$$

Upon integrating, we get $T v^{\gamma - 1} = constant$. This is the equation that describes the adiabatic, fully resisted process undergone by the air in cylinder B. The final pressure in B may now be calculated using

$$P_{2,B} = P_{1,B} \left(\frac{V_{1,B}}{V_{2,B}} \right)^{\gamma}$$

Since $V_{2,B} = V_{1,B} - 2\, A_{p,B} = 0.03194$ m³, we finally get $P_{2,B} = 442.84$ kPa.

(g) Final temperature in B is given as

$$T_{2,B} = \frac{P_{2,B} V_{2,B}}{m_B R} = \frac{442.84 \times 10^3 \times 0.03194}{0.15 \times 288.68} = 327\,\text{K}$$

(h) If we apply first law to the air in cylinder B, we get

$$\Delta E = \Delta U = \overset{0}{\cancel{Q}} - W \Rightarrow W_B = m_B\, C_v \left(T_{2,B} - T_{1,B} \right)$$

Therefore, W_B = -(0.15)(721.7)(327-300) = -2923 J.

$$\left(\frac{A_{p,A}}{A_{p,A}} \right) \quad A_{p,A} \quad A_{p,A}$$

Hence, the final pressure in A, $P_{2,A} = 258.7$ kPa.

(j) Final volume of the air in cylinder A, $V_{2,A} = V_{1,A} + 2 A_{p,A} = 0.10212$ m³.
Final temperature in A may now be calculated as

$$T_{2,A} = \frac{P_{2,A} V_{2,A}}{m_A R} = \frac{258.7 \times 10^3 \times 0.10212}{0.2 \times 288.68} = 458 \, \text{K}$$

(k) If we apply first law to the air in A, we get

$$
\begin{aligned}
\Delta E = \Delta U \;&=\; Q - W \\
\Rightarrow Q_A \;&=\; \Delta U_A + W_A \\
&=\; \Delta U_A - (W_B + W_{piston} + W_{atm}) \\
&=\; m_A C_v (T_{2,A} - T_{1,A}) - (W_B + W_{piston} + W_{atm})
\end{aligned}
$$

where we have used the fact that $W_A + W_B + W_{piston} + W_{atm} = 0$. Upon substituting the values, we finally get $Q_A = 26.714$ kJ.

■ **EXAMPLE 6.5**

Air and N_2 are contained in an insulated piston cylinder apparatus as shown.

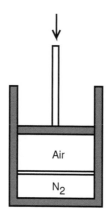

The thin rigid partition that separates the two chambers is perfectly thermally conducting. Initially, the air is at 500 kPa and 473 K and N_2 is at 1000 kPa and they each occupy 0.015 m³. The air is now compressed slowly until the pressure of

The mass of air and N_2 may be determined from the respective equations of state using the given information. Since the partition in the middle is perfectly thermally conducting, the air and N_2 are always in thermal equilibrium. Thus

$$m_{air} = \left. \frac{P_1 V_1}{R T_1} \right|_{air} = \frac{500 \times 10^3 \times 0.015}{288.68 \times 473} = 0.055 \, \text{kg}$$

and

$$m_{N_2} = \left. \frac{P_1 V_1}{R T_1} \right|_{N_2} = \frac{1000 \times 10^3 \times 0.015}{296.93 \times 473} = 0.107 \, \text{kg}$$

N_2 undergoes a constant volume process. Since the final pressure is known, the final temperature may be evaluated from

$$T_2 = \left. \frac{P_2 V_2}{m R} \right|_{N_2} = \frac{1100 \times 10^3 \times 0.015}{0.107 \times 296.93} = 519 \, \text{K}$$

The first law for a system comprising the air, N_2 and the partition may be written as

$$\Delta E = \Delta U = \overset{0}{\cancel{Q}} - W = 0$$
$$\Rightarrow \quad \left(m_{air} C_{v,air} + m_{N_2} C_{v,N_2} \right) (T_2 - T_1) = -W$$

Note that $Q = 0$, since the vessel is insulated. Upon substituting the numerical values, we get

$$W = -(0.055 \times 721.7 + 0.107 \times 742.32)(519 - 473) = -5479.61 \, \text{J}$$

The work interaction for air is -5479.61 J, since the work interaction for N_2 is zero.

If we apply first law to a system that contains the air alone, we get

$$\Delta E = \Delta U = Q - W = 0$$
$$m_{air} C_{v,air} (T_2 - T_1) = Q - W$$

Note that the Q that appears in this expression is the heat interaction between the air and N_2 across the thermally conducting partition. Hence

$$Q = W + m_{air} C_{v,air} (T_2 - T_1) = -3653.71 \, \text{J}$$

Alternatively, if we apply first law to a system that contains the N_2 alone, we get

$$\Delta E = \Delta U = Q - \overset{0, \, V \, = \, const}{\cancel{W}} = 0$$
$$\Rightarrow \quad m_{N_2} C_{v,N_2} (T_2 - T_1) = Q$$
$$\Rightarrow \quad Q = (0.107)(742.32)(519 - 473) = 3653.71 \, \text{kJ}$$

Since, $Q_{air} = -Q_{N_2}$, we get the same value for Q_{air} as before.

kPa and the other side contains 64 kg of O_2 (Mol. Wt = 32 kg/kmol, $\gamma = 1.4$) at 425 K, 500 kPa. The partition is now removed and the gases are allowed to mix while 1000 kJ of energy in the form of electrical work is supplied. Determine the final temperature and the final pressure.

Solution : For CO_2, $R = 8314 \div 44 = 188.95$ J/kg.K and $C_v = R \div (\gamma - 1) = 472.39$ J/kg.K and for O_2, $R = 8314 \div 32 = 259.81$ J/kg.K and $C_v = R \div (\gamma - 1) = 649.53$ J/kg.K.

Initial volume occupied by both the gases may be evaluated as

$$V_{1,CO_2} = \frac{mRT_1}{P_1}\bigg|_{CO_2} = \frac{44 \times 188.95 \times 300}{200 \times 10^3} = 12.4707 \, \text{m}^3$$

and

$$V_{1,O_2} = \frac{mRT_1}{P_1}\bigg|_{O_2} = \frac{64 \times 259.81 \times 425}{500 \times 10^3} = 14.1337 \, \text{m}^3$$

The volume of the vessel is therefore $12.4707 + 14.1337 = 26.6044$ m^3.

If we take the entire contents of the vessel as the system and apply first law, we get

$$\Delta E = \Delta U = \overset{0}{\cancel{Q}} - W$$
$$\Delta U_{CO_2} + \Delta U_{O_2} = -(-1000 \times 10^3) = 1000 \times 10^3$$

where we have set Q to be zero since the vessel is insulated and W = -1000 kJ. Displacement work is zero since the system boundary does not deform during the process. After mixing, the final temperature of the two gases are the same. Hence,

$$m_{CO_2} C_{v,CO_2} (T_2 - T_{1,CO_2}) + m_{O_2} C_{v,O_2} (T_2 - T_{1,O_2}) = 1000 \times 10^3$$

This may be rearranged to give

$$T_2 = \frac{1000 \times 10^3 + m_{CO_2} C_{v,CO_2} T_{1,CO_2} + m_{O_2} C_{v,O_2} T_{1,O_2}}{m_{CO_2} C_{v,CO_2} + m_{O_2} C_{v,O_2}}$$

Upon substituting the known values, we can get the final temperature, $T_2 = 399$ K.

The final mixture pressure is the sum of the partial pressures of CO_2 and O_2. Thus,

$$P_2 = \left(\frac{mRT_2}{V_2}\right)_{CO_2} + \left(\frac{mRT_2}{V_2}\right)_{O_2}$$

$$= \quad 374.062\,\text{kPa}$$

It should be noted that this process is very unlikely to be a quasi-equilibrium process. Fortunately, this does not cause any difficulty since the analysis requires property values at the end state only, which, in any case is an equilibrium state.

▮ EXAMPLE 6.7

Redo the previous problem assuming that the partition is perfectly thermally conducting and free to move without friction inside the vessel. In addition, the electrical work supplied may be taken as zero.

Solution : The initial volumes occupied by CO_2, $V_{1,CO_2} = 12.4707$ m^3 and by O_2, $V_{1,O_2} = 14.1337$ m^3. The total volume occupied by the gases together is thus 26.6044 m^3.

If we take the entire contents of the vessel as the system and apply first law, we get

$$\Delta E = \Delta U = \overset{0}{\cancel{Q}} - \overset{0}{\cancel{W}} = 0 \quad \Rightarrow \quad \Delta U_{CO_2} + \Delta U_{O_2} = 0$$

where we have set Q to be zero since the vessel is insulated and $W = 0$ since the electrical work in this case is given to be zero. As already mentioned, displacement work is zero since the system boundary does not deform during the process. The final temperature of the two gases are the same as the partition is perfectly thermally conducting. Hence

$$T_2 = \frac{m_{CO_2}\, C_{v,CO_2}\, T_{1,CO_2} + m_{O_2}\, C_{v,O_2}\, T_{1,O_2}}{m_{CO_2}\, C_{v,CO_2} + m_{O_2}\, C_{v,O_2}} = 383\,\text{K}$$

When the final equilibrium state is attained, the pressure is the same on both sides of the partition. Since the total volume occupied by the gases also remains the same, we may write $V_{2,CO_2} + V_{2,O_2} = 26.6044$ m^3. If we rewrite this expression using the equation of state, we get

$$\frac{T_2}{P_2}\,(m_{CO_2} R_{CO_2} + m_{O_2} R_{O_2}) = 26.6044$$

Therefore, the final pressure is given as

$$P_2 = \frac{T_2\,(m_{CO_2} R_{CO_2} + m_{O_2} R_{O_2})}{26.6044} = 359.06\,\text{kPa}$$

In this example, the process is a partially resisted process and hence the intermediate states are indeterminate. However, the end states alone are required for the analysis, as in the previous example.

at 1 MPa and 500 K. The right chamber (labelled B) is initially filled with air at 100 kPa, 300 K. The valve connecting the two chambers is now opened allowing the air to flow from the left to the right chamber. The valve is closed when the pressures in both the chambers are equal. Determine (a) the mass of air initially present in each chamber (b) the final pressure and (c) the final temperature of the air in each chamber. What assumptions (if any) have to be made to determine the final temperatures?

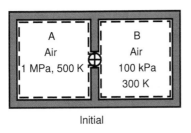

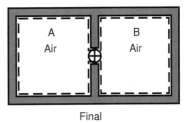

Initial   Final

Solution : (a) The mass of air initially present in each compartment may be determined as

$$m_{1,A} = \left.\frac{P_1 V}{R T_1}\right|_A = \frac{10^6 \times 0.2}{288.68 \times 500} = 1.3856 \text{ kg}$$

and

$$m_{1,B} = \left.\frac{P_1 V}{R T_1}\right|_B = \frac{10^5 \times 0.2}{288.68 \times 300} = 0.2309 \text{ kg}$$

(b) Applying first law to the system shown in the figure, we get

$$\Delta E = \Delta U = \cancel{Q}^{0} - \cancel{W}^{0} = 0$$

where we have set $Q = 0$ since the cylinder is insulated and $W = 0$ since all forms of work including displacement work are absent. Upon expanding this expression, we can get

$$m_{2,A} T_{2,A} + m_{2,B} T_{2,B} = m_{1,A} T_{1,A} + m_{1,B} T_{1,B}$$

If we now use the equation of state, this may be simplified as

$$P_2 = \frac{P_{1,A} V_A + P_{2,A} V_B}{V_A + V_B}$$

$$\frac{i_{2,A} \, v_A}{RT_{2,A}} + \frac{i_{2,B} \, v_B}{RT_{2,B}} = \frac{i_{1,A} \, v_A}{RT_{1,A}} + \frac{i_{1,B} \, v_B}{RT_{1,B}}$$

Upon simplification, this may be written as

$$\frac{1}{T_{2,A}} + \frac{1}{T_{2,B}} = \frac{P_{1,A}}{P_2} \frac{1}{T_{1,A}} + \frac{P_{1,B}}{P_2} \frac{1}{T_{1,B}}$$

where we have used $V_A = V_B$. As we have one equation with two unknowns, it is clear that additional information must be drawn from a closer examination of the process.

Given the large pressure difference between the compartments during the initial stage, it may be inferred that the air that enters compartment B undergoes a partially resisted and hence an out-of-equilibrium process. However, the air that finally remains in compartment A, may be safely assumed to have undergone an adiabatic, fully resisted process owing to the presence of the valve. Hence, the process obeys $Pv^\gamma = $ constant. The final temperature of the air in compartment A may thus be evaluated as

$$T_{2,A} = T_{1,A} \left(\frac{P_2}{P_{1,A}} \right)^{(\gamma-1)/\gamma} = 421 \, \text{K}$$

The final temperature of the air in compartment B may be evaluated from the above expression as 536 K.

■ **EXAMPLE 6.9**

Air is contained in an insulated rigid tank of volume 1000 L. The air is initially at 1 MPa, 300 K. A valve on the top of the tank is now opened and the air is allowed to escape slowly into the atmosphere until the pressure in the vessel reaches 100 kPa at which point, the valve is closed. Determine (a) the final temperature of the air inside the tank and (b) the amount of air that escapes.

Solution : (a) A system appropriate for this problem is shown in Fig. 2.5. If we apply first law to this system at an intermediate instant, we get

$$dE = dU = \delta\cancel{Q}^{\,0} - \delta W = -P \, dV$$

where we have set $\delta Q = 0$, since the system and the surrounding air are always at the same temperature and the vessel is insulated. This expression may be written as

$$m \, C_v \, dT = -m \, P \, dv$$
$$C_v \, dT = -\frac{RT}{v} \, dv$$
$$\frac{dT}{T} = -(\gamma - 1) \frac{dv}{v}$$

Note that the mass of the system itself is not required for this analysis.

(b) Mass contained in the tank initially may be calculated as $m_1 = P_1 V \div (R T_1)$ = 11.547 kg. The mass in the tank finally is given as $m_2 = P_2 V \div (R T_2) = 2.235$ kg (this is also the mass of the system considered in part a). Therefore, the mass that escapes is 11.547 - 2.235 = 9.312 kg.

■ EXAMPLE 6.10

Consider the piston cylinder arrangement shown in the figure in which the piston is spring loaded. The top surface of the piston is exposed to the atmosphere at 100 kPa. Initially, the cylinder contains a certain quantity of N_2 at 200 kPa, 27°C and occupies 0.1 m³. At this position, the spring just touches the piston but does not exert any force. The cross-sectional area of the frictionless piston is 0.05 m² and the spring constant is 15000 N/m. The valve is now opened and the N_2 from the line at 800 kPa, 27°C flows into the cylinder until the pressure inside the cylinder reaches the line pressure at which point, the valve is closed. Determine the final temperature of the N_2 inside the cylinder and the amount of N_2 that enters the cylinder.

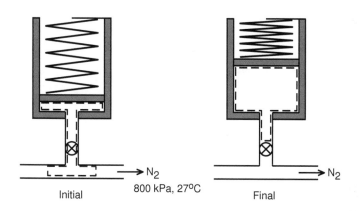

Initial   800 kPa, 27°C   Final

Solution : For N_2, $M = 28$ kg/kmol and $\gamma = 7/5$. Therefore, $R = 8314 \div 28 = 296.93$ J/kg.K and $C_v = R \div (\gamma - 1) = 742.33$ J/kg.K.

The mass of N_2 initially in the cylinder may be evaluated as

$$m_1 = \frac{P_1 V_1}{R T_1} = \frac{200 \times 10^3 \times 0.1}{296.93 \times 300} = 0.2245 \text{ kg}$$

where x is the spring deflection. Therefore, when the pressure finally reaches 800 kPa, the volume occupied by N_2 may be evaluated as

$$V_2 = V_1 + (800 - 200) \times 10^3 \times \frac{A_p^2}{k} = 0.2\,\text{m}^3$$

The thermodynamic system appropriate for this problem is shown in the figure using a dashed line. With respect to this system, let the mass of N_2 initially in the line be m_{line} kg. This is also the mass that enters the cylinder. At the final state,

$$(m_1 + m_{line})\, R\, T_2 = P_2\, V_2 \tag{1}$$

If we apply first law to the system shown, we get

$$\Delta E = \Delta U = \overset{0}{\cancel{Q}} - W$$
$$m_1\, C_v\, (T_2 - T_1) + m_{line}\, C_v\, (T_2 - T_{line}) = -W$$

Here, we have set $Q = 0$ since the cylinder and the piston are both insulated. The displacement work W is the sum of the displacement work in that part of the system that is inside the line and that part next to the piston. Hence,

$$
\begin{aligned}
W &= P_{line}\,(0 - V_{1,line}) + \int_1^2 P\,dV \\
&= -m_{line}\, R\, T_{line} + \int_1^2 \left(200 \times 10^3 + \frac{k}{A_p^2}\,(V - V_1)\right) dV \\
&= -m_{line}\, R\, T_{line} + 200 \times 10^3 \times (0.2 - 0.1) + \frac{15000}{0.05^2}\,\frac{0.1^2 - 0}{2} \\
&= -R\, T_{line}\, m_{line} + 50000
\end{aligned}
$$

Therefore,

$$m_1\, C_v\, (T_2 - T_1) + m_{line}\, C_v\, (T_2 - T_{line}) = -50000 + R\, T_{line}\, m_{line}$$

This may be rearranged to give

$$m_1\, C_v\, T_2 + m_{line}\, C_v\, [T_2 - T_{line} - (\gamma - 1)\, T_{line}] = m_1\, C_v\, T_1 - 50000$$

Or

$$\frac{1}{\gamma - 1}\, \underbrace{R\, T_2\,(m_1 + m_{line})}_{=P_2\,V_2} - \gamma\, m_{line}\, C_v\, T_{line} = m_1\, C_v\, T_1 - 50000$$

Upon substituting the known values, we get m_{line} = 1.283 kg. From Eqn. (1), we may get T_2 = 357 K.

■ EXAMPLE 6.11

A saturated mixture of R134a initially at 100 kPa is contained in a rigid vessel of volume 0.02 m³. The mixture is heated until it attains the critical state. Determine the mass of the mixture and the heat transferred.

Solution : The mixture undergoes a constant volume process. Since the final state is the critical state, the property values are known. From Table E, we can retrieve v_2 = 0.00197 m³/kg and u_2 = 233.9 kJ/kg.

Since $v_1 = v_2$ = 0.00197 m³/kg, the mass of the mixture, $m = V \div v_1$ = 10.15 kg. Taking the contents of the vessel to be the system, if we apply first law, we get,

$$\Delta E = \Delta U = Q - \overset{0}{\cancel{W}} = Q$$

Here, we have set W = 0 since the vessel is rigid and hence the displacement work is zero. Thus,

$$Q = \Delta U = m \left(u_2 - u_1 \right)$$

The initial state is a saturated mixture at 100 kPa and we may retrieve the following property values from Table E: $v_f = 0.726 \times 10^{-3}$ m³/kg, v_g = 0.19255 m³/kg, u_f = 17.19 kJ/kg and u_g = 215.21 kJ/kg. The dryness fraction at the initial state, x_1, can now be evaluated.

$$x_1 = \frac{v_1 - v_f}{v_g - v_f} = 0.0065$$

The specific internal energy at the initial state may be calculated as

$$u_1 = u_f + x_1 \left(u_g - u_f \right) = 18.47713 \, kJ/kg$$

Therefore, the heat supplied, $Q = m \left(u_2 - u_1 \right) = 2186.54$ kJ.

■ EXAMPLE 6.12

A piston cylinder device contains 5 kg of saturated R134a vapor initially at 25°C. The vapor is compressed slowly until the volume is reduced by two-thirds of the initial volume. Heat transfer during the process with the ambient at 25°C keeps the temperature of the contents of the cylinder constant. Determine the final pressure, work done and the heat transferred.

Solution : Initial state is saturated vapor at T_1 = 25°C and mass, m = 5 kg. From Table D, we can retrieve the following property values: specific volume, $v_1 = v_g$ =

compression is accommodated by the reduction in the volume occupied on account of the condensation and the pressure thus remains the same until all the vapor has condensed.

Final specific volume, $v_2 = v_1 \div 3 = 0.0103$ m³/kg. As this is greater than v_f corresponding to 25°C, condensation is not complete and so the pressure remains at the saturation pressure, 665.8 kPa.

The dryness fraction at the final state may be evaluated as

$$x_2 = \frac{v_2 - v_f}{v_g - v_f} = 0.3148$$

where $v_f = 0.8286 \times 10^{-3}$ m³/kg, from Table D. The specific internal energy at the initial state may be calculated as

$$u_2 = u_f + x_1 (u_g - u_f) = 135.552 \text{ kJ/kg}$$

where $u_f = 85.85$ kJ/kg, from Table D.

Since the pressure remains constant during the process, the displacement work may be calculated as

$$W = P_{sat} (V_2 - V_1) = P_{sat}\, m\, (v_2 - v_1) = -68.644 \text{ kJ}$$

If we apply first law to the system consisting of the R134a substance, we get

$$\Delta E = \Delta U = Q - W$$
$$\Rightarrow \quad Q = \Delta U + W = m(u_2 - u_1) + W$$

Upon substituting numerical values into this expression, $Q = -609.084$ kJ. Heat has to be removed to maintain the system at the same temperature.

■ EXAMPLE 6.13

A piston cylinder device contains 2.6 kg of saturated R134a mixture initially at 500 kPa and occupying 0.1 m³. The mixture expands according to the relation $PV = constant$ until it becomes a saturated vapor. Determine the final pressure, work done and heat transferred.

Solution : Given a saturated mixture of R134a at $P_1 = 500$ kPa and $V_1 = 0.1$ m³, and mass, $m = 2.6$ kg. The specific volume at the initial state is $v_1 = V_1 \div m = 0.0385$

where $v_f = 0.806 \times 10^{-3}$ m^3/kg and $v_g = 0.04117$ m^3/kg, from Table E. The specific internal energy at the initial state may be calculated as

$$u_1 = u_f + x_1 (u_g - u_f) = 227.824 \, \text{kJ/kg}$$

where $u_f = 72.92$ kJ/kg and $u_g = 238.77$ kJ/kg, from Table E.

Since the mixture expands according to $PV = constant$, we may write $P_2 v_2 = P_1 v_1$, as the mass remains the same. At first sight, it may appear that we have two unknowns, namely, P_2 and v_2, in this expression. This is not so, since $v_2 = v_g(P_2)$. The final pressure may be obtained by trial and error as $P_2 = 100$ kPa.

The specific volume at the final state is then $v_2 = 0.19255$ m^3/kg and the specific internal energy is $u_2 = 215.21$ kJ/kg.

The work interaction for the process may be obtained from

$$W = \int_1^2 P \, dV = \int_1^2 m P \, dv = m P_1 v_1 \ln \left(\frac{v_2}{v_1} \right) = P_1 V_1 \ln 5$$

Upon substituting the values, we get $W = 80.472$ kJ.

Apply first law to the system consisting of R134a to get,

$$\Delta E = \Delta U = Q - W \Rightarrow Q = m(u_2 - u_1) + W$$

Upon substituting the values, we get $Q = 47.676$ kJ.

It should be noted that, although the process obeys $PV = constant$, the temperature does not remain constant, since the pure substance under consideration is not an ideal gas.

EXAMPLE 6.14

A vertical piston-cylinder arrangement contains 2 kg of water initially at 1 bar. The piston initially rests on stops and will move when the pressure reaches 3 bar. The enclosed volume when the piston is on the stops is 250 L. An amount of heat equal to 4500 kJ is now transferred to the water. Determine the final state and the work done.

Solution : The specific volume at the initial state, $v_1 = V_1 \div m = 0.125$ m^3/kg. From Table B, corresponding to 1 bar, we can retrieve, $v_f = 1.0431 \times 10^{-3}$ m^3/kg and $v_g = 1.6958$ m^3/kg, $u_f = 417.3$ kJ/kg and $u_g = 2505.5$ kJ/kg. Since $v_f < v_1 < v_g$, the

Q

water is initially a saturated mixture.

The dryness fraction at the initial state may be evaluated as

$$x_1 = \frac{v_1 - v_f}{v_g - v_f} = 0.07314$$

The specific internal energy at the initial state may be calculated as

$$u_1 = u_f + x_1(u_g - u_f) = 570.031 \text{ kJ/kg}$$

The heat required to raise the pressure to 3 bar may be calculated by applying first law to a system that consists of the water substance. This gives,

$$\Delta E = \Delta U = Q_{1-2} - \cancel{W}_{1-2}^{\;0}$$
$$\Rightarrow \quad Q_{1-2} = m(u_2 - u_1)$$

where the displacement work has been set to zero since the pressure increase occurs at a constant volume. Hence, at state 2, $P = 3$ bar and $v = 0.125 \text{ m}^3/\text{kg}$. From Table B, corresponding to 3 bar, we can retrieve, $v_f = 1.0731 \times 10^{-3} \text{ m}^3/\text{kg}$ and $v_g = 0.6063 \text{ m}^3/\text{kg}$, $u_f = 561.19 \text{ kJ/kg}$ and $u_g = 2543.4 \text{ kJ/kg}$. Since $v_f < v_2 < v_g$, state 2 is also a saturated mixture state.

The dryness fraction at state 2 may be evaluated as

$$x_2 = \frac{v_2 - v_f}{v_g - v_f} = 0.2048$$

The specific internal energy at state 2 may be calculated as

$$u_2 = u_f + x_2(u_g - u_f) = 967.15 \text{ kJ/kg}$$

Therefore, $Q_{1-2} = 794.238$ kJ. It thus becomes clear that the remainder, namely, 4500-794.2384 = 3705.762 kJ of heat is further added to the system. Note that the pressure remains constant during this process (2-3), since the pressure is entirely due to atmospheric pressure and the weight of the piston, both of which do not change

$$\Delta E = \Delta U = Q_{2-3} - W_{2-3}$$
$$m\,(u_3 - u_2) = Q_{2-3} - m\,P_2\,(v_3 - v_2)$$

If we rearrange this expression, we can get

$$h_3 = u_2 + P_2\,v_2 + \frac{Q_{2-3}}{m}$$

where we have used the fact that $P_3 = P_2$ and the definition of specific enthalpy, $h = u + Pv$. Therefore, $h_3 = 2857.53$ kJ/kg. Since $h_g(3$ bar$) < h_3$, state 3 is a superheated state.

From Table C, the specific volume at state 3 may be obtained by interpolation as 0.7103 m^3/kg. Therefore, the work done

$$W_{1-3} = W_{2-3} = m\,P_2\,(v_3 - v_2) = 351.18\,\text{kJ}$$

■ EXAMPLE 6.15

An insulated rigid vessel of volume 500 L, which is initially evacuated is connected to a line in which steam at 20 bar, 400°C flows, through a valve. The valve is opened and steam is allowed to slowly flow into the vessel until the pressure inside is the same as the line pressure at which point, the valve is closed. Determine (a) the final temperature inside the vessel and (b) the mass that enters. Neglect heat loss and KE and PE changes.

Solution : (a) The appropriate system for this problem is shown in Fig. 2.4. If we apply first law to this system, we get

$$\Delta E = \Delta U = \overset{0}{\cancel{Q}} - W$$
$$m(u_2 - u_1) = -P_{line}\,(V_2 - V_1)_{line}$$
$$= -P_{line}\,m_{line}\,(0 - v_{line})$$

Here, m is the mass in the system and for the system under consideration, $m = m_{line}$. Also $u_1 = u_{line}$. Therefore,

$$m_{line}\,(u_2 - u_{line} - P_{line}\,v_{line}) = 0$$
$$\Rightarrow u_2 = u_{line} + P_{line}\,v_{line} = h_{line}$$

From Table C, corresponding to 20 bar, 400°C, we can get $h_{line} = 3247.5$ kJ/kg. Therefore, $u_2 = 3247.5$ kJ/kg. From Table B, since u_g at 20 bar is 2599.5 kJ/kg,

to the specific enthalpy of the steam in the line - the flow work term in the specific enthalpy is a part of the specific internal energy at the final state.

(b) Final mass in the vessel may be evaluated as $m = V \div v_2 = 0.5 \div 0.1937 = 2.58$ kg. This is the mass that enters the vessel from the line.

■ EXAMPLE 6.16

Consider the frictionless piston cylinder arrangement shown. The cylinder is divided into two compartments A and B by a rigid metal partition which ensures

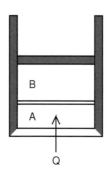

that the temperature is always the same in the two compartments. The piston and all sides of the cylinder except the bottom are well insulated. The volume of compartment A is 0.2023 m³ and it initially contains steam at 5 bar, 180°C. Compartment B initially has a volume of 0.2 m³ and contains 2 kg of a saturated mixture of liquid water and water vapor. Heat is now added to compartment A until the pressure becomes equal in both the compartments. Determine (a) the initial pressure in compartment B, (b) the dryness fraction of the mixture initially in compartment B (c) the mass contained in compartment A and (d) the final pressure and temperature. Calculate the heat transferred.

Solution : Given mass of steam in compartment B, $m_B = 2$ kg and volume, $V_{1,B} = 0.2\,\mathrm{m}^3$.

(a) The initial specific volume in compartment B is $v_{1,B} = V_{1,B} \div m_B = 0.1\,\mathrm{m}^3/\mathrm{kg}$. Furthermore, $T_{1,B} = T_{1,A} = 180°C$. For this value of temperature and specific volume, the substance is a saturated mixture. The initial pressure in compartment B is thus equal to the saturation pressure corresponding to 180°C, namely, 10.02 bar,

(b) The initial dryness fraction may be calculated as

$$x_{1,B} = \frac{v_{1,B} - v_f}{v_g - v_f} = 0.513$$

(c) Given $P_1 = 5$ bar and $T_1 = 180°C$ in compartment A. For this value of pressure and temperature, the state is superheated. Accordingly, the following property values may be retrieved from Table C: $v_{1,A} = 0.4045$ m³/kg, $u_{1,A} = 2609.5$ kJ/kg.

Since V_A is given to be 0.2023 m³, mass contained in compartment A, $m_A = V_A \div v_{1,A} = 0.5$ kg.

(d) The pressure in compartment B at any instant is the sum of the atmospheric pressure and the pressure due to the weight of the piston. Since these two remain the same, the pressure in compartment B remains constant. Hence, $P_{2,B} = P_{1,B} = 10.02$ bar. Further, it is given that the final pressure is the same in both the compartments. Hence, $P_{2,A} = P_{2,B} = 10.02$ bar.

Since the volume of compartment A remains the same, $v_{2,A} = v_{1,A} = 0.4045$ m³/kg.

The final pressure and specific volume in compartment B are known and hence the final state in compartment B is known. From Table C, it can be determined that the final state in B is superheated. The final temperature in compartment B, $T_{2,B} = 607°C$, from Table C.

The final temperature in compartment A is the same as in B. Note that, since the pressure and temperature in both the compartments are the same, the final states are also the same.

Taking compartments A and B together as the system, application of first law gives

$$\Delta E = \Delta U = Q - W$$

Here, the work interaction is displacement work in compartment B and this may be evaluated as

$$W = P_{1,B}(V_{2,B} - V_{1,B}) = P_{1,B} m_B (v_{2,B} - v_{1,B}) = 610.218 \text{ kJ}$$

The internal energy change in compartments A and B may be evaluated using

$$\Delta U = m_A(u_{2,A} - u_{1,A}) + m_B(u_{2,B} - u_{1,B})$$

The specific internal energy at the final state is, $u_{2,B} = u_{2,A} = 3309.8$ kJ/kg. Hence, $\Delta U = 3576.87$ kJ. Therefore, heat transferred, $Q = \Delta U + W = 4187.088$ kJ.

EXAMPLE 6.17

An insulated vertical cylinder is divided into two parts by a frictionless, insulated piston as shown in the figure. The upper part contains N_2 initially at 100 kPa, 300 K and occupying a volume of 1 m³. The piston initially rests at the bottom of the cylinder. The lower part connected through a valve to a steam line in which steam at 7 bar, 300°C flows. The valve is opened and the steam is allowed to slowly flow into the lower part of the cylinder until the pressure becomes 7 bar in both parts of the cylinder. Determine the final temperature of N_2 and the steam in the cylinder and the mass of steam that enters. Gravity effect may be neglected.

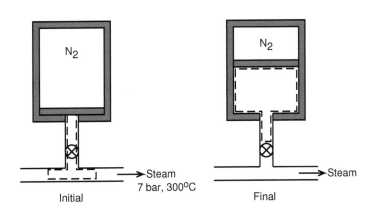

Solution : For N_2, $M = 28$ kg/kmol and $\gamma = 7/5$. Therefore, $R = 8314 \div 28 = 296.93$ J/kg.K and $C_v = R \div (\gamma - 1) = 742.32$ J/kg.K.

The mass of N_2 initially in the cylinder may be evaluated as

$$m_{N_2} = \left.\frac{P_1 V_1}{R T_1}\right|_{N_2} = \frac{100 \times 10^3 \times 1}{296.93 \times 300} = 1.1225 \text{ kg}$$

N_2 undergoes a fully resisted, adiabatic compression process. It has already been shown in Example 6.4 that a perfect gas undergoing such a process obeys the relation $PV^\gamma = constant$. Since the final pressure of N_2 is known, the final volume

The final temperature of N_2 is given by

$$T_{2,N_2} = \left. \frac{P_2 V_2}{m R} \right|_{N_2} = 525 \text{ K}$$

Considering N_2 as the system, if we apply first law, we get

$$\Delta E = \Delta U = \cancelto{0}{Q} - W$$
$$\Rightarrow \quad W_{N_2} = -m C_v (T_2 - T_1) = -187.482 \text{ kJ}$$

Note that work is done on the N_2 by the steam.

Property values for the steam flowing in the line may be retrieved from Table C: $v_{line} = 0.3714$ m^3 and $u_{line} = 2798.6$ kJ/kg. For the steam, the system appropriate for this problem is shown in the figure using a dashed line. Let m_{line} be the mass of steam initially in the line. This is also the mass of steam that enters the cylinder. If we apply first law to this system, we get

$$\Delta E = \Delta U = \cancelto{0}{Q} - W = -W$$

Displacement work occurs at two parts of the system boundary – in the line and next to the piston. Thus,

$$
\begin{aligned}
W &= P_{line} (V_2 - V_1)_{line} + (-W_{N_2}) \\
&= P_{line} (0 - m_{line} v_{line}) + 187.482 \\
&= -259.98 \, m_{line} + 187.482
\end{aligned}
$$

Therefore,

$$m_{line} (u_2 - u_1) = 259.98 \, m_{line} - 187.482$$

Upon substituting for u_1 (since $u_1 = u_{line}$) and rearranging this expression, we get

$$3058.58 \, m_{line} - m_{line} \, u_2 = 187.482 \qquad (1)$$

In the final state, since the mass of the steam in the cylinder is m_{line}, we may write,

$$m_{line} = \frac{V_{steam}}{v_2} = \frac{V_{1,N_2} - V_{2,N_2}}{v_2} = \frac{0.75}{v_2}$$

If we substitute this into Eqn. 1 above, we finally get

$$\frac{(3058.58)(0.75)}{v_2} - \frac{0.75 \, u_2}{v_2} = 187.482$$

$$\frac{3058.58 - u_2}{v_2} = 249.976$$

The final temperature of the steam in the cylinder is 393°C.

Mass of steam that enters the cylinder may be obtained as $0.75 \text{ m}^3 \div 0.435 \text{ m}^3/\text{kg} = 1.7241$ kg.

FIRST LAW ANALYSIS OF CONTROL VOLUMES

In this chapter, first law analysis of steady and unsteady flow processes is presented. The steady flow analysis involves flow devices that are used extensively in practical applications. As these are steady flow devices, the steady flow energy equation, namely, Eqn. 4.7 or 4.8 is used for the analysis. The process occurring in each one of these devices may be viewed as an interplay between different terms of this equation. This is brought out in the the following classification of these devices based on their heat and work interactions.

Devices in the first category convert enthalpy to KE (nozzle), KE to enthalpy (diffuser), transfer enthalpy from one stream to another (mixing chamber, heat exchanger) or cause a change of state without a change in enthalpy (throttling).

Devices in the second category convert enthalpy to useful work (steam and gas turbine), PE to useful work (hydraulic turbine), work to enthalpy (compressor, pump), work to flow work (fan, blower) or work to PE (domestic pump).

Devices in the third category convert heat (energy contained in fuels such as coal and oil) into enthalpy (boiler, combustor) or extract enthalpy as heat (condenser).

$\dot{W}_x = 0, \ \dot{Q}_x \approx 0:$

Nozzle, Diffuser

Mixing chamber, Heat exchanger

Throttling

$\dot{W}_x \neq 0, \ \dot{Q}_x \approx 0:$

$\dot{W}_x > 0:$	Turbine - steam, gas and water
$\dot{W}_x < 0:$	Compressor, Fan, Blower and Pump (liquid)

$\dot{W}_x = 0, \ \dot{Q}_x \neq 0:$

$\dot{Q}_x > 0:$	Boiler, Combustor
$\dot{Q}_x < 0:$	Condenser

7.1 Steady flow analysis

The steady flow energy equation, Eqn. 4.7 is reproduced here for the sake of convenience:

$$\dot{Q} - \dot{W}_x + \dot{m}\left[\left(h_i + \frac{V_i^2}{2} + gz_i\right) - \left(h_e + \frac{V_e^2}{2} + gz_e\right)\right] = 0.$$

It is customary in steady flow analysis to denote the inlet and exit states *1* and *2* respectively, instead of *i* and *e*.

7.1.1 Nozzle

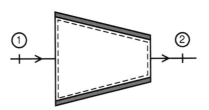

Figure 7.1: Schematic representation of a nozzle

Nozzles are quite extensively used in propulsion applications for generating thrust. A convergent nozzle is used in aircraft engines whereas a convergent-divergent nozzle is used in rocket engines. A convergent nozzle is shown in Fig. 7.1. It is well known from high school physics that when a fluid such as water is forced

flow rate at any section along the length of the nozzle, given as $\dot{m} = \rho A V$, has to be the same, a reduction in the area causes the velocity to increase (density being a constant for water). The function of a nozzle is to convert the enthalpy of the fluid into kinetic energy. This can be understood by applying the SFEE to the CV shown in Fig. 7.1:

$$\left(h_1 + \frac{V_1^2}{2}\right) - \left(h_2 + \frac{V_2^2}{2}\right) = 0$$

Here, we have used that the fact that there is no work interaction and assumed that the heat loss to the ambient and PE change are negligibly small. Since $h = u + Pv$ and the change in internal energy of the fluid (in this case, water) is negligible as the flow remains isothermal, and the density remains constant as the fluid is incompressible, the above expression may be written as

$$\frac{P_1 - P_2}{\rho} + \frac{V_1^2 - V_2^2}{2} = 0$$

Given P_1, V_1, A_1 (or $\dot{m}$) and P_2, this expression may be used to evaluate V_2. The exit area A_2 may be evaluated using the given mass flow rate or by using the fact that the mass flow rate is the same at inlet and outlet.

In the case of a compressible fluid such as a gas or steam or R134a vapor, the decrease in pressure along the length of the nozzle causes the fluid to expand, in addition. This results in a decrease in the density. Since $\dot{m} = \rho A V$, this reduction in density causes the fluid to accelerate even more in order to pass the same mass flow rate at each section[†]. This is the reason why the exit velocity in such cases can become equal to or exceed the local speed of sound.

▣ EXAMPLE 7.1

Dry, saturated steam at 5 bar enters an adiabatic nozzle with negligible velocity. It leaves at 3 bar with a dryness fraction of 0.97. Assuming steady state operation and neglecting PE changes, determine (a) the exit velocity and (b) the exit area if the mass flow rate is 2 kg/s.

Solution : The SFEE in this case reduces to (with $\dot{Q} = 0$, $\dot{W}_x = 0$, and PE changes neglected),

$$\dot{m}\left[h_1 - \left(h_2 + \frac{V_2^2}{2}\right)\right] = 0$$

[†] In case the decrease in density is more than the increase in the velocity, then the area actually has to *increase* to accommodate the same mass flow rate. This happens when the velocity of the fluid is supersonic *i.e.*, exceeds the local speed of sound.

At state 1, $h_1 = 2748.6$ kJ/kg, from Table B. At state 2, $v_f = 1.0731 \times 10^{-3}$ m^3/kg, $v_g = 0.6063$ m^3/kg, $h_f = 561.51$ kJ/kg and $h_g = 2725.2$ kJ/kg, from Table B. Since $x_2 = 0.97$,

$$h_2 = h_f + x_2 (h_g - h_f) = 2660.29 \text{ kJ/kg}$$

and

$$v_2 = v_f + x_2 (v_g - v_f) = 0.5881 \text{ m}^3/\text{kg}$$

(a) Therefore, $V_2 = 420$ m/s.

(b) The mass flow rate at the exit, $\dot{m} = A_2 V_2/v_2$. Upon substituting the numerical values, we get the exit area A_2 to be 28 cm^2.

■ EXAMPLE 7.2

Steam at 7 bar, 260°C enters an adiabatic nozzle with negligible velocity. The mass flow rate is 0.076 kg/s. Assume steady state operation and neglect PE change. Determine the diameter of the nozzle at (a) a location where the pressure is 3.8 bar and specific volume is 0.55 m^3/kg and (b) the exit where the pressure is 1 bar and the dryness fraction is 0.96.

Solution : At state 1, $h_1 = 2974.2$ kJ/kg, from Table C. It can be established that the steam is superheated at the intermediate location denoted i where $P_i = 3.8$ bar and $v_i = 0.55$ m^3/kg. From Table C, we can get $h_i = 2840$ kJ/kg by interpolation.

(a) From SFEE, the velocity $V_i = \sqrt{2(h_1 - h_i)} = 518$ m/s. The area at this location may be calculated from $\dot{m} = A_i V_i \div v_i$. Substituting the numerical values, we get $A_i = 8.0695 \times 10^{-5}$ m^2. The diameter is thus 10.14 mm.

(b) We can calculate the exit velocity V_2 in a similar manner using the expression $V_2 = \sqrt{2(h_1 - h_2)}$. At the exit, $v_f = 1.0431 \times 10^{-3}$ m^3/kg, $v_g = 1.6958$ m^3/kg, $h_f = 417.4$ kJ/kg and $h_g = 2675.1$ kJ/kg, from Table B. Since $x_2 = 0.96$,

$$h_2 = h_f + x_2 (h_g - h_f) = 2584.79 \text{ kJ/kg}$$

and

$$v_2 = v_f + x_2 (v_g - v_f) = 1.628 \text{ m}^3/\text{kg}$$

Therefore, $V_2 = 882.51$ m/s. The exit area comes out to be 1.40122×10^{-4} m^2 and the exit diameter is 13.36 mm. Note that the exit diameter in this case is actually greater than the diameter at the intermediate location, demonstrating the nozzle to be diverging beyond this location!

C (M=44 kg/kmol, γ=4/3) by volume expands adiabatically in a nozzle from 700 kPa, 727°C, negligible velocity to 350 kPa. Assume steady state operation and neglect PE changes. Determine (a) the exit temperature given that the compression process obeys $Pv^\gamma = constant$, where γ is the ratio of the specific heats of the mixture (b) the exit velocity and (c) the exit area, if the mass flow rate is 5 kg/s.

Solution : Since the mole fractions of the individual components are given, the mixture molecular weight may be evaluated using Eqn. 5.7 as

$$M = \sum_i y_i M_i = 0.3 \times 40 + 0.4 \times 28 + 0.3 \times 44 = 36.4 \, \text{kg/kmol}$$

The mixture gas constant, $R = \mathcal{R}/M = 228.41$ J/kg.K.

The mixture molar specific heat may be evaluated using

$$
\begin{aligned}
\bar{C}_p = \sum_i y_i \bar{C}_{p,i} &= \sum_i y_i \frac{\gamma}{\gamma - 1} \mathcal{R} \\
&= \left(0.3 \times \frac{5}{2} + 0.4 \times \frac{7}{2} + 0.3 \times \frac{4}{1} \right) \mathcal{R} \\
&= 3.35 \, \mathcal{R} \, \text{J/kmol.K}
\end{aligned}
$$

Hence, $\gamma/(\gamma - 1) = 3.35$ from which $\gamma = 1.4255$ for the mixture. Also, the mixture C_p is given as $\bar{C}_p/M = 765.162$ J/kg.K.

(a) Since the process obeys $Pv^\gamma = $ constant, we have

$$\frac{T_2^\gamma}{P_2^{\gamma-1}} = \frac{T_1^\gamma}{P_1^{\gamma-1}}$$

from which

$$T_2 = T_1 \left(\frac{P_2}{P_1} \right)^{(\gamma-1)/\gamma} = 813 \, \text{K}$$

(b) Application of SFEE for this case gives (with $\dot{Q} = 0$, $\dot{W}_x = 0$, and PE changes neglected),

$$V_2 = \sqrt{2\,(h_1 - h_2)} = \sqrt{2\,C_p\,(T_1 - T_2)} = 535 \, \text{m/s}$$

(c) The mass flow rate at the exit,

$$\dot{m} = \rho_2 A_2 V_2 = \frac{P_2}{R\,T_2} A_2 V_2$$

from which the exit area A_2 may be evaluated as 49.6 cm².

Figure 7.2: Schematic representation of a diffuser

Diffusers (Fig. 7.2) are quite extensively used in the intake of engines for aircrafts and missiles that fly at supersonic speeds. The function of the diffuser is to decelerate the air that enters the engine and convert the KE of the incoming air stream into enthalpy (or equivalently, the momentum of the air into a pressure rise). The diffuser acts like a nozzle in reverse, in that, the air is forced to flow through the device by its momentum and the diverging passage causes it to decelerate resulting in a pressure increase. At supersonic flight speeds, a substantial amount of compression can be accomplished in this manner. Consequently, the work required for further compression (if any) can be minimized or eliminated altogether. However, diffusion (as a result of deceleration) is quite difficult to accomplish in practice in a stable manner owing to the adverse pressure gradient.

◼ EXAMPLE 7.4

While cruising at an altitude of 24000 m and a flight speed of 972 m/s, which is 3.2 times the local speed of sound, the diffuser in the engine of a SR-71 Blackbird captures the outside ambient air which is at 2.55 kPa, -51°C. The air is decelerated in the diffuser to accomplish a pressure ratio of 39. Assume steady state operation. Neglect PE changes and any heat loss. Determine (a) the final temperature assuming that the compression obeys $Pv^\gamma = constant$ and (b) the final speed.

Solution : For air, M = 28.8 kg/kmol and γ = 7/5. Therefore, R = 8314÷28.8 = 288.68 J/kg.K and $C_p = \gamma R \div (\gamma - 1)$ = 1010.38 J/kg.K.

(a) Since the process obeys Pv^γ = constant, we have

$$\frac{T_2^\gamma}{P_2^{\gamma-1}} = \frac{T_1^\gamma}{P_1^{\gamma-1}}$$

from which

$$T_2 = T_1 \left(\frac{P_2}{P_1}\right)^{(\gamma-1)/\gamma} = 632\,\mathrm{K}$$

Or

$$\dot{m}\left[\left(C_p T_1 + \frac{V_1^2}{2}\right) - \left(C_p T_2 + \frac{V_2^2}{2}\right)\right] = 0$$

Hence, the final speed is given as

$$V_2 = \sqrt{V_1^2 + 2\,C_p\,(T_1 - T_2)} = 341\,\mathrm{m/s}$$

In the actual engine, the deceleration is accomplished quite differently on account of weight constraint which demands as short an intake as possible. The deceleration process is much more rapid and effective though not as efficient as the process that obeys $Pv^\gamma = constant$.

7.1.3   Mixing chamber

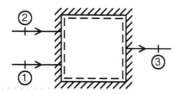

Figure 7.3: Schematic representation of a mixing chamber

A mixing chamber, as the name suggests, is a chamber in which a high enthalpy and a low enthalpy stream are mixed to get an intermediate enthalpy stream. Since the streams are physically mixed, they are both at the same pressure and so the exiting stream is also at the same pressure. Mixing chambers are used extensively in engineering applications for waste heat recovery. They are also used in steam power plants under the name of open feedwater heaters where a part of the steam undergoing expansion in a turbine is extracted and mixed with water going to the boiler. The exiting stream is them pumped into the boiler. Although the extraction of a part of the steam from the turbine decreases the power produced (see Example 7.8 below), the heating of the feedwater in the mixing chamber reduces the heat required to be added in the boiler, resulting in an increase in the efficiency. This strategy is called regenerative heating and is discussed in detail later.

determine the ratio of mass flow rates of the incoming streams. Neglect KE and PE changes.

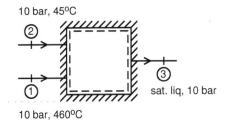

10 bar, 45°C

sat. liq, 10 bar

10 bar, 460°C

Solution : Equation 4.8 when applied the CV shown in the figure, simplifies as follows:

$$\dot{m}_1\, h_1 + \dot{m}_2\, h_2 - \dot{m}_3\, h_3 = 0$$

From Eqn. 4.9, we get $\dot{m}_3 = \dot{m}_1 + \dot{m}_2$. Hence

$$\dot{m}_1\, h_1 + \dot{m}_2\, h_2 - (\dot{m}_1 + \dot{m}_2)\, h_3 = 0$$

Upon rearranging, we get

$$\frac{\dot{m}_1}{\dot{m}_2} = \frac{h_3 - h_2}{h_1 - h_3}$$

From Table C, we can retrieve, $h_1 = 3392.2$ kJ/kg. State 2 is a compressed liquid state. Hence, $h_2(10\,\text{bar}, 45°\text{C}) = u_f(45°\text{C}) + 1000\ v_f(45°\text{C}) = 189.42$ kJ/kg, from Table A. State 3 is a saturated liquid state at 10 bar and so $h_3 = h_f(10\,\text{bar}) = 762.8$ kJ/kg, from Table B. Therefore,

$$\frac{\dot{m}_1}{\dot{m}_2} = 0.2181$$

7.1.4 Throttling

Throttling is a process in which a pure substance at a high pressure and temperature is very quickly converted to a low pressure, low temperature state with enthalpy almost remaining constant. Usually, a porous plug or a partially open valve or a capillary tube is used for this purpose. The process is illustrated in Fig. 7.4.

Application of the SFEE to the CV shown in Fig. 7.4 gives (with $\dot{W}_x = 0$)

$$0 = \dot{m}\left(h_1 + \frac{V_1^2}{2} - h_2 - \frac{V_2^2}{2}\right)$$

Porous plug

Figure 7.4: Schematic representation of throttling using a porous plug

where heat loss to the ambient and PE changes have been neglected. In general, the change in specific volume as a result of throttling is assumed to be small. This, coupled with the fact that the mass flow rate and the cross-sectional area are the same at stations 1 and 2, results in $V_2 \approx V_1$. Consequently, $h_2 = h_1$, suggesting that throttling is an isenthalpic process. In case the change in specific volume is not small, then KE changes cannot be neglected. The SFEE has to be supplemented with the mass conservation equation, $\dot{m} = A_1 V_1 \div v_1 = A_2 V_2 \div v_2$, to determine the exit state. Throttling is employed quite extensively in refrigerators and domestic air-conditioners using capillary tubes as the required change of state can quite easily be accomplished within a small space.

■ EXAMPLE 7.6

Saturated liquid R134a at $30°C$ is throttled to 200 kPa inside a domestic refrigerator. Heat loss to the ambient amounts to 2 kW per unit mass flow rate of refrigerant. Determine the final temperature and dryness fraction (if saturated) of the refrigerant. KE and PE changes may be neglected.

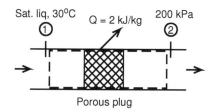

Porous plug

Solution : The specific enthalpy at the inlet may be retrieved from Table D as 93.58 kJ/kg.

The SFEE for this problem simplifies to $h_2 = \dot{Q} + h_1 = $ -2 + 93.58 = 91.58 kJ/kg.

From Table E, corresponding to 200 kPa, h_f = 38.41 kJ/kg and h_g = 244.5 kJ/kg. Since $h_f < h_2 < h_g$, the exit state after throttling is a saturated mixture. Hence, the final temperature is $T_2 = T_{sat}(200\,\text{kPa}) = -10°C$. The dryness fraction may be

Throttling is also used as a means to determine the dryness fraction of a saturated mixture in a device called the throttling calorimeter. Here, a two phase mixture is throttled so that it becomes superheated. By measuring the pressure and temperature of the superheated vapor (both of which are easily measurable), and using the fact that the enthalpy of the two states is the same, the dryness fraction of the initial two phase mixture can be determined. This is illustrated in the following example.

■ EXAMPLE 7.7

A saturated mixture of water at 20 bar is throttled to 1 bar, 120°C in a throttling calorimeter. Determine the dryness fraction of the mixture.

Solution : After throttling, the mixture becomes superheated. From Table C, the specific enthalpy at this state is 2716.3 kJ/kg.

The SFEE applied to this problem gives $h_1 = h_2 = 2716.3$ kJ/kg. From Table B, corresponding to 20 bar, $h_f = 908.62$ kJ/kg and $h_g = 2798.7$ kJ/kg. Hence, the dryness fraction at the initial state is

$$x_1 = \frac{h_1 - h_f}{h_g - h_f} = 0.9564$$

The limitation of this method is that it can be used only in cases where the dryness fraction is high, as otherwise, the mixture remains in the two phase region even after throttling. It should be noted that perfect gases cannot be throttled since their enthalpy depends on temperature alone.

7.1.5 Turbine

The thermodynamic process that occurs in a turbine is the same as in a nozzle, namely, expansion[†]. However, in the case of the turbine, the enthalpy of the fluid is converted into work. There are several different types of turbines depending on how this conversion is accomplished. In the case of a compressible fluid, the axial flow turbine illustrated in Fig. 7.5 is almost exclusively utilized. It is called an axial turbine because the fluid flows along the axial direction from the inlet to the exit. This turbine utilizes two sets of blades - one set (called rotor) attached to the shaft and another set (called stator) attached to the casing. The former set of blades rotate along with the shaft while the latter set of blades is stationary. Conversion of enthalpy

---

[†]This is true only in the case of a compressible fluid. In the case of hydraulic turbines, where water is the working fluid, the decrease in pressure is not accompanied by expansion of the fluid, since water is incompressible.

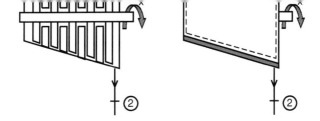

Figure 7.5: Schematic representation of an axial flow turbine

into work takes place in the rotor, whose blades are shaped appropriately so that as the fluid expands in the passage between the blades, the flow pushes against the rotor while being simultaneously turned by the blade surface, thereby doing work. In the case of the turbine, the cross-sectional area increases to accommodate the expanding flow as shown in Fig. 7.5, so that the velocity in the direction of the flow (*i.e.*, along the axis) remains more or less constant. In fact, the passage between the blade surfaces of a rotor is convergent just like a nozzle. Each set of rotor blades is followed by a set of stator blades whose purpose is to undo the turning induced in the flow by the rotor blades and direct the flow smoothly on to the next set of rotor blades. For the present purpose, it is sufficient to know that the fluid does work as it expands inside the turbine.

■ EXAMPLE 7.8

Steam at 60 bar, 680°C enters an insulated turbine that is operating at steady state. Part of the steam is extracted from the turbine at 10 bar, 460°C. The rest is expanded completely and leaves at 45°C, 94 percent dryness fraction. Determine the power developed if the extracted mass flow rate is 15.25 percent of the total. KE and PE changes may be neglected. Also, determine the power developed in the absence of extraction.

Solution : SFEE applied to the CV shown in Fig. 7.5 reduces to (with $\dot{Q} = 0$ and KE, PE changes neglected),

$$0 = -\dot{W}_x + \dot{m}_1\, h_1 - \dot{m}_2\, h_2 - \dot{m}_3\, h_3$$

Here, subscripts 1 and 2 denote inlet and exit as usual and 3 denotes the intermediate outlet. From Table C, we can get $h_1 = 3846.3$ kJ/kg and $h_3 = 3392.2$ kJ/kg. From Table A, corresponding to 45°C, $h_f = 188.42$ kJ/kg and $h_g = 2582.3$ kJ/kg. With $x_2 = 0.94$,

$$h_2 = h_f + x_2\,(h_g - h_f) = 2438.67\,\text{kJ/kg}$$

per unit inlet mass flow rate.

In the absence of extraction, the power developed is given as

$$\dot{W}_x = \dot{m} \left(h_1 - h_2 \right) = 1407.63 \, \text{kW}$$

It can thus be seen that, in the absence of extraction, the power produced is more.

7.1.6 Compressor and pump

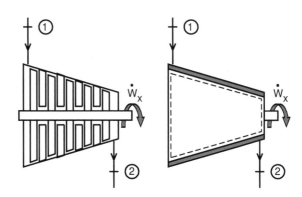

Figure 7.6: Schematic representation of an axial flow compressor

Compressors, as the name suggests, are used for increasing the pressure of a compressible working fluid using external work. Here, the work is used to increase the enthalpy of the fluid. Compressors in wide use today are reciprocating, centrifugal and axial compressors. The choice of which one to use in a particular application depends upon the required pressure ratio and mass flow rate as well as the size. Functionally, these differ significantly in the manner in which they accomplish the compression. A reciprocating compressor (Fig. 2.6) accomplishes this by taking in a fixed quantity of the working fluid and *physically* compressing it using the external work supplied. A centrifugal compressor uses external work to move the fluid continuously in a radially outward direction, in a rotor thereby increasing its pressure. An axial compressor (illustrated in Fig. 7.6 and also in the bottom in Fig. 2.10) uses the diffusion process mentioned earlier to accomplish the compression. The axial compressor also has a rotor and a stator, just like a turbine. However, in this case, the input work is used to diffuse the flow while turning it inside the passage between the appropriately shaped blades, which are such that each rotor blade passage acts just like a diffuser. In the case of the axial compressor, as the fluid

utilized to increase the enthalpy of the fluid.

■ EXAMPLE 7.9

Air at 100 kPa, 27°C enters an insulated compressor steadily at a rate of 1 m³/min, where it is compressed to 1 MPa. Assuming the compression process to obey $Pv^\gamma = constant$, determine the required compressor power. KE and PE changes may be neglected.

Solution : Application of SFEE to the CV shown in Fig. 7.6 leads to (with $\dot{Q} = 0$ and KE, PE changes neglected),

$$0 = -\dot{W}_x + \dot{m}\,(h_1 - h_2)$$

Therefore

$$\dot{W}_x = \dot{m}\,C_p\,(T_1 - T_2)$$

The mass flow rate is the product of the volume flow rate at the inlet and the density at the inlet. Hence,

$$\dot{m} = \rho_1 \times \frac{1}{60} = \frac{P_1}{RT_1}\frac{1}{60} = 0.01924\,\mathrm{kg/s}$$

Since the process obeys $Pv^\gamma = constant$,

$$T_2 = T_1\left(\frac{P_2}{P_1}\right)^{(\gamma-1)/\gamma} = 579\,\mathrm{K}$$

The compressor power required is 5423 W.

■ EXAMPLE 7.10

Refrigerant 134a is compressed steadily in a compressor from 140 kPa, -10°C to 700 kPa, 120°C. The volumetric flow rate at the inlet is 0.06 m³/min. If the heat loss to the ambient is 7 percent of the input power, determine the power required. KE and PE changes may be neglected.

Solution : SFEE for this case reduces to (with KE, PE changes neglected),

$$0 = \dot{Q} - \dot{W}_x + \dot{m}\,(h_1 - h_2)$$

From Table F, we can get $v_1 = 0.1461$ m³/kg, $h_1 = 246.4$ kJ/kg and $h_2 = 358.9$ kJ/kg. The mass flow rate may be evaluated as

$$\dot{m} = \frac{0.06}{v_1} = 0.4107\,\mathrm{kg/min}$$

The power required by the compressor, $|\dot{W}_x| = 49.68$ kJ/min.

▆ EXAMPLE 7.11

Saturated liquid at 45°C is pumped to 20 bar pressure steadily by a pump. Neglecting heat losses, KE and PE changes, determine the power required per unit mass flow rate.

Solution : In this case, the SFEE when applied to a CV that encloses the pump reduces to (with $\dot{Q} = 0$ and KE, PE changes neglected),

$$0 = -\dot{W}_x + \dot{m}(h_1 - h_2)$$

Or

$$\dot{W}_x = \dot{m}(u_1 + P_1 v_1 - u_2 - P_2 v_2)$$

Since the inlet state is a saturated liquid state and the outlet state is a compressed liquid state, the above expression may be simplified further. If we assume that the increase in temperature of the fluid across the pump to be small and in addition, use the fact that water is incompressible and hence $v_2 = v_1$, the above expression becomes

$$\dot{W}_x = \dot{m} v_1 (P_1 - P_2)$$

From Table A, $v_1 = v_f(45°C) = 1.0099 \times 10^{-3}$ m³/kg and $P_1 = P_{sat}(45°C) = 0.09593$ bar. Hence, the required pump power per unit mass flow rate is 2010.11 W per unit mas flow rate.

▆ EXAMPLE 7.12

A pump is to be used for lifting water in a sump to an overhead tank as shown in the figure. The height to which the water is to be raised is 20 m above the pump outlet. The depth from which the water is drawn is 3 m and the surface of the water in the sump is 1 m below the pump inlet. If the volume flow rate is 10 litres/s, determine the power required to run the pump. The inlet and outlet pipe diameters may be assumed to be the same.

Solution : The CV for this problem encloses the inlet pipe, pump and outlet pipe between stations labelled *1* and *2* in the figure. SFEE applied to this problem reduces to ($\dot{Q} = 0$, $V_2 = V_1$, since the pipe diameters are the same and $u_2 = u_1$, since the temperature change is negligible),

$$0 = -\dot{W}_x + \dot{m}\left(\frac{P_1}{\rho} + g z_1 - \frac{P_2}{\rho} - g z_2\right)$$

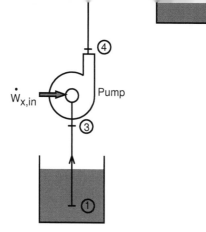

It can be seen from the illustration that $P_2 = P_{atm}$ and $P_1 = P_{atm} + \rho g (3 - 1)$, since the height of the water column above the pipe inlet is 2 m. Hence, the above expression may be simplified to yield (with $\dot{m} = \rho \dot{V}$)

$$\dot{W}_x = \rho \dot{V} g \left[2 + (-3) - 20 \right]$$

where z is measured with respect to the pump. Hence, the pump work is 2.06 kW.

Note that the control volume in the previous problem enclosed just the pump and used the information given at the pump inlet and outlet (stations 3 and 4 in the figure).

7.1.7 Heat exchanger

A heat exchanger is a device that transfers enthalpy from one stream to another *without mixing*. Since there is no mixing, the streams can be at different pressures. Heat exchangers are used extensively for heating a stream as well as for cooling. Different types of heat exchangers, such as parallel flow, counter flow and cross flow, depending on the relative direction of the flow of the streams, are available. A counterflow exchanger is depicted in the following example with air as the stream being cooled and water as the coolant stream. Heat losses from the exchanger itself to the ambient as well as pressure losses within the streams are usually neglected in a thermodynamic analysis.

316.22 kPa to 1000 kPa in the second stage. The air exiting from the first stage is cooled to 27°C in a heat exchanger before entering the second stage. Water at 30°C enters the heat exchanger and leaves at 60°C. Neglecting heat losses to the ambient and KE, PE changes, determine the required mass flow rate of water. Also, determine the total compression work and compare with the value obtained before.

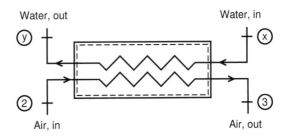

Solution : Let the inlet and exit states of the first compressor stage be denoted *1* and *2*, while those of the second stage be denoted *3* and *4*. We assume that there is no loss of pressure in the heat exchanger (so that $P_3 = P_2$).

Application of SFEE to the first stage compressor leads to (with $\dot{Q} = 0$ and KE, PE changes neglected),
$$0 = -\dot{W}_x + \dot{m}\,(h_1 - h_2)$$

Therefore
$$\dot{W}_x = \dot{m}\,C_p\,(T_1 - T_2)$$

The mass flow rate of air is the same as before, hence $\dot{m} = 0.01924$ kg/s.

Since the process obeys $Pv^\gamma = constant$,
$$T_2 = T_1\left(\frac{P_2}{P_1}\right)^{(\gamma-1)/\gamma} = (300)\,(3.1622)^{2/7} = 417\,\mathrm{K}$$

The compressor power required for the first stage is thus 2274.45 W. The power required for the second stage is the same since the pressure ratio $P_4 \div P_3 = P_2 \div P_1$ and $T_3 = T_1$. Hence, the total power required is $2 \times 2274.45 = 4548.9$ W. This represents a 16 percent *reduction* when compared with single stage compression.

SFEE applied to the heat exchanger (for the CV shown in the figure) reduces to (with $\dot{Q} = 0$, $\dot{W}_x = 0$, and KE, PE changes neglected),
$$0 = \dot{m}_{water}\,(h_x - h_y) + \dot{m}_{air}\,(h_2 - h_3)$$

EXAMPLE 7.14

The exiting stream from the mixing chamber in Example 7.5 is pumped to 60 bar pressure, and then sent to a boiler. It leaves the boiler at 60 bar, 680°C. Determine the heat added per unit mass flow rate.

Solution : The SFEE reduces to, in this case (with $\dot{W}_x = 0$ and KE, PE changes neglected),

$$0 = \dot{Q} + \dot{m}\,(h_1 - h_2).$$

From Table C, the specific enthalpy of the steam at the boiler exit, $h_2 = 3846.3$ kJ/kg. The enthalpy of the compressed liquid stream entering the boiler may be evaluated as,

$$
\begin{aligned}
h_1 &= h_f(10\,\text{bar}) + v_f(10\,\text{bar})(60 - 10) \times 100 \\
&= 762.8 + 1.1272 \times 10^{-3} \times 50 \times 100 \\
&= 768.436\,\text{kJ/kg}
\end{aligned}
$$

Hence, $\dot{Q} = (1)(3846.3 - 768.436) = 3077.864$ kW per unit mass flow rate.

EXAMPLE 7.15

The saturated mixture leaving the turbine in Example 7.8 enters a condenser and leaves as a saturated liquid. Neglecting pressure losses, determine the heat removed. Neglect KE and PE changes.

Solution : The SFEE reduces to, in this case (with $\dot{W}_x = 0$ and KE, PE changes neglected),

$$0 = \dot{Q} + \dot{m}\,(h_1 - h_2)$$

The enthalpy of the saturated mixture leaving the turbine has already been evaluated in Example 7.8 to be 2438.67 kJ/kg. Since the water leaves as a saturated liquid at 45°C, the specific enthalpy of the stream exiting from the condenser is $h_f(45°C)$ = 188.42 kJ/kg, from Table A. Thus, we can evaluate the heat interaction in the condenser to be $\dot{Q} = (1\text{-}0.1525)(188.42 - 2438.67) = -1907.1$ kW per unit mass flow rate entering the turbine.

EXAMPLE 7.16

The saturated liquid leaving the condenser in the previous example is pumped to 60 bar pressure and then sent to the boiler described in Example 7.14. Determine the heat added per unit mass flow rate.

From Table C, the specific enthalpy of the steam leaving the boiler, h_2 = 3846.3 kJ/kg. The enthalpy of the compressed liquid stream entering the boiler may be evaluated as,

$$\begin{aligned} h_1 &= h_f(45°\text{C}) + \text{v}_\text{f}(45°\text{C})(60 - \text{P}_\text{sat}(45°\text{C})) \times 100 \\ &= 188.42 + 1.0099 \times 10^{-3} \times (60 - 0.09593) \times 100 \\ &= 194.47\,\text{kJ/kg} \end{aligned}$$

Hence, $\dot{Q}$ = (1) (3846.3 - 194.47) = 3651.83 kW per unit mass flow rate.

We are now in a position to examine the claim made earlier that extraction from the turbine for use in the mixing chamber results in an increase in the efficiency. The values calculated in Examples. 7.8, 7.14 and 7.16 for the power and heat added for the same mass flow rate, may be summarized as follows:

With extraction : Power = 1262.22 kW; heat added = 3077.864 kW

Without extraction : Power = 1407.63 kW; heat added = 3651.83 kW

It maybe ascertained that, indeed, more power is produced per unit kW of input heat with extraction than without extraction †.

7.2 Unsteady analysis

In this section, examples that require an unsteady analysis are worked out. A few more examples are given later in Chapter 9. It may be recalled that an unsteady analysis is required in cases where, either the mass or the energy or both, inside the control volume change with time during the process.

EXAMPLE 7.17

Repeat Example 6.15 using an appropriate control volume:
An insulated rigid vessel of volume 500 L, which is initially evacuated is connected to a line in which steam at 20 bar, 400°C flows, through a valve. The valve is opened and steam is allowed to slowly flow into the vessel until the pressure inside is the same as the line pressure at which point, the valve is closed. Determine (a) the final temperature inside the vessel and (b) the mass that enters. Neglect heat loss and KE and PE changes.

† Here we have neglected the pump work as it is very small

$$\frac{dE_{CV}}{dt} = \frac{dU_{CV}}{dt} = \dot{m}_i\, h_i$$

The unsteady mass balance equation, Eqn. 4.5, reduces to

$$\frac{dm_{CV}}{dt} = \dot{m}_i$$

Upon combining the two expressions above, we get

$$\frac{dU_{CV}}{dt} = h_i\, \frac{dm_{CV}}{dt}$$

Or

$$\frac{d}{dt}(mu)_{CV} = h_{line}\, \frac{dm_{CV}}{dt}$$

Upon integrating this equation, we get,

$$(m_2\, u_2 - m_1\, u_1) = h_{line}\, (m_2 - m_1)$$

The subscript CV has been dropped as there is no danger of ambiguity. This can be simplified further after noting that the vessel is initially evacuated and so $m_1 = 0$. Thus $u_2 = h_{line}$, which is the same as what we obtained in Example 6.15. The rest of the calculation is identical to that in the earlier example and is not repeated here.

■ EXAMPLE 7.18

Air is contained in an insulated rigid tank of volume 1000 L. The air is initially at 1 MPa, 300 K. A valve on the top of the tank is now opened and the air is allowed to escape slowly into the atmosphere until the pressure in the vessel reaches 100 kPa at which point, the valve is closed. The temperature of the air inside the tank is maintained constant during this process by an electric resistance heater placed inside the tank. Determine (a) the mass of air that escapes and (b) the electrical work done during the process. KE and PE changes may be neglected.

Solution :
(a) Mass of air that escapes may be calculated from

$$m_1 - m_2 = \frac{V}{RT}\, (P_1 - P_2) = 10.4\,\mathrm{kg}$$

(b) The control volume for this case is shown in the figure. For this CV, electrical work and *not heat* crosses the control surface. Accordingly, the unsteady energy

300 K

CV

equation applied to this CV reduces to (with $\dot{Q} = 0$, $\dot{m}_i = 0$ and after neglecting KE and PE terms)

$$\frac{dE_{CV}}{dt} = \frac{dU_{CV}}{dt} = -\dot{W}_x - \dot{m}_e \, h_e$$

The unsteady mass balance equation, Eqn. 4.5, reduces to

$$\frac{dm_{CV}}{dt} = -\dot{m}_e$$

Upon combining the two expressions above, we get

$$\frac{dU_{CV}}{dt} = -\dot{W}_x + h_e \frac{dm_{CV}}{dt}$$

Since $h_e = h_{CV} = C_p T_{CV}$ and the temperature inside the CV is held constant, h_e does not vary with time. Hence, we may integrate the above expression to get

$$m_2 \, u_2 - m_1 \, u_1 = -W_x + C_p \, T \, (m_2 - m_1)$$

Or

$$W_x = C_p T \left(\frac{P_2 \, V}{RT} - \frac{P_1 \, V}{RT} \right) - \left(\frac{P_2 \, V}{RT} C_v \, T - \frac{P_1 \, V}{RT} C_v \, T \right).$$

This may be simplified to give

$$W_x = V \, (P_2 - P_1) = -900 \, \text{kJ}$$

■ **EXAMPLE 7.19**

A rigid vessel initially contains 1 kg of saturated R134a vapor and 15 kg of saturated R134a liquid at 900 kPa. A relief valve is provided on the top of the vessel in order to maintain the pressure inside the vessel constant at 900 kPa by allowing saturated vapor to escape. Heat is now added to the vessel until all the liquid evaporates. Determine (a) the mass of vapor that escapes and (b) the heat supplied.

Solution :

(a) At the initial state, the dryness fraction is given by $x_1 = m_g \div (m_f + m_g) = $

Since saturated vapor fills the entire vessel at the final state and the pressure remains constant at 900 kPa, $v_2 = v_g = 0.0227$ m³/kg and $u_2 = u_g = 248.88$ kJ/kg. Hence, $m_2 = V \div v_2 = 0.0356 \div 0.0227 = 1.57$ kg. The mass of vapor that escapes is $m_1 - m_2 = 16 - 1.57 = 14.43$ kg.

(b) The control volume for this case is as shown in Fig. 2.8(c). The unsteady energy equation, Eqn. 4.4, in this case reduces to (with $\dot{W}_x = 0$, $\dot{m}_i = 0$ and after neglecting KE and PE terms)

$$\frac{dE_{CV}}{dt} = \frac{dU_{CV}}{dt} = \dot{Q} - \dot{m}_e \, h_e$$

The unsteady mass balance equation, Eqn. 4.5, reduces to

$$\frac{dm_{CV}}{dt} = -\dot{m}_e$$

Upon combining the two expressions above, we get

$$\frac{dU_{CV}}{dt} = \dot{Q} + h_e \frac{dm_{CV}}{dt}$$

Since saturated vapor alone is allowed to escape, $h_e = h_g$ and so

$$\frac{d}{dt}(mu)_{CV} = \dot{Q} + h_g \frac{dm_{CV}}{dt}$$

Upon integrating this equation (note that h_g remains constant during the process since pressure remains constant), we get,

$$(m_2 \, u_2 - m_1 \, u_1) = Q + h_g \, (m_2 - m_1)$$

The subscript CV has been dropped as there is no danger of ambiguity.

From Table E, we have $h_g = 269.31$ kJ/kg. If we substitute the numerical values into the above expression, we get $Q = 2515.285$ kJ.

■ EXAMPLE 7.20

A rigid vessel of volume 600 L contains 8 kg of water at 20 bar. Liquid is now allowed to escape slowly from the bottom of the vessel. Heat is transferred to the vessel so as to maintain the pressure constant. The process is stopped when no more liquid remains in the vessel. Determine (a) the mass that escapes and (b) heat transferred.

$$x_1 = \frac{v_1 - v_f}{v_g - v_f} = 0.75$$

From Table B, we have $u_f = 906.27$ kJ/kg and $u_g = 2599.5$ kJ/kg. Thus, $u_1 = u_f + x_1 (u_g - u_f) = 2176.2$ kJ/kg.

Since saturated vapor fills the entire vessel at the final state, $v_2 = v_g = 0.0996$ m³/kg and $u_2 = u_g = 2599.5$ kJ/kg. Hence, $m_2 = V \div v_2 = 0.6 \div 0.0996 = 6.024$ kg. The mass of vapor that escapes is $m_1 - m_2 = 8 - 6.024 = 1.976$ kg.

(b) The control volume for this case is the entire vessel. Proceeding in the same manner as the previous example, we finally arrive at

$$\frac{dU_{CV}}{dt} = \dot{Q} + h_e \frac{dm_{CV}}{dt}$$

Since saturated liquid alone is allowed to escape, $h_e = h_f$ and so

$$\frac{d}{dt}(mu)_{CV} = \dot{Q} + h_f \frac{dm_{CV}}{dt}$$

Upon integrating this equation (note that h_f remains constant during the process since pressure remains constant), we get,

$$(m_2 u_2 - m_1 u_1) = Q + h_f (m_2 - m_1)$$

The subscript CV has been dropped as there is no danger of ambiguity.

From Table B, we have $h_f = 908.62$ kJ/kg. If we substitute the numerical values into the above expression, we get $Q = 45.22$ kJ.

■ EXAMPLE 7.21

A rigid vessel of volume 500 L contains air initially at a pressure of 1 bar. The vessel is connected to a vacuum pump that can extract air at a constant rate of 0.01 m³/min. During the process, heat exchange occurs between the vessel and the surroundings so as to maintain the temperature of the air within the vessel constant. Determine (a) the time taken to reduce the pressure in the vessel to one-fourth of the initial value and (b) the magnitude and sign of the heat interaction between the vessel and the surroundings.

Solution : The control volume in this case may be taken as the vessel. The unsteady energy equation, Eqn. 4.4, in this case reduces to (with $\dot{W}_x = 0$, $\dot{m}_i = 0$ and after

The unsteady mass balance equation, Eqn. 4.5, reduces to

$$\frac{dm_{CV}}{dt} = -\dot{m}_e \qquad (2)$$

(a) This expression may be rewritten as

$$\frac{d}{dt}\left(\frac{PV}{RT}\right) = -\dot{V}_e\,\rho_e = -\dot{V}_e\,\frac{P}{RT}$$

where we have used the fact that the density of the exiting stream at any instant is the same as the density inside the control volume. This simplifies to

$$\frac{dP}{dt} = -\frac{\dot{V}_e}{V}P$$

This may be integrated to give,

$$P = P_1\,e^{-0.02\,t} = e^{-0.02\,t}$$

where t is in minutes and P is in bar.

Time taken to reduce the pressure to one-fourth of the initial value is thus, -50ln 0.25 = 69.31 mins.

(b) If we combine Eqns. (1) and (2) above, we get, after noting that since $h_e = h_{CV} = C_p T_{CV}$ and the temperature inside the CV is constant, h_e does not vary with time,

$$\dot{Q} = \frac{dU_{CV}}{dt} - h_e\,\frac{dm_{CV}}{dt}$$

$$= \frac{d}{dt}\left(\frac{PV}{RT}C_v\,T\right) - C_p T\,\frac{d}{dt}\left(\frac{PV}{RT}\right)$$

$$= \frac{V}{\gamma-1}\frac{dP}{dt} - \frac{\gamma V}{\gamma-1}\frac{dP}{dt}$$

$$= -V\,\frac{dP}{dt}$$

We may integrate the above expression to get $Q = -V\,(P_2 - P_1) = 37.5$ kW.

Air is contained in a rigid tank of volume V initially at a pressure and temperature of P_1 and T_1 respectively. The air is discharged slowly through an insulated turbine (see figure) into the atmosphere that is at a pressure P_{atm}. It may be assumed that the air is always expanded to the atmospheric pressure. The mass of air in the turbine at any instant and its energy may be neglected. Develop an expression for the work developed by the turbine for each of the following cases: (a) the tank is insulated and (b) the air inside the tank remains at the same temperature.

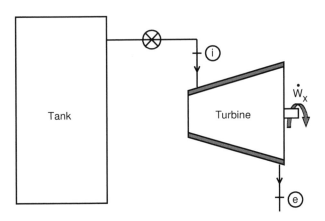

Solution :
(a) Considering the tank and turbine together as the control volume, the unsteady energy and mass balance equation reduce to (with $\dot{Q} = 0$, $\dot{m}_i = 0$ and after neglecting KE and PE terms)

$$\frac{dU_{CV}}{dt} = -\dot{W}_x - \dot{m}_e h_e$$

$$\frac{dm_{CV}}{dt} = -\dot{m}_e$$

Upon combining these two expressions, we get,

$$\frac{d(mu)_{tank}}{dt} = -\dot{W}_x + \frac{dm_{tank}}{dt} h_e$$

where the mass inside the turbine has been neglected.

where we have used the fact that the final pressure in the tank is P_{atm} and thus the final temperature in the tank $T_2 = T_e$, in writing the last equality. Note that, in this case, the exit temperature T_e does not change during the process.

We may now integrate the above expression for work to get

$$W_x = C_p T_e (m_2 - m_1) - C_v (m_2 T_2 - m_1 T_1)$$

where $m = PV/(RT)$ is the mass in the tank at any instant. This may be simplified to yield

$$W_x = \frac{\gamma}{\gamma - 1} P_{atm} V \left(1 - \frac{T_e}{T_1} \frac{P_1}{P_{atm}}\right) - \frac{1}{\gamma - 1} P_{atm} V \left(1 - \frac{P_1}{P_{atm}}\right)$$

Or

$$W_x = \frac{\gamma}{\gamma - 1} P_{atm} V \left[1 - \left(\frac{P_1}{P_{atm}}\right)^{1/\gamma}\right] - \frac{1}{\gamma - 1} P_{atm} V \left(1 - \frac{P_1}{P_{atm}}\right)$$

If the initial pressure and temperature in the tank are assumed to be, say, 200 bar and 300 K respectively, then the final temperature of the air in the tank comes out to be an incredibly low value of 66 K! This motivates the constant temperature case that we discuss next.

(b) Considering the turbine as the control volume, the unsteady energy and mass balance equation reduce to (with $\dot{Q} = 0$ and after neglecting KE and PE terms)

$$0 = -\dot{W}_x + \dot{m}_i h_i - \dot{m}_e h_e$$

$$0 = \dot{m}_i - \dot{m}_e$$

where the mass in the turbine has been neglected. If we let $\dot{m}_i = \dot{m}_e = \dot{m}$, then

$$\dot{W}_x = \dot{m} (h_i - h_e)$$

The expansion process in the turbine may be assumed to follow $Pv^\gamma = constant$. Hence,

$$T_e = T_i \left(\frac{P_e}{P_i}\right)^{(\gamma-1)/\gamma} = T_1 \left(\frac{P_{atm}}{P_i}\right)^{(\gamma-1)/\gamma}$$

Note that, in this case, the exit temperature varies during the process.

$$W_x = \dot{m}\, C_p\, T_1 \left[1 - \left(\frac{P_{atm}}{P_i}\right)\right]$$

The mass flow rate that enters the turbine is equal to the mass flow rate that escapes from the tank. Hence, $\dot{m} = -dm_{tank}/dt$, where $m_{tank} = P_{tank}V \div (RT_{tank})$ is the mass inside the tank at any instant. The negative sign is required on account of the fact that dm_{tank}/dt is negative. Upon substituting this into the expression for work, we are led to

$$\dot{W}_x = -\frac{V}{R}\frac{dP_i}{dt}\, C_p \left[1 - \left(\frac{P_{atm}}{P_i}\right)^{(\gamma-1)/\gamma}\right]$$

This may be simplified as

$$\dot{W}_x = -\frac{\gamma V}{\gamma - 1}\frac{dP_i}{dt}\left[1 - \left(\frac{P_{atm}}{P_i}\right)^{(\gamma-1)/\gamma}\right]$$

Or

$$\dot{W}_x = -\frac{\gamma V}{\gamma - 1}\left[\frac{dP_i}{dt} - P_{atm}^{(\gamma-1)/\gamma}\, P_i^{-(\gamma-1)/\gamma}\frac{dP_i}{dt}\right]$$

This may now be integrated to give

$$W_x = \frac{\gamma}{\gamma - 1}\, P_{atm}\, V \left[-1 + \frac{P_1}{P_{atm}} + \gamma - \gamma\left(\frac{P_1}{P_{atm}}\right)^{1/\gamma}\right]$$

EXAMPLE 7.23

Repeat Example 6.10 using an appropriate control volume:
Consider the piston cylinder arrangement shown in the figure in which the piston is spring loaded. The top surface of the piston is exposed to the atmosphere at 100 kPa. Initially, the cylinder contains a certain quantity of N_2 at 200 kPa, 27°C and occupies 0.1 m³. At this position, the spring just touches the piston but does not exert any force. The cross-sectional area of the frictionless piston is 0.05 m² and the spring constant is 15000 N/m. The valve is now opened and the N_2 from the line at 800 kPa, 27°C flows into the cylinder until the pressure inside the cylinder reaches the line pressure at which point, the valve is closed. Determine the final temperature of the N_2 inside the cylinder and the amount of N_2 that enters the cylinder.

Solution : The control volume appropriate for the analysis is shown in the figure using a dashed line. It is important to note that the control surface adjacent to the

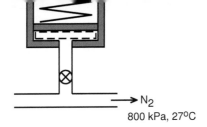

N₂
800 kPa, 27°C

piston *deforms* during the process and hence the work interaction in this case is *displacement work*.

The unsteady energy equation, Eqn. 4.4, in this case reduces to (with $\dot{Q}_x = 0$, $\dot{m}_e = 0$ and after neglecting KE and PE terms)

$$\frac{dE_{CV}}{dt} = \frac{dU_{CV}}{dt} = -\dot{W}_{disp} + \dot{m}_i \, h_i$$

The unsteady mass balance equation, Eqn. 4.5, reduces to

$$\frac{dm_{CV}}{dt} = \dot{m}_i$$

Upon combining the two expressions above, we get

$$\frac{dU_{CV}}{dt} = -\dot{W}_{disp} + h_i \frac{dm_{CV}}{dt}$$

Or

$$\frac{d}{dt}(mu)_{CV} = -\dot{W}_{disp} + h_{line} \frac{dm_{CV}}{dt}$$

Let the piston move up by a distance dx in a time interval dt. The displacement work is thus $\delta W_{disp} = P \, A \, dx = P \, dV$ and the rate at which displacement work is being done is,

$$\dot{W}_{disp} = \frac{\delta W_{disp}}{dt} = \frac{P \, dV}{dt}$$

Therefore,

$$\frac{d}{dt}(mu)_{CV} = -\frac{P \, dV}{dt} + h_{line} \frac{dm_{CV}}{dt}$$

$$=50000 \, J$$

As indicated above, the displacement work has already been shown to be equal to 50000 J in Example 6.10. If we denote the mass that enters the cylinder from the line as m_{line}, then $m_2 = m_1 + m_{line}$. The above expression then simplifies to

$$\gamma \, m_{line} \, T_{line} = -m_1 \, C_v \, T_1 + 50000 + \frac{P_2 \, V_2}{\gamma - 1}$$

which is identical to the expression derived in Example 6.10.

SECOND LAW OF THERMODYNAMICS

In this chapter, the second law of thermodynamics is introduced from the perspective of the performance of devices (engines) that execute a cyclic process. Two different statements of the law are given and discussed. Most importantly, their equivalence is established. Limits imposed by the second law on how best these devices can perform are also discussed.

8.1 Need for the second law

Consider again the example discussed in section 4.1, where 100 J of energy is transferred to the system in three different ways to be converted to work. It was said that when heat is supplied to a system, it is utilized to increase the energy associated with molecular motion, which is disordered. Hence, it cannot be converted entirely into work. In an engineering context, years of efforts to improve the performance of engines that convert heat into work, and run continuously, led to the realization that there is fundamentally a limit on how efficient such engines can be, even under ideal circumstances. Interestingly enough, the first law, Eqn. 4.1, seemed to place no such restriction - the heat supplied could entirely be converted to work without violating

own, however, once mixed, the constituents *never* have been seen to separate out on their own. Air in a high pressure vessel escapes most easily on its own but the reverse does not happen spontaneously. Most, if not all, spontaneous processes[†] in nature always take place in one direction on their own but not in the opposite direction, although the first law does not forbid this from happening.

The above considerations clearly bring out the need for a second law. The interested reader is referred to the books by Peter Atkins in which, many, many more interesting, every day examples for motivating the need for the second law are discussed.

8.2 Heat engines

We start the development of the second law with a description of cyclic devices. Right from the beginning, inventors focussed on developing engines that operate in a cycle and convert heat into work. The emphasis on cyclic operation stemmed from the fact that, such engines can operate continuously and forever (in other words, perpetually). Hence, a lot of effort was put into evaluating novel "engines" that could operate perpetually while seemingly converting a high percentage of the heat input into work. Early inventors realized from practical experience that a collection of devices, each member of which operated at or close to the best possible efficiency, when operated in a cycle, performed at an efficiency far below the efficiency of the individual devices. Thus, the price that had to be paid for cyclic (and hence perpetual) operation was quite high. It became clear that this was inescapable and a fundamental limitation which could not be overcome by improving the performance of the individual devices.

In this context, a proper framework is necessary to determine what exactly the highest efficiency allowed in nature is. In this chapter, answers to the following two questions are sought: (1) what this efficiency cannot be and (2) what it can be. It may seem at first sight that the answer to the second question obviates the need to answer the first question. This is not true, since it will be established that, in reality, the answers to these questions are far apart and remain so even under ideal circumstances.

Following Spalding and Cole, we define a heat engine as a *continuously* operating *thermodynamic system* which has heat and work interactions with the surroundings. A heat engine itself may be power producing (direct engine) or power absorbing

---

[†] These will be called irrversible processes, later on

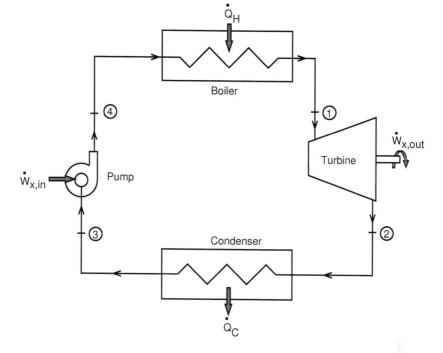

Figure 8.1: Illustration of a basic steam power plant

The block diagram of a basic steam power cycle is illustrated in Fig. 8.1. Water is the working substance in this cycle. Superheated steam at high pressure and high temperature (state *1*) enters the turbine, where it expands and produces work. After expansion, the steam leaves as a two phase mixture (state *2*) at a low pressure and low temperature (typically the ambient temperature) and enters the condenser. Here, it is cooled by water from a river or lake until it becomes a saturated liquid at the same pressure and temperature (state *3*). It is then pumped to a high pressure in the pump and leaves as a compressed liquid (state *4*) and enters the boiler. Heat is added at constant pressure in the boiler until the water becomes high pressure, high temperature superheated steam (state *1*) and the sequence is repeated. This cycle is called the basic Rankine cycle.

It is quite clear that this arrangement operates continuously (since it executes a cyclic process) and is a thermodynamic system since the mass of the working substance inside the device remains the same. Hence, it may be classified as a heat engine and in addition, as a direct engine since the net power production is positive.

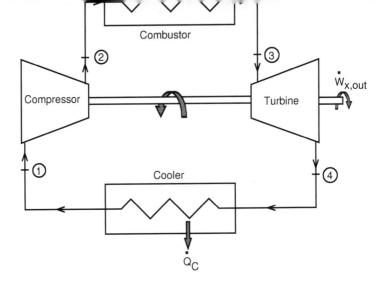

Figure 8.2: Illustration of a basic gas turbine power plant

The block diagram of a basic gas turbine power plant (similar to the one in the bottom in Fig. 2.10) is illustrated in Fig. 8.2. Here, air is the working fluid. Air enters the compressor usually at ambient pressure and temperature (state *1*), where it is compressed to a high pressure (accompanied by an increase in its temperature). The air (state *2*) then enters a combustor where heat is added at constant pressure thereby increasing the temperature of the air[†]. The air (state *3*) undergoes an expansion process in the turbine next. The power produced by the turbine is partly utilized to run the compressor and the remainder is usually used to run a generator to produce electricity. After expansion in the turbine, the low pressure, low temperature air (state *4*) is further cooled at constant pressure in the cooler (which is a heat exchanger) by the ambient air. The air then enters the compressor and the sequence is repeated. This cycle is called the basic Brayton cycle.

[†]This implies that the combustor is actually a heat exchanger. In reality, however, the fuel is mixed with the air and burnt. The combustion products then undergo expansion in the turbine. Since the air is now contaminated with combustion products, it cannot be used again and hence is usually exhausted into the atmosphere. Such an arrangement does not execute a cyclic process and is also not a thermodynamic system since the mass of the working substance increases due to the addition of fuel in the combustor. Hence, it is not a heat engine.

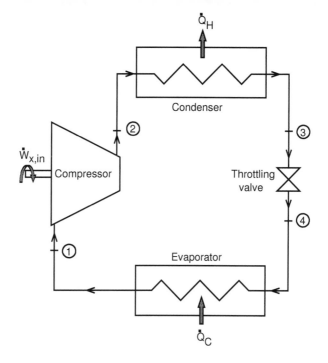

Figure 8.3: Illustration of a basic vapor compression refrigeration cycle

The last example is a power absorbing cycle illustrated schematically in Fig. 8.3. The working substance is a refrigerant such as R134a. Saturated vapor at a low temperature and pressure (state *1*) enters the compressor from the refrigerated space (called evaporator). It is compressed to high pressure and consequently, high temperature. The superheated vapor (state *2*) enters the condenser where it is cooled by the ambient air and exits as a saturated liquid at the same pressure (state *3*). The saturated liquid (at a temperature just above the ambient temperature) is throttled to a low pressure, low temperature saturated mixture state (state *4*). This mixture then enters the evaporator where it absorbs the heat from the refrigerated space. Consequently, it becomes a saturated vapor at the same pressure and enters the compressor next. This sequence of processes is then repeated.

Once again, it is clear that this arrangement operates continuously and is a thermodynamic system as the mass of the refrigerant remains constant. Hence, it is a heat engine. Since the only work interaction is the supply of power to the compressor, it is a reverse heat engine.

in which heat is rejected and a turbine (except the vapor compression cycle) which produces more power than what is required to run the pump or compressor. The vapor compression cycle has a throttling valve in place of the turbine, since the scope for developing work is very limited in this case. Consequently, installing a turbine is not worthwhile, given the increased complexity/size and cost.

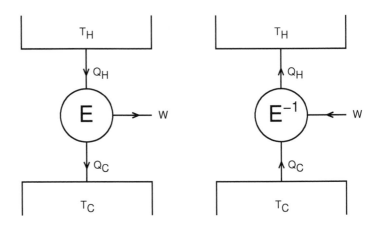

Figure 8.4: Heat engines, direct (left) and reverse (right)

The aforementioned similarities suggest that a heat engine, whether direct or reverse, can be represented simply as shown in Fig. 8.4. The engine interacts with two thermal reservoirs, operates in a cycle and absorbs or produces work. It is customary to have W, Q_H and Q_C as positive values and affix the appropriate sign based on the sign convention that we have adopted. For instance, application of first law to the cyclic process executed by a direct engine would give $W = Q_H - Q_C$, whereas the same for a reverse heat engine would yield $-W = Q_C - Q_H$.

For now, we shall assume that the reservoirs are infinite, so that irrespective of the amount of heat supplied by them or rejected to them, their temperature remains the same. For the three cycles shown here, heat is invariably rejected to the ambient, which can easily be seen as an infinite reservoir[†]. Towards the end of this chapter, examples illustrating the use of finite reservoirs are worked out.

[†]Alas, only until recently! Now the temperature of even this seemingly infinite reservoir has started to increase as a result of human activity.

$$\eta = \frac{\text{Net work output}}{\text{Heat input}} = \frac{W_{net}}{Q_H} = 1 - \frac{Q_C}{Q_H} \tag{8.1}$$

where the last equality comes from the application of the first law for a cyclic process ($W_{net} = Q_{net} = Q_H - Q_C$). The first equality is applicable to any device (whether it executes a cyclic process or not), whereas the second equality is applicable only to heat engines.

Reverse heat engines are used in real life as refrigerators or heat pumps. In the former case, the low temperature reservoir, the refrigerated space, is at a sub-zero temperature whereas the high temperature reservoir, the ambient air, is at, say, 27°C. The objective in this case is to remove as much heat as possible from the refrigerated space for every unit of input work. On the other hand, in the latter application, the low temperature reservoir is the cold ambient at a sub-zero temperature and the high temperature reservoir is a house or dwelling which is to be maintained at, say, 27°C. The objective in this case is to maximize the amount of heat transferred to the house for every unit of input work.

For a reverse engine, the performance metric, called the Coefficient of Performance (COP), is defined as follows:

$$\text{COP}_{\text{refrigerator}} = \frac{\text{Heat removed}}{\text{Work input}} = \frac{Q_C}{W} = \frac{1}{\frac{Q_H}{Q_C} - 1} \tag{8.2}$$

and

$$\text{COP}_{\text{heat pump}} = \frac{\text{Heat supplied}}{\text{Work input}} = \frac{Q_H}{W} = \frac{1}{1 - \frac{Q_C}{Q_H}} \tag{8.3}$$

As before, the second equality in these two expressions follows from application of first law for a cyclic process. It should be noted that, as defined above, the maximum permissible value for η is 1 and that for the COP is ∞.

8.3 Kelvin-Planck and Clausius statements

The Kelvin-Planck statement of the second law may formally be written as follows: "It is impossible to construct a device that operates in a *cycle* and produces no effect other than the *raising of a weight* and exchange of heat with a *single reservoir*".

In other words, the efficiency of a direct heat engine cannot be 100 percent, or, equivalently, some amount of heat must be rejected in order for a direct heat engine

is impossible to construct a device that operates in a *cycle* and produces *no effect other than* the *transfer of heat from a cold to a hot body*".

In other words, the COP of a reverse heat engine cannot be infinity. An informal and quite popular version of this statement is: A refrigerator does not work until it is turned on.

Both the Kelvin-Planck and the Clausius statements emphasize three conditions to be met for the law to be applicable and these are shown above in italics. Even if one of these conditions is not met, then the law is silent on the viability of such a device. Let us examine this through the arrangement shown in Fig. 3.1. The desired effect in this case is the lifting of the weight and the input given is electrical work from the battery and so an efficiency relating these two may easily be defined. Under ideal circumstances, it is not difficult to realize that the efficiency of this arrangement can be quite close to 100 percent. The arrangement has access to a single reservoir (the ambient) and produces useful work (raising of a weight). Does this violate the Kelvin-Planck statement? The answer to this question is that Kelvin- Planck statement is not applicable in this case, since the arrangement does not execute a cyclic process.

One important point about these two statements of the second law as well as the statement of the first law, Eqn. 4.1, is that they are based on extensive observations alone and without any proof. However, no instance of a violation of any of these statements has been reported so far[†].

8.3.1 Equivalence of the two statements

Since the Kelvin-Planck statement is applicable to direct heat engines and the Clausius statement is applicable to reverse heat engines, the possibility that a device may violate one but not the other, if real, would be a vexing dilemma. We now proceed to determine if this can be a real possibility.

[†]There have been instances in the past when it was thought the first and second law were in jeopardy. When nuclear fission was realized, it appeared as if first law was being violated. This was reconciled after taking into account the famous Einstein's mass-energy equivalence, $E = mc^2$. In the second instance, it was shown in 1972, by Hawking and co-workers that, unless black holes emitted radiation, they violated the second law. The dilemma was that, since black holes were "black", nothing, even light, could escape them. However, in 1973, Hawking himself showed that such a radiation was possible owing to quantum mechanical effects and rescued second law. This radiation, appropriately enough, is called the Hawking radiation. Interested readers may see Chapter 7 of the book *A Brief History of Time* by Stephen Hawking. For a more formal discussion of the connection between thermodynamics and black holes, readers may

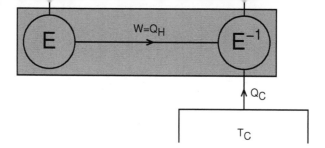

Figure 8.5: Equivalence of the Kelvin-Planck and the Clausius statement. The direct engine E violates the Kelvin-Planck statement. The reverse engine E^{-1} is a legitimate reverse engine.

Figure 8.5 depicts a direct engine on the left that violates the Kelvin-Planck statement (since the heat rejected is zero and hence its efficiency is 100 percent) and a legitimate reverse engine on the right. Both the engines use the same hot reservoir. If we allow the work produced by the former to run the latter, then the net interaction with the hot reservoir during each cycle is Q_C. It can be ascertained that the combined device inside the shaded box, (a) executes a cyclic process (b) requires no work input and (c) transfers an amount of heat Q_C from the cold reservoir to the hot reservoir during each cycle. The combined device thus violates the Clausius statement. This is possible only because a direct engine that violates the Kelvin-Planck statement is available.

Figure 8.6 depicts a reverse heat engine on the right which violates the Clausius statement (since the work input is zero and hence the COP is infinity) and a legitimate direct heat engine on the left. Both of them use the same hot and cold reservoirs. The direct heat engine rejects the exact same amount of heat Q_C to the cold reservoir that the reverse engine removes during each cycle. It is evident that, during each cycle, the combined device inside the shaded box receives a net amount of heat $Q_H - Q_C$ from the hot reservoir and produces an amount of work equal to $Q_H - Q_C$ and hence has an efficiency of 100 percent. This is in violation of the Kelvin-Planck statement and is possible only because a reverse engine that violates the Clausius statement is available.

---

see the article *Thermodynamics of black holes* by P. C. W. Davies that appeared in Reports on Progress in Physics, Vol. 41, 1978, pp. 1314-1355.

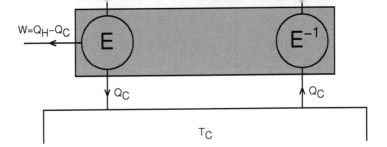

Figure 8.6: Equivalence of the Clausius and the Kelvin-Planck statement. The reverse engine E^{-1} violates the Clausius statement. The direct engine E is a legitimate direct engine.

The equivalence of the two statements is thus established and so the aforementioned dilemma, quite reassuringly, does not arise. The Kelvin-Planck and the Clausius statements answer the first of the two questions posed in the beginning of section 8.2. We now turn to finding the answer to the second question, which is, if the efficiency cannot be 100 percent and the COP cannot be infinity, what can they be?

8.4 Reversible processes and the Carnot cycle

It may be recalled that the notion of a fully resisted or quasi-equilibrium process was discussed in detail in sections 2.3 and 3.2.1. It is quite easy to see that, so long as the departure from mechanical equilibrium in a process is small, the work developed during expansion is the highest possible and the work required in the case of a compression process is the smallest possible. In other words, processes with a non-zero work interaction in a cycle must be quasi-equilibrium or reversible processes.

It follows logically that any process with a non-zero heat interaction must be very close to being in thermal equilibrium in order for it to be reversible. This requires that the temperature difference between the reservoir and the system be infinitesimally small and remain so during the process. In other words, it is desirable to accomplish any heat interaction by means of a reversible isothermal process. This also requires then that all other processes in the cycle be adiabatic. Since the work interaction in an isothermal process is non-zero, it must also be a reversible process as argued in the previous paragraph.

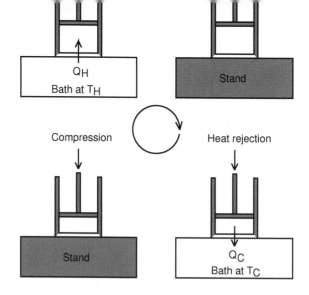

Figure 8.7: Illustration of the various processes in the Carnot cycle

We thus arrive at the following conclusion: the best possible cycle (one with the highest efficiency or COP) must be composed entirely of reversible isothermal and reversible adiabatic processes. Such a cycle is called a Carnot cycle. This is described next.

Consider a piston cylinder apparatus that is completely insulated except for the bottom surface as shown in Fig. 8.7. The cylinder contains a fixed quantity of the working substance that is made to execute a Carnot cycle. In what follows, for the sake of simplicity and without any loss of generality, we assume the working substance to be an ideal gas. The processes undergone by the working substance in the Carnot cycle are shown in Fig. 8.8. Processes *1-2* and *3-4* are reversible isothermal processes and *2-3* and *4-1* are reversible adiabatic processes.

To begin with, the gas in the cylinder is a temperature infinitesimally less than T_H and the bottom surface of the cylinder is brought into contact with the hot reservoir at T_H, as shown in Fig. 8.7. As the gas absorbs the heat, the piston moves outward causing the gas to expand slowly, and hence remain at the temperature T_H (process *1-2* in Fig. 8.8). Once the desired amount of heat, Q_H, has been transferred, the cylinder is kept on an insulated stand, where it undergoes further expansion, until

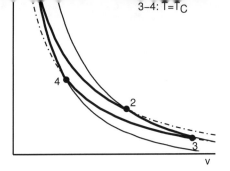

Figure 8.8: Carnot cycle in $P - v$ coordinates

its temperature decreases to the temperature of the cold reservoir, T_C (process *2-3* in Fig. 8.8). The cylinder is then brought into contact with the cold reservoir in the same manner as before. The piston now starts moving inward in such as a manner as to compress the gas slowly so that the temperature remains constant (process *3-4* in Fig. 8.8). Once the required amount of heat, Q_C, has been rejected to the cold reservoir, the cylinder is kept on the insulated stand. The gas is now compressed slowly until its temperature increases to that of the hot reservoir (process *4-1* in Fig. 8.8 and the entire sequence is repeated.

The Carnot engine described above is an appropriate idealization of heat engines in which the working substance undergoes non-flow processes. For instance, a single cylinder in the multi-cylinder engine shown in the top in Fig. 2.10 could be such a heat engine. However, the working substance undergoes steady flow processes in the heat engines described in section 8.2. A Carnot engine that is appropriate for such a situation will employ a reversible isothermal turbine, reversible adiabatic turbine, reversible isothermal compressor and a reversible adiabatic compressor to accomplish processes *1-2*, *2-3*, *3-4* and *4-1* in Fig. 8.8 respectively. The processes remain the same, as they should, since the arguments made at the beginning of this section in this connection are quite general. The realization of the device alone differs, depending on whether the processes that comprise the cycle are required to be flow or non-flow processes.

If the direct Carnot cycle shown in Fig. 8.8 is executed in reverse, then the signs of the heat and work interaction are changed but their magnitudes remain the same.

It was argued earlier that the best possible cycle must be composed entirely of reversible isothermal and reversible adiabatic processes. Strictly speaking, it is sufficient if the processes in the cycle are reversible and they need not be isothermal and adiabatic, in addition. These additional constraints make the the development above easy to understand, and, as will be shown in the next chapter, make the analysis easier as well. Since the Carnot engine is itself an ideal construct, it is not difficult to imagine an engine that executes a reversible cycle that is different from the Carnot cycle. However, we must establish whether such an engine will have the same or a better efficiency than a Carnot engine operating between the same reservoirs.

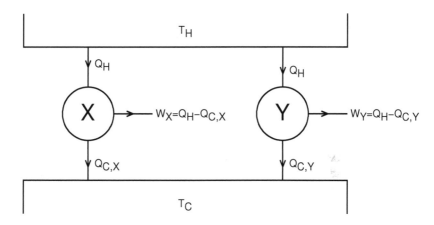

Figure 8.9: X and Y are two reversible engines. Engine X executes a Carnot cycle while engine Y executes a reversible cycle different from a Carnot cycle.

Consider the two engines labelled X and Y shown in Fig. 8.9, that operate between the same reservoirs. Let engine X execute a Carnot cycle and engine Y execute a reversible cycle that is not a Carnot cycle. Let us assume that the efficiency of engine Y is higher than that of X. Since the heat supplied to both the engines is the same, this would require $W_Y > W_X$ or $Q_{C,X} > Q_{C,Y}$. Since X is a reversible engine, let it be run as a reverse engine and let the work required be supplied by engine Y. It is quite easy to establish that in each cycle, reverse engine X and engine Y together extract a net amount of heat equal to $Q_{C,X} - Q_{C,Y}$ from the cold reservoir and develop a net amount of work equal to $Q_{C,X} - Q_{C,Y}$. This is in violation of the Kelvin-Planck statement. It may thus be concluded that *all* reversible engines operating between the same reservoirs have the same efficiency.

is the highest possible. In real life, there are many factors which cause actual engines to depart from this ideal engine and hence suffer a degradation in the performance. These factors are, friction, dissipation, mechanical (partially resisted) and thermal disequilibrium (heat transfer across a finite temperature difference), mixing and chemical reactions, to name a few. Consequently, these factors render a real heat engine irreversible. Given the same input and heat reservoirs, irreversibilities cause a decrease in the desired effect, when compared to a reversible engine.

If the same amount of heat Q_H from a hot reservoir at temperature T_H is given to a reversible and an irreversible direct engine, and they both reject heat to a cold reservoir at temperature T_C, the work developed by a reversible and irreversible engine may be written as

$$W_{rev} = Q_H - Q_{C,rev}; \quad W_{irr} = Q_H - Q_{C,irr}$$

Since $W_{irr} < W_{rev}$, it follows that $Q_{C,irr} > Q_{c,rev}$. In other words, the irreversible direct engine rejects more of the given heat than the reversible one and consequently is less efficient.

The factors (irreversibilities) mentioned above that cause an engine to be irreversible, may be broadly classified into internal and external irreversibilities, depending upon whether a particular irreversibility is inside the engine (system) or outside (surroundings). Accordingly, friction, dissipation, mixing and chemical reactions are internal irreversibilities whereas, partial resistance and heat transfer across a finite temperature difference are external irreversibilities. Thus, a heat engine may be internally reversible but externally irreversible or *vice versa*.

8.5 Thermodynamic scale of temperature

The efficiency of any heat engine is given by Eqn. 8.1, and it depends on Q_H and Q_C. An examination of the Carnot cycle in Fig. 8.8 shows that the *only* parameters that control the cycle are the reservoir temperatures, T_H and T_C. Therefore, the efficiency of a Carnot engine depends only on the reservoir temperatures, *viz.*,

$$\eta_{Carnot} = 1 - \frac{Q_C}{Q_H} = \psi(T_H, T_C),$$

where ψ is an unknown function. It follows then that

$$\frac{Q_C}{Q_H} = 1 - \psi(T_H, T_C) = \phi(T_H, T_C).$$

Consider three Carnot engines labelled I, II and III as shown in Fig. 8.10. Engine I

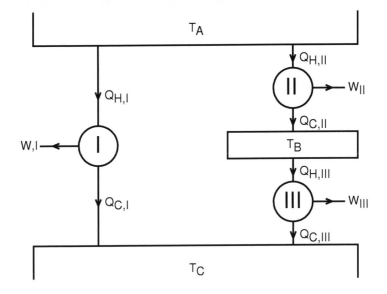

Figure 8.10: Derivation of the absolute or thermodynamic temperature scale

engines I and III reject the same amount of heat to the reservoir at T_C. Engine III receives the same amount of heat from the reservoir at T_B that is rejected by engine II. For each of these engines, we may write,

$$\frac{Q_{C,I}}{Q_{H,I}} = \phi(T_A, T_C); \quad \frac{Q_{C,II}}{Q_{H,II}} = \phi(T_A, T_B); \quad \frac{Q_{C,III}}{Q_{H,III}} = \phi(T_B, T_C)$$

Since, $Q_{H,I} = Q_{H,II}$, $Q_{C,I} = Q_{C,III}$ and $Q_{H,III} = Q_{C,II}$, the above expression may be written as

$$\frac{Q_{C,I}}{Q_{H,I}} = \frac{Q_{C,III}}{Q_{H,III}} \frac{Q_{C,II}}{Q_{H,II}} \quad \Rightarrow \quad \phi(T_C, T_A) = \phi(T_C, T_B) \phi(T_B, T_A)$$

This expression can be satisfied only if $\phi(T_C, T_H)$ is of the form

$$\phi(T_C, T_H) = \frac{\xi(T_C)}{\xi(T_H)}$$

where $\xi(T)$ is an unknown function. Lord Kelvin initially proposed $\xi(T) = e^T$, a temperature scale in which the temperature values range from $-\infty$ to ∞. He later proposed an even simpler functional form for ξ, namely, $\xi(T) = T$. Temperatures in this scale range from 0 to ∞. This has come to be known as the Kelvin or

0 K corresponds to -273.15° C – an assertion that has withstood the test of time!

The efficiency of a Carnot engine, (Eqn. 8.1), may now be written as

$$\eta_{\text{Carnot}} = 1 - \frac{Q_C}{Q_H} = 1 - \frac{T_C}{T_H} \tag{8.4}$$

In the same manner, for a reverse Carnot engine, we may write Eqns. 8.2 and 8.3 as

$$\text{COP}_{\text{refrigerator, Carnot}} = \frac{1}{\frac{Q_H}{Q_C} - 1} = \frac{1}{\frac{T_H}{T_C} - 1} \tag{8.5}$$

and

$$\text{COP}_{\text{heat pump, Carnot}} = \frac{1}{1 - \frac{Q_C}{Q_H}} = \frac{1}{1 - \frac{T_C}{T_H}} \tag{8.6}$$

The above expressions provide the answer to the second question posed in section 8.2, namely, what the maximum possible efficiency of a direct heat engine or COP of a reverse heat engine can be. For instance, if a direct heat engine operates between reservoirs at 1200 K and 300 K, then the efficiency of such an engine cannot exceed 75 percent. This is a profound assertion, since the Kelvin-Planck statement forbids the efficiency of this engine from being 100 percent and it emerges now that the same cannot even be higher than 75 percent! Since the Carnot engine is itself an ideal construct, it is not possible to improve upon its efficiency and approach 100 percent (with the same reservoir temperatures). In other words, even the Carnot engine, which is composed of the best possible device(s) can only achieve an efficiency which is far below 100 percent. From Eqn. 8.4, it may be inferred that a temperature of 0 K is not attainable, for then we could construct a Carnot engine that will have an efficiency of 100 percent, violating the Kelvin-Planck statement[†].

It may be recalled that, towards the end of section 2.4, it was mentioned that there is a fundamental difficulty in constructing a thermometer to measure and give values for temperature that are independent of the thermometric substance or calibration relation. Equation 8.4 suggests that the Carnot engine may be thought of as a thermometer that does not suffer from these shortcomings, since its efficiency does not depend on the working substance. By measuring the efficiency of a Carnot engine that uses the body whose temperature is to be measured as one reservoir and another reservoir at a known temperature, the unknown temperature may be evaluated using Eqn. 8.4.

[†] The current record for the lowest temperature achieved is 500 picoKelvin by a team of scientists from MIT in September 2003.

Which of the following two choices is the best way to increase the thermal efficiency of a reversible engine operating between T_H and T_C? (a) Increase T_H while keeping T_C constant, or (b) Decrease T_C while keeping T_H constant?

Solution : The efficiency of the reversible engine is given as

$$\eta = 1 - \frac{T_C}{T_H}$$

If we take the total derivative of this expression, we get

$$
\begin{aligned}
\Delta\eta &= \frac{\partial\eta}{\partial T_H}\Delta T_H + \frac{\partial\eta}{\partial T_C}\Delta T_C \\
&= \frac{T_C}{T_H^2}\Delta T_H - \frac{1}{T_H}\Delta T_C \\
&= \frac{1}{T_H}\frac{T_C}{T_H}\Delta T_H - \frac{1}{T_H}\Delta T_C
\end{aligned}
$$

Each term on the right hand side of this expression gives the change in the efficiency due to a given change in the reservoir temperature. For an increase in the efficiency ($\Delta\eta > 0$), it is clear that T_H must increase ($\Delta T_H > 0$) or T_C must decrease ($\Delta T_C < 0$). For the same change in temperature of the reservoir, $\Delta T_H = \Delta T_C$, it can be inferred from the coefficients in the above expression that decreasing the cold reservoir temperature is the better strategy, since $T_C < T_H$.

In reality, however, the lowest possible value for T_C is the ambient temperature and so it is not possible to adopt the above strategy. Instead, the trend has been to increase T_H as much as metallurgical limitations would allow.

■ EXAMPLE 8.2

A house is to be maintained at 27°C when the outside ambient temperature is -5°C. Two strategies are being considered: (a) operate a 1 kW space heater for 1 hour and (b) supply the same work to a reversible heat pump that operates between the ambient and the house. In each case, determine the amount of heat that is transferred to the house.

Solution :
(a) When the 1 kW heater is operated for an hour, the amount of heat transferred to the house is simply 3600 kJ. The COP of the heater is thus equal to 1, which is quite poor.

$$1 - \frac{T_C}{T_H} \qquad 1 - \frac{\ldots}{27+273}$$

Hence, the heat transferred to the house, $Q_H = 9.375 \times 3600 = 33750$ kJ!

■ EXAMPLE 8.3

A seawater desalination plant using solar energy is being proposed. The plant collects 9.9×10^9 J of solar energy during a 10 hour period at an average temperature of $100°C$ and rejects heat to the ambient at $25°C$. It is claimed that the plant can deliver anywhere between 1000 L/sec to 1000 L/min of fresh water. Evaluate this claim. What is the maximum possible yield for this plant? It is known that, in an ideal process, seawater has to be pumped through a pressure difference of 27 bar, to obtain fresh water.

Solution : From Example 7.11, the work required is given by,

$$W = \dot{m} v \, \Delta P = \dot{V} \, \Delta P$$

where $\dot{V}$ is the volume flow rate and ΔP is the pressure difference, which, in this case is equal to 27 bar.

If we assume that a Carnot engine is used, then the work produced by the engine is given as

$$W = \eta \, Q_H = 9.9 \times 10^9 \left(1 - \frac{T_C}{T_H} \right) = 9.9 \times 10^9 \left(1 - \frac{25+273}{100+273} \right) = 1.991 \times 10^9 \text{ J}$$

Therefore, the maximum possible volume flow rate of desalinated water that can be delivered is given as

$$\dot{V} = \frac{W}{\Delta P} = \frac{1.991 \times 10^9}{27 \times 10^5} = 737.27 \, \frac{\text{m}^3}{\text{day}} = 0.512 \, \frac{\text{m}^3}{\text{min}} = 512 \, \frac{\text{L}}{\text{min}}$$

It can thus be seen that, even an ideal reversible engine can produce only 512 L/min of fresh water.

■ EXAMPLE 8.4

Two reversible power cycles, labelled I and II, are arranged in series. The first cycle receives energy from a reservoir at T_H and rejects energy to a reservoir at an intermediate temperature T. The second cycle receives the energy rejected by the first cycle to the reservoir at T and rejects energy to a reservoir at a temperature T_C lower than T. Derive an expression for the intermediate temperature T in terms of T_H and T_C, when (a) the net work of the two cycles are equal; (b) the thermal efficiencies of the two cycles are equal.

Or

$$Q_{H,I} = Q_{C,I}\frac{T_H}{T}; \quad Q_{C,II} = Q_{H,II}\frac{T_C}{T}$$

(a) From first law, the net work produced by each cycle is given as $W_I = Q_{H,I} - Q_{C,I}$ and $W_{II} = Q_{H,II} - Q_{C,II}$. As these are given to be equal, we may write

$$Q_{C,I}\left(\frac{T_H}{T} - 1\right) = Q_{H,II}\left(1 - \frac{T_C}{T}\right)$$

Since it is given that $Q_{C,I} = Q_{H,II}$, the above expression leads to

$$T = \frac{T_H + T_C}{2}$$

The temperature of the intermediate reservoir must be the arithmetic mean of the temperatures of the other two reservoirs in order for the work produced by the two power cycles to be equal.

(b) Since the cycles are reversible, and their efficiencies are given to the equal, we have

$$1 - \frac{T}{T_H} = 1 - \frac{T_C}{T}$$

It follows then, that, for the efficiencies to be the same, the temperature of the intermediate reservoir must be geometric mean of the temperatures of the other two reservoirs, i.e., $T = \sqrt{T_H T_C}$.

■ EXAMPLE 8.5

A reversible engine operates between reservoirs at 1000 K and 500 K. The entire work delivered by this engine is used to operate a reversible heat pump that operates between the reservoir at 500 K and one at a temperature T (1000 K $> T > 500$ K). The heat removed by the pump and that received by the engine are equal. Determine the temperature T.

Solution : Let the reservoirs be respectively labelled A, B and C in descending order of temperature. For the engine, since it is reversible,

$$\frac{Q_A}{1000} = \frac{Q_C}{500} \Rightarrow Q_C = \frac{1}{2}Q_A$$

The work developed by the engine per cycle is given as, $W = Q_A - Q_C = Q_A / 2$.

Since the work input for the heat pump is equal to the work developed by the engine, we may write

$$Q_B - Q_C' = \frac{Q_A}{2} \implies \left(\frac{T}{500} - 1\right) Q_C' = \frac{Q_A}{2}$$

Since it is given that $Q_A = Q_C'$, it follows that $T = 750$ K.

▣ EXAMPLE 8.6

An irreversible heat engine, X, takes heat from a reservoir at 1000 K and rejects heat to the ambient at 300 K. A part of the work output from the engine is used to drive an irreversible refrigerator, Y, which operates between a reservoir at 280 K and the ambient. The refrigerator removes 240 kJ of heat per cycle from the reservoir at 280 K and the total heat input to the ambient is 800 kJ per cycle. The efficiency of the heat engine and the COP of the refrigerator are respectively 75 percent of the values of a reversible engine and a reversible refrigerator operating between the same temperatures. Determine (a) work input required for the refrigerator per cycle and (b) the heat input to the engine.

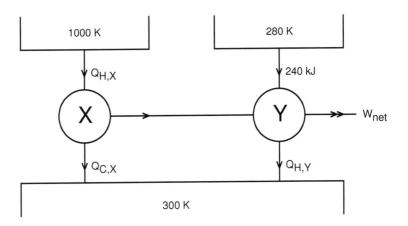

Solution : The schematic of the combined engine-refrigerator is shown in the figure. It is given that the efficiency of the heat engine and the COP of the refrigerator are respectively 75 percent of the values of a reversible engine and a reversible refrigerator operating between the same temperatures. Hence,

$$\eta_X = 1 - \frac{Q_{C,X}}{Q_{H,X}} = \frac{3}{4}\left(1 - \frac{300}{1000}\right) = 0.525 \implies Q_{H,X} = \frac{Q_{C,X}}{0.475}$$

Since $Q_{C,Y}$ is given to be 240 kJ, we can get $Q_{H,Y} = 262.857$ kJ.

(a) Therefore, the work input required for the refrigerator per cycle is $W_Y = Q_{H,Y} - Q_{C,Y} = 22.857$ kJ.

(b) Since $Q_{C,X} + Q_{H,Y}$ is given to be equal to 800 kJ, we can get $Q_{C,X} = 800 - 262.857 = 537.143$ kJ. From the expression above, we can get $Q_{H,X} = 537.143 \div 0.475 = 1130.827$ kJ.

■ EXAMPLE 8.7

An internally reversible engine operates between a reservoir at 1000 K and the ambient at 300 K. If the engine receives and rejects heat across a 20 K temperature difference with the reservoirs, determine its efficiency.

Solution : Let the engine receive heat Q_H from the reservoir at T_H and reject heat Q_C to the reservoir at T_C. Also let the temperature difference between the engine and the hot and cold reservoirs be ΔT_H and ΔT_C respectively.

Since the engine is internally reversible, we have,

$$\frac{Q_H}{T_H - \Delta T_H} = \frac{Q_C}{T_C + \Delta T_C}$$

Therefore, the efficiency of the engine, $\eta = 1 - Q_C \div Q_H$ may be written as

$$\eta = 1 - \frac{T_C + \Delta T_C}{T_H - \Delta T_H}$$

Since both ΔT_H and ΔT_C are positive, the efficiency decreases as a result of the external irreversibility. Upon substituting the values, we get $\eta = 0.6735$, instead of 0.7.

■ EXAMPLE 8.8

A reversible heat engine receives 2000 kJ heat from a source at 1000 K and 3000 kJ heat from another source at 750 K. It rejects heat to the ambient at 300 K. All other processes of the cycle are adiabatic. Determine the work output of the engine and the overall thermal efficiency.

working substance to be an ideal gas). As indicated in this figure, we can imagine the given engine to be composed of two Carnot engines, labelled I and II, executing

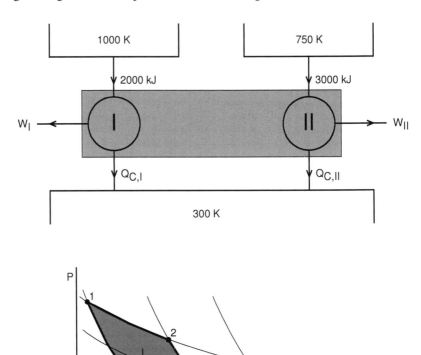

cyclic processes *1-2-3-x-6-1* and *3-4-5-x-3*, respectively. It is important to note that this decomposition does not alter the net heat or work interaction of the original cycle, since the additional process *3-x* is an adiabatic process and the positive work produced in this process in cycle I is negated entirely by the same amount of work absorbed in this process in cycle II. Hence, we may write for each Carnot engine,

Heat rejected to the cold reservoir, Q_C is the sum of the heat rejected by each Carnot engine, and so $Q_C = Q_{C,I} + Q_{C,II}$. From the above expression, we may write

$$Q_C = Q_{C,I} + Q_{C,II} = T_C \left(\frac{Q_{H,I}}{T_{H,I}} + \frac{Q_{H,II}}{T_{H,II}} \right)$$

Upon substituting the numerical values, we get, $Q_C = 1800$ kJ.

The work output of the engine is, $W = Q_{H,I} + Q_{H,II} - Q_C$, from first law. Hence, $W = 3200$ kJ. The efficiency of the engine is, $\eta = W \div (Q_{H,I} + Q_{H,II}) = 64$ percent.

■ EXAMPLE 8.9

A reversible engine, while executing a cyclic process consisting of reversible isotherms and reversible adiabats, receives 2000 kJ of heat from a reservoir at 900 K and in addition, receives an unknown amount of heat from a reservoir at 700 K. It rejects 500 kJ of heat to the ambient at 300 K and in addition, rejects an unknown amount of heat to a reservoir at 500 K. If the efficiency of the engine is 0.6, determine the unknown heat interactions.

Solution : A schematic illustration of the engine and the cyclic process executed by the engine are shown in the following figures. The engine may be decomposed into four Carnot engines labelled I, II, III and IV that execute cyclic processes *1-2-13-14-1, 2-3-4-10-11-12-13-2, 4-5-9-10-4* and *5-6-7-8-9-5* respectively (assuming the working substance to be an ideal gas), as shown in the figure.

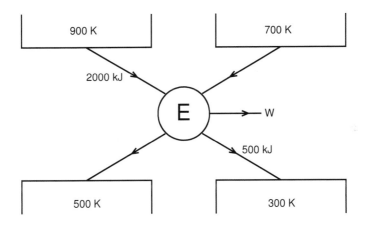

For each engine, we may write

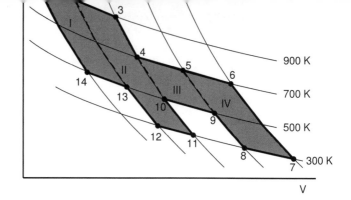

$$\frac{Q_{H,I}}{900} = \frac{Q_{C,I}}{500}; \quad \frac{Q_{H,II}}{900} = \frac{Q_{C,II}}{300}; \quad \frac{Q_{H,III}}{700} = \frac{Q_{C,III}}{500}; \quad \frac{Q_{H,IV}}{700} = \frac{Q_{C,IV}}{300}$$

It is given that $Q_{H,I} + Q_{H,II} = 2000$ kJ $= Q'_H$, say, and $Q_{C,II} + Q_{C,IV} = 500$ kJ $= Q'_C$. The first expression may also be written as

$$\frac{9}{5} Q_{C,I} + 3 Q_{C,II} = Q'_H$$

The efficiency of the engine is given as

$$\eta = 1 - \frac{\sum_{i=I}^{IV} Q_{C,i}}{\sum_{i=I}^{IV} Q_{H,i}} = 0.6$$

If we expand the summation in the numerator and the denominator and use the given information, we get

$$\frac{Q'_C + Q_{C,I} + Q_{C,III}}{Q'_H + Q_{H,III} + Q_{H,IV}} = 0.4$$

This may be written as

$$
\begin{aligned}
Q_{C,I} + Q_{C,III} &= 0.4 Q'_H - Q'_C + (0.4)\frac{7}{5} Q_{C,III} + (0.4)\frac{7}{3} Q_{C,IV} \\
&= 0.4 Q'_H - Q'_C + (0.4)\frac{7}{5} Q_{C,III} + (0.4)\frac{7}{3} (Q'_C - Q_{C,II}) \\
&= 0.4 Q'_H - \frac{0.2}{3} Q'_C + (0.4)\frac{7}{5} Q_{C,III}
\end{aligned}
$$

Therefore, $Q_{C,I} + Q_{C,III} = 328.2828$ kJ. It follows that $Q_{H,III} + Q_{H,IV} = 70.70707$ kJ. These are the required heat interactions of the engine with the reservoirs at 500 K and 700 K respectively.

8.6 Finite reservoirs

As already mentioned, in real life, reservoirs are finite. Consequently, when a finite amount of heat is supplied by or rejected to such a reservoir, its temperature changes by a finite amount. This poses a challenge in the operation of a reversible engine, since the heat addition or rejection processes are required to be reversible and isothermal. This difficulty may be circumvented by limiting the magnitude of the heat interaction to an infinitesimally small amount δQ during a cycle and executing an infinite number of such cycles, rather than executing a single cycle with a finite amount of heat interaction. The temperature of the reservoir thus changes only by a differential amount dT as a result of heat addition or rejection in each cycle and hence this process may be treated as reversible and isothermal. Examples illustrating this concept are discussed next.

■ **EXAMPLE 8.10**

A block of mass m and specific heat capacity c is initially at a temperature T_1. A heat engine operates between the block and the ambient at a temperature T_0. Determine the maximum amount of work that can be extracted.

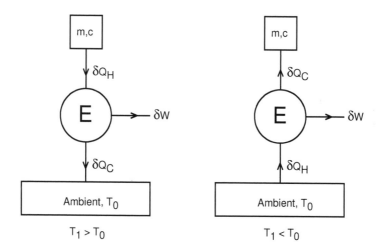

and the ambient as the cold reservoir. This will deliver the maximum amount of work. The engine will operate until the temperature of the block becomes equal to the ambient temperature.

Let the temperature of the block at the beginning of a cycle at some intermediate instant be T. During the cycle, an infinitesimal amount of heat δQ_H is transferred from the block to the engine, the engine delivers work δW while rejecting an amount of heat δQ_C to the ambient, as shown on the left in the figure. Since the cycle is reversible, we may write,

$$\frac{\delta Q_H}{T} = \frac{\delta Q_C}{T_0} \Rightarrow \delta Q_C = \frac{T_0}{T}\delta Q_H$$

If we apply first law to the block, we get $dE = dU = \delta Q - \delta W$, which may be simplified to give $\delta Q = m\,c\,dT$, since $\delta W = 0$. This is a negative quantity (since dT is negative), as it should be, since heat is rejected by the block. Hence, the heat received by the engine is $\delta Q_H = -m\,c\,dT$. Therefore,

$$\delta Q_C = -T_0\,\frac{m\,c\,dT}{T}$$

Upon integrating this, we get

$$Q_C = -m\,c\,T_0 \int_{T_1}^{T_0} \frac{dT}{T} = m\,c\,T_0\,\ln\frac{T_1}{T_0}$$

The total work delivered is given as

$$
\begin{aligned}
W &= Q_H - Q_C \\
&= m\,c\,(T_1 - T_0) - m\,c\,T_0\,\ln\frac{T_1}{T_0} \\
&= m\,c\,T_0\left(\frac{T_1}{T_0} - 1 - \ln\frac{T_1}{T_0}\right)
\end{aligned}
$$

In the second case, a reversible heat engine operates with the ambient as the hot reservoir and the block as the cold reservoir. The disposition of the engine at the beginning of a cycle at an intermediate instant is shown on the right in the figure. Proceeding as before,

$$\frac{\delta Q_H}{T_0} = \frac{\delta Q_C}{T} \Rightarrow \delta Q_H = \frac{T_0}{T}\delta Q_C$$

If we apply first law to the block, we get $\delta Q_C = m\,c\,dT$. If we repeat the same steps as before, we get

$$Q_H = m\,c\,T_0\,\ln\frac{T_0}{T_1}$$

The *reversible work* developed is positive in both the cases and is defined as the *exergy* of the block.

◼ EXAMPLE 8.11

A piston cylinder device (see figure) contains m kg of a perfect gas initially at a temperature T_1. A heat engine operates between the vessel and the ambient at a temperature T_0 ($< T_1$). Determine the maximum amount of work that can be extracted.

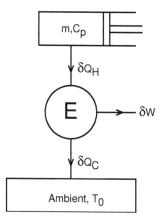

Solution : In this case, the pressure inside the cylinder remains constant while heat is supplied to the engine. The solution proceeds along the same lines as before but with this modification. Accordingly,

$$\delta Q_C = \frac{T_0}{T} \delta Q_H$$

If we apply first law to the gas inside the piston cylinder device, we get $dE = dU = \delta Q - \delta W$. The displacement work is non-zero in this case and is equal to $P\,dV$, with P being constant. Hence $\delta Q = dU + P\,dV = dH = m\,C_p\,dT$. The heat received by the engine is $\delta Q_H = -m\,C_p\,dT$. Proceeding as before, we get,

$$Q_C = m\,C_p\,T_0 \ln \frac{T_1}{T_0}$$

and

$$W = m\,C_p\,T_0 \left(\frac{T_1}{T_0} - 1 - \ln \frac{T_1}{T_0} \right)$$

A block of mass 15 kg, specific heat capacity 800 J/kg.K and initially at 600 K is available. Two strategies are being considered for maintaining the house at 27°C:

(a) allow the block (finite reservoir) to cool inside the house and

(b) run a reversible heat engine between the block (finite reservoir) and the house and supply the work produced to a reversible heat pump that operates between the ambient and the house.

In each case, determine the amount of heat that is transferred to the house.

Solution :

(a) In this case, the reservoir cools down from 600 K to 300 K. The heat transferred to the house is simply $Q_H = m c (T_1 - T_2) = 3600$ kJ.

(b) In this case, a reversible direct engine operates with the block and the house as reservoirs, until the the temperature of the former becomes equal to that of the latter. The heat supplied to the house by the direct engine is (from Example 8.10),

$$Q_C = m c T_0 \ln \frac{T_1}{T_0} = 2495.33 \text{ kJ}$$

The work produced by this engine is (from Example 8.10),

$$W = m c T_0 \left(\frac{T_1}{T_0} - 1 - \ln \frac{T_1}{T_0} \right) = 1104.67 \text{ kJ}$$

The COP of the reversible heat pump is, from Eqn. 8.6,

$$\text{COP} = \frac{1}{1 - \frac{T_C}{T_H}} = \frac{1}{1 - \frac{-5+273}{27+273}} = 9.375$$

The heat supplied to the house by the heat pump may be calculated using Eqn. 8.3, as

$$Q_H = W \times \text{COP} = 1104.67 \times 9.375 = 10356.28 \text{ kJ}$$

Therefore, the total heat supplied to the house is $2495.33 + 10356.28 = 12851.61$ kJ. This is 3.57 times more than the energy that is available in the block.

Two important observations can be made from this example and Example 8.2:

- A heat pump provides a multiplication factor equal to its COP on the input work
- If the input provided is work rather than heat, then the overall output is higher

required, assuming that there is no heat loss. If the same final state is to be attained in a reversible manner, determine the work required.

Solution : Considering the air as the system, application of first law to the stirring process denoted *1-2* gives

$$\Delta E = \Delta U = \cancel{Q}^{\,0} - W \;\Rightarrow\; W = -m\,C_v\,(T_2 - T_1)$$

where Q has been set to zero as there is no heat loss. If we substitute the numerical values, we get the stirring work to be equal to 288.681 kJ.

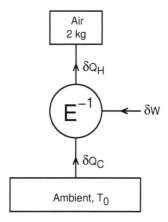

In order to reach the same end state in a reversible manner, a reversible heat pump that operates between the ambient at a temperature T_0 and the vessel is utilized, as shown in the figure. The heat pump is stopped when the temperature of the air in the reservoir reaches the desired final temperature.

As before, for any intermediate cycle, we may write,

$$\frac{\delta Q_H}{T} = \frac{\delta Q_C}{T_0} \;\Rightarrow\; \delta Q_C = \frac{T_0}{T}\,\delta Q_H$$

Considering the air as the system, application of first law gives $dE = dU = \delta Q - \delta W$, which may be simplified to give $\delta Q = m\,C_v\,dT$, since $\delta W = 0$. Hence, the heat supplied by the heat pump is $\delta Q_H = m\,C_v\,dT$. Therefore,

$$\delta Q_C = T_0\,\frac{m\,C_v\,dT}{T}$$

The total work input is given as

$$W = Q_H - Q_C$$

$$= m\,C_v\,(T_2 - T_1) - m\,C_v\,T_0\,\ln\frac{T_2}{T_1}$$

$$= m\,C_v\,T_0\left(\frac{T_2}{T_0} - 1 - \ln\frac{T_2}{T_1}\right)$$

Upon substituting numerical values, we get the reversible work input to be equal to 67.482 kJ, which is only 24 percent of the stirring work. This demonstrates how highly irreversible the stirring process is. Students will learn in a later chapter that this diference is referred to as "exergy loss".

▐ EXAMPLE 8.14

A rigid vessel is divided into two chambers A and B by a thin partition. Chamber A contains 2 kg of air initially at 100 kPa, 300 K. Chamber B contains 3 kg of air initially at 200 kPa, 400 K. The partition is removed and the air from the two chambers are allowed to mix and attain a final equilibrium state. Determine the final pressure and temperature of the air, neglecting any heat exchange with the ambient. If the same final state is to be attained in a reversible manner, determine the amount of work required/produced.

Solution : The volume occupied initially by the air in chamber A may be calculated using

$$V_{1,A} = \frac{m_A\,R\,T_{1,A}}{P_{1,A}} = 1.732\,\text{m}^3$$

Similarly, the volume initially occupied by the air in chamber B, $V_{1,B}$, may be evaluated to be 1.732 m³.

Considering the air in both the chambers as the system, application of first law to the system gives

$$\Delta E = \Delta U = \cancel{Q}^{0} - \cancel{W}^{0}$$

$$\Delta U = m_A\,C_v\,(T_3 - T_{1,A}) + m_B\,C_v\,(T_3 - T_{1,B})$$

$$= 0$$

Here, the work interaction is zero since the volume of the system remains constant and the heat interaction is given to be negligible. If we substitute the numerical

$$V_{1,A} + V_{1,B}$$

In order to attain the same end state reversibly, we let the air in chamber A and chamber B separately undergo a constant volume process *1-2* until the pressure becomes 150 kPa and then a constant pressure process *2-3* until the final temperature becomes 360 K. Both these processes are carried out reversibly by using the air as a finite reservoir for a reversible heat engine as shown in Examples 8.11 and 8.13.

For the air in chamber A, $P_2 = 150$ kPa and $V_2 = V_{1,A} = 1.732$ m^3. Therefore, the temperature at the end of this process, $T_{2,A} = P_2 V_2/(m_A R) = 450$ K. It is clear that heat is added to the air in chamber A during process *1-2* and removed during process *2-3*. Hence, it is used as a hot reservoir at constant volume for a reversible heat pump and as a hot reservoir at constant pressure for reversible direct engine during process *2-3*.

For the air in chamber B, $P_2 = 150$ kPa and $V_2 = V_{1,B} = 1.732$ m^3. Therefore, the temperature at the end of this process, $T_{2,B} = P_2 V_2/(m_B R) = 300$ K. In this case, heat is removed during process *1-2* and added during process *2-3*. Hence, the air in chamber B is used as a hot reservoir at constant volume for a reversible direct engine and as a hot reservoir at constant pressure for reversible heat pump during process *2-3*.

In both cases, the ambient at 300 K is used as the cold reservoir. The work interactions for the individual processes can easily be evaluated using the expressions derived and methods developed in the previous examples and so the details are not given here. The final values are

$$W_{1-2,A} = m_A C_v T_0 \left(\frac{T_{2,A}}{T_0} - \frac{T_{1,A}}{T_0} - \ln \frac{T_{2,A}}{T_{1,A}} \right) = 40.936 \text{ kJ (in)}$$

$$W_{2-3,A} = m_A C_p T_0 \left(\frac{T_{2,A}}{T_0} - \frac{T_3}{T_0} - \ln \frac{T_{2,A}}{T_3} \right) = 46.593 \text{ kJ (out)}$$

$$W_{1-2,B} = m_B C_v T_0 \left(\frac{T_{1,B}}{T_0} - \frac{T_{2,B}}{T_0} - \ln \frac{T_{1,B}}{T_{2,B}} \right) = 29.652 \text{ kJ (out)}$$

$$W_{2-3,B} = m_B C_p T_0 \left(\frac{T_3}{T_0} - \frac{T_{2,B}}{T_0} - \ln \frac{T_3}{T_{2,B}} \right) = 16.072 \text{ kJ (in)}$$

The net work interaction is -40.936 + 46.593 + 29.652 - 16.076 = +19.233 kJ. Hence, if the same final state had been attained in a reversible manner, it would have resulted in a positive work output of 19.233 kJ as opposed to zero. It is evident that mixing is more irreversible than stirring and the loss of this work is usually referred to as "exergy loss".

A reversible heat engine operates with the blocks as reservoirs. Determine the final temperature and the work produced.

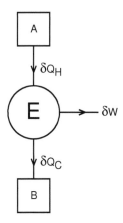

Solution : The disposition of the engine at an intermediate instant is shown in the figure. Since the engine is reversible,

$$\frac{\delta Q_H}{T_H} = \frac{\delta Q_C}{T_C}$$

If we apply first law to blocks A and B, we get $\delta Q_H = -m_A\, c_A\, dT_H$ and $\delta Q_C = m_B\, c_B\, dT_C$. If we substitute these into the above expression, we get

$$-\frac{m_A\, c_A\, dT_H}{T_H} = \frac{m_B\, c_B\, dT_C}{T_C}$$

Upon integrating this expression, and noting that the engine stops working when the temperature of both blocks become equal, we get

$$-m_A\, c_A \int_{T_{1,A}}^{T_2} \frac{dT_H}{T_H} = m_B\, c_B \int_{T_{1,B}}^{T_2} \frac{dT_C}{T_C}$$

and so, the final temperature may be obtained as

$$T_2 = \left(T_{1,A}\, T_{1,B}^{\kappa}\right)^{1/(1+\kappa)}.$$

where $\kappa = (m_B \div m_A)\,(c_B \div c_A)$. The work produced by the engine is given as

$$
\begin{aligned}
W &= m_A\, c_A\,(T_{1,A} - T_2) - m_B\, c_B\,(T_2 - T_{1,B}) \\
&= m_A\, c_A\, T_{1,A} + m_B\, c_B\, T_{1,B} - T_2\,(m_A\, c_A + m_B\, c_B)
\end{aligned}
$$

temperature to be T_f. Derive an expression for the final temperature of the blocks, the work produced and the overall efficiency. Also, determine the work lost as a result of the reservoirs being finite.

Solution : With $m_A = m_B$ and $c_A = c_B$, the final temperature of the blocks, $T_f = \sqrt{T_1 T_2}$. The work developed is given as

$$W = mc \left(T_1 + T_2 - 2\sqrt{T_1 T_2} \right)$$

The overall efficiency of the engine is given as

$$\eta = \frac{W}{Q_H} = \frac{mc \left(T_1 + T_2 - 2 \left(T_1 T_2 \right)^{1/2} \right)}{mc(T_1 - \sqrt{T_1 T_2})}$$

which may be simplified to give

$$\eta = 1 - \sqrt{\frac{T_2}{T_1}}$$

If the same amount of heat Q_H had been given to a reversible engine with infinite reservoirs at temperatures T_1 and T_2, the work produced, W', would have been $Q_H \left(1 - T_2/T_1 \right)$, according to Eqn. 8.4. Hence, the work lost due to the reservoirs being finite is

$$
\begin{aligned}
W' - W &= Q_H \left(1 - \frac{T_2}{T_1} - 1 + \sqrt{\frac{T_2}{T_1}} \right) \\
&= Q_H \sqrt{\frac{T_2}{T_1}} \left(1 - \sqrt{\frac{T_2}{T_1}} \right) \\
&= \frac{\sqrt{T_2}}{\sqrt{T_1} + \sqrt{T_2}} W'
\end{aligned}
$$

ENTROPY

In the previous chapter, the focus was primarily on cyclic processes and their efficiencies. In this chapter, the focus is on individual processes with the objective of assessing their performance. As mentioned earlier, real world processes are irreversible and they cause a degradation in the performance of not only the cycle, but the individual processes that comprise the cycle as well. In this chapter, a metric is developed for assessing the amount of this degradation. This metric utilizes the change in a new property called entropy. By evaluating the change in entropy of an ideal process and the actual process, a comparison of their performance can be carried out and means to improve the performance of the latter may be proposed. Calculation of the change of entropy for a system as well as the rate of change of entropy in a control volume are discussed in detail. A very fundamental law involving entropy, namely, the principle of increase of entropy is also given. This is regarded universally as a profound statement on the evolution of the universe and everything within. Hence, this forms the starting point for most non-engineering thermodynamic treatises.

process by a sequence of processes, namely, A-1, 1-2 and 2-B (Fig. 9.1, right). The first and the last processes in this sequence are reversible adiabatic processes and are unique since only one reversible adiabat each passes through state points A and B. The middle process in this sequence is a reversible isothermal process and is not unique since state points 1 and 2 are, as yet, not known. We determine this isotherm by requiring that the net work and heat interaction for process A-B and A-1-2-B be the same. This is illustrated graphically in Figs. 9.1 and 9.2.

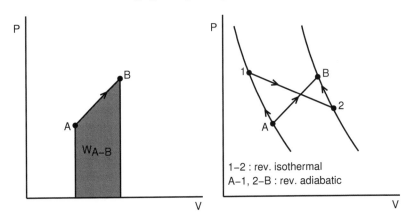

Figure 9.1: The reversible process A-B executed by a system is to be replaced by a sequence of reversible isothermal and reversible adiabatic processes as shown in the right.

The work interaction for process A-B is the shaded area in Fig. 9.1, as it is a reversible process. The work interaction along with the sign for processes A-1, 1-2 and 2-B are shown shaded in Figs. 9.2 (a), (b) and (c) respectively. The algebraic sum of these three areas is shown in Fig. 9.2 (d). Point Q in Fig. 9.2 is the point of intersection of the original process A-B and the reversible isotherm 1-2. If we choose the reversible isotherm 1-2 in such a manner that the geometric location of point Q results in Area A1Q = Area Q2B, then, the net area under the curve for A-1-2-B becomes identical to the shaded area shown in Fig. 9.1 which is equal to W_{A-B}. Hence, $W_{A-1-2-B} = W_{A-B}$.

Let us now apply first law to these processes separately, to get

$$\Delta E|_{A-B} = Q_{A-B} - W_{A-B}$$

and

$$\Delta E|_{A-1-2-B} = Q_{A-1-2-B} - W_{A-1-2-B}$$

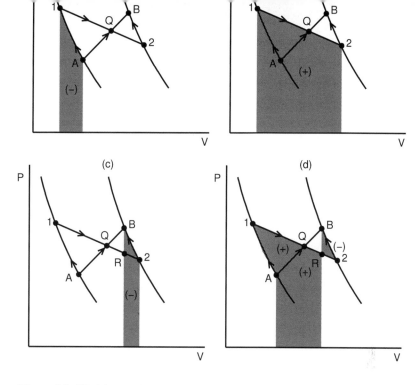

Figure 9.2: Work interaction for process (a) A-1, (b) 1-2, (c) 2-B and (d) A-1-2-B

Since E is a property, the term on the left hand side of these two expressions are identical. The work interactions in these two expressions have also been proven to be equal. Therefore, $Q_{A-1-2-B} = Q_{A-B}$. Hence, we may state that any internally reversible process may be replaced entirely by a sequence of reversible adiabatic and reversible isothermal processes.

Consider the internally reversible cyclic process shown in Fig. 9.3 executed by a system. Reversible adiabats (dashed lines) and reversible isotherms (solid lines) are shown overlaid on this cycle in this figure. A small segment in the upper portion of the cyclic process, denoted a-b, may be replaced by the sequence a-p, p-q and q-b in the manner discussed before. Similarly, the corresponding segment, c-d, in the lower portion of the cyclic process may be replaced by the sequence c-s, s-r and r-d. Note that, states a and d lie on the same reversible adiabat and states b and c lie on the same reversible adiabat. It is evident that p-q-c-s-r-a-p is a Carnot cycle (the

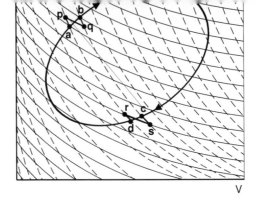

Figure 9.3: A reversible cyclic process executed by a system overlaid with reversible isotherms (solid lines) and reversible adiabats (dashed lines)

remaining segments q-b and r-d are part of the adjacent Carnot cycles). Therefore, the entire reversible cycle may be replaced by an infinite number of infinitesimally small Carnot cycles. For each infinitesimal Carnot cycle, we know that

$$\frac{\delta Q_H}{T_H} = \frac{\delta Q_C}{T_C} \quad \Rightarrow \quad \frac{\delta Q_H}{T_H} - \frac{\delta Q_C}{T_C} = 0$$

It is important to note that T_H and T_C are the temperatures at which the engine receives and rejects heat, and not necessarily the temperatures of the hot and cold reservoirs. In other words, it is sufficient for this development if the cycle is internally reversible. For the entire cycle, we may write

$$\sum \left(\frac{\delta Q_H}{T_H} - \frac{\delta Q_C}{T_C} \right) = 0$$

This may be written more compactly as

$$\oint \frac{\delta Q}{T} = 0$$

where we have used the fact that the sign convention adopted here will automatically change the sign of δQ in the top and bottom portion of the cycle.

If the engine is internally irreversible, then for the same δQ_H, T_H and T_C, the heat rejected in each infinitesimal cycle is more. That is, $\delta Q_{C,irr} > \delta Q_C$ and hence the sum (and the integral) above becomes negative in this case. Hence, we may write

$$\oint \frac{\delta Q_{irr}}{T} < 0$$

is applicable otherwise. This is the well-known Clausius inequality. This may be written in a convenient form in which the inequality is absent. Thus

$$\oint \frac{\delta Q}{T} = -\sigma_{int} \qquad (9.2)$$

where σ_{int} refers to the internal irreversiblities in the cycle and is zero or positive:

$$\sigma_{int} = \begin{cases} 0 \ (reversible) \\ > 0 \ (irreversible) \end{cases}$$

It is left as an exercise to the reader to rework Examples 8.8 and 8.9 using the Clausius inequality.

9.2 Entropy change of a system during a process

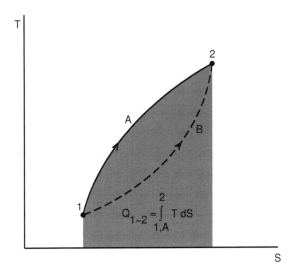

Figure 9.4: Entropy change of a system for an internally reversible (A) and irreversible (B) process between the same states.

Similar to the first law for a cyclic process executed by a system in which the cyclic integral of the quantity $\delta Q - \delta W$ was zero, now, the cyclic integral of the quantity $\delta Q/T$ equals zero for an internally reversible process executed by a system.

The property S is called the entropy of the system under consideration. If we integrate the above equation between states 1 and 2 along an internally reversible process labelled A as shown in Fig. 9.4, we get

$$\Delta S = S_2 - S_1 = \int_1^2 \frac{\delta Q}{T}\bigg|_{rev} \tag{9.4}$$

Since entropy is a property, its value at a state, by definition, does not depend on the process by which the system arrived at that state. Consequently, entropy change for the irreversible process B shown in Fig. 9.4 is the same as it is for the reversible process A since both occur between the same states.

Let us now run process A in reverse, as it is an internally reversible process and denote this A^{-1}. If we apply Eqn. 9.2 to the cyclic process 1-2 along B and 2-1 along A^{-1}, we get

$$\int_1^2 \frac{\delta Q}{T}\bigg|_B + \int_2^1 \frac{\delta Q}{T}\bigg|_{A^{-1}} = -\sigma_{int}$$

If we use Eqn. 9.4, this expression may be written as

$$\int_1^2 \frac{\delta Q}{T}\bigg|_B + (S_1 - S_2) = -\sigma_{int}$$

This may be rearranged to give

$$\Delta S = S_2 - S_1 = \int_1^2 \frac{\delta Q}{T}\bigg|_B + \sigma_{int}$$

or, in a general form,

$$\Delta S = S_2 - S_1 = \int_1^2 \frac{\delta Q}{T} + \sigma_{int} \tag{9.5}$$

for *any* process - reversible or irreversible, between states 1 and 2. Note that, Eqn. 9.5 reduces to Eqn. 9.4 in the case of a reversible process, since $\sigma_{int} = 0$ in this case.

Equation 9.5 may be written in words as:

$$\begin{pmatrix} \text{Entropy} \\ \text{change of} \\ \text{system} \end{pmatrix} = \begin{pmatrix} \text{Entropy} \\ \text{transfer} \end{pmatrix} + \begin{pmatrix} \text{Entropy} \\ \text{generation due} \\ \text{to internal} \\ \text{irreversibilities} \end{pmatrix} \tag{9.6}$$

the system or positive in case heat is supplied to the system and zero otherwise. In general, the entropy of a system may increase, decrease or remain the same depending upon the interplay between these terms. A few examples are used to illustrate this.

Mixing : As already mentioned, mixing is a highly irreversible process. The irreversibility in this case is an internal irreversibility. When two fluids at different initial thermodynamic states (pressure, temperature and composition) are mixed, the entropy increases since the entropy generated is quite large. Even if there is heat loss to the surroundings and the attendant reduction in entropy, the net entropy change is positive.

Although work transfer does not appear in Eqn. 9.5, it can still influence the entropy change through the σ_{int} term.

Stirring work : When a system is stirred using a paddle wheel (Fig. 3.5), the process is highly irreversible, again due to internal irreversibility. In this case also, the entropy increases irrespective of whether the process is carried out adiabatically or not.

Electrical work : When electrical work is transferred to an embedded resistor (Fig. 3.6), the entropy increases. When the system is chosen as the one in the top in Fig. 3.6, the second term in Eqn. 9.5 is zero and the first term is positive. On the other hand, when the system shown in the bottom in Fig. 3.6 is used, then the first term in Eqn. 9.5 is zero and the second term is non-zero and hence positive.

Displacement work : In the absence of friction and any other internal irreversibility ($\sigma_{int} = 0$), the entropy of the system will increase or decrease depending upon whether it is being compressed or undergoing expansion and the heat interaction with the surroundings. If the process is carried out adiabatically, then the entropy remains constant, since the first term in Eqn. 9.5 is zero as well.

9.2.1 Isentropic process

For an adiabatic process, the first term in the right hand side of Eqn. 9.5 is identically zero. Since $\sigma_{int} \geq 0$, entropy can increase or remain the same in an adiabatic process but cannot decrease. For an adiabatic, reversible process, both terms in the right hand side of Eqn. 9.5 are zero and hence $s_2 = s_1$. Furthermore, from Eqn. 9.3, $ds = 0$, from which it follows that, $s = constant$ for the reversible adiabatic process. Consequently, the reversible adiabatic process is also an *isentropic* process.

reversible adiabatic process which is also an isentropic process is very special.

Perhaps, the most popular statement regarding entropy is that it is a measure of the disorder in the system. This interpretation comes from a microscopic perspective[†] and for that reason, it is not possible to corroborate this statement within the framework of the macroscopic approach. In addition, in the latter approach, it is the *entropy change* and not the entropy itself that is required and evaluated. Notwithstanding this, it is still possible to make a connection between disorder and entropy in the macroscopic approach as follows. Equation 9.5 shows that the entropy of a system increases as a result of heat addition. It was mentioned in section 4.1 that, addition of heat directly causes the internal energy of the system to increase and that internal energy is a disordered form of energy since it is associated with molecular motion. Hence, we may conclude that an increase in entropy due to heat addition is associated with an increase in internal energy with an attendant increase in the molecular motion.

9.3 $T\,ds$ relations

For a fully resisted (and hence reversible) process and after neglecting KE and PE changes, the first law for a system undergoing a non-cyclic process, Eqn. 4.2 may be written as

$$dE = dU = \delta Q - P\,dV$$

or, in intensive form as

$$du = \delta q - P\,dv$$

If we substitute for δq from Eqn. 9.3, we get

$$T\,ds = du + P\,dv \qquad (9.7)$$

The above equation may be expressed in terms of enthalpy after using $dh = du + v\,dP + P\,dv$ to eliminate du. Thus,

$$T\,ds = dh - v\,dP \qquad (9.8)$$

Equations 9.7 and 9.8 are called the $T\,ds$ relations and are applicable to any simple, compressible substance. Note that they are written here in intensive form and may easily be written in extensive form as well.

[†] In fact, this comes directly from the Boltzmann equation, $S = k_B \ln w$, where w is the number of available microscopic states.

role in the development of fundamental ideas such as the Carnot cycle, absolute temperature scale and the Clausius inequality. Hence, depicting processes using $T - s$ coordinates will better bring out aspects concerning irreversibilities than using $P - V$ coordinates.

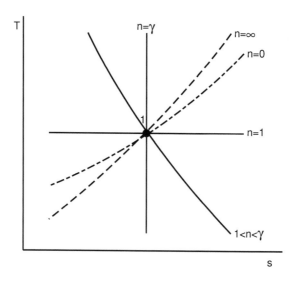

Figure 9.5: Polytropic processes for an ideal gas corresponding to various polytropic indices in $T - s$ coordinates

Equations 9.7 and 9.8 may be written for a perfect gas as

$$ds = \left(C_v \frac{dT}{T} + R \frac{dv}{v} \right) \tag{9.9}$$

$$= \left(C_p \frac{dT}{T} - R \frac{dP}{P} \right) \tag{9.10}$$

We may also write

$$ds = \left(C_v \frac{dP}{P} + C_p \frac{dv}{v} \right) \tag{9.11}$$

after using the equation of state.

Integration of Eqn. 9.11, after setting $ds = 0$, leads to the familiar $Pv^\gamma = constant$ as the relation obeyed by an isentropic process. If we set $dP = 0$ in Eqn. 9.10, we

Similarly, we can get the slope of an isochor to be (after setting $dv = 0$ in Eqn. 9.9)

$$\left.\frac{dT}{ds}\right|_{v=constant} = \frac{T}{C_v}$$

Since the slope of the isobar and the isochor are positive, they lie in the first and the third quadrant with respect to a given initial state as shown in Fig. 9.5. Also, since $C_p > C_v$, the isochors are steeper than the isobars. For a general polytropic process that obeys $Pv^n = constant$, we can get from Eqn. 9.9,

$$\left.\frac{dT}{ds}\right|_{Pv^n=constant} = \frac{n-1}{n-\gamma}\frac{T}{C_v}$$

The slope is negative for $1 < n < \gamma$ and hence the process line $PV^n = constant$ lies in the second and the fourth quadrant with respect to the initial state, as shown in Fig. 9.5.

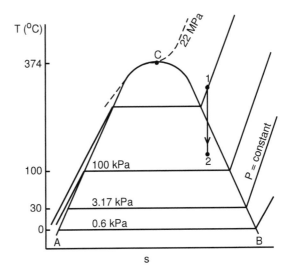

Figure 9.6: $T - s$ diagram for liquid water-water vapor mixture. Not to scale.

The $P - v$ diagram for water in Fig. 5.2 is shown in $T - s$ coordinates in Fig. 9.6. For the sake of clarity and since they are also the most widely used, isobars alone are shown in this diagram.

the area under the curve for an internally reversible process gives the heat interaction for the system, as can be ascertained from Eqn. 9.3, *viz.,*

$$\frac{Q_{1-2}}{m} = \int_1^2 T\, ds$$

This is illustrated in Fig. 9.4. It must be recalled that the area under the curve for a reversible process in $P - v$ coordinates gives the work interaction.

■ **EXAMPLE 9.1**

Consider the two internally reversible cycles I and II between the same temperature and entropy limits, executed by a system as shown in the figure. Determine if they are power absorbing or power producing and which one is more efficient. For each cycle, determine the temperature limits of an internally reversible Carnot cycle between the same entropy limits.

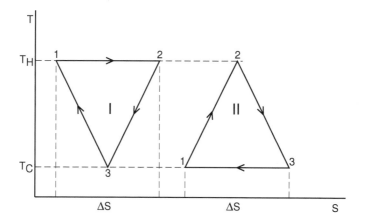

Solution : Since the cycles are internally reversible, the area under the process curve is the net heat interaction during each cycle. In the case of cycles I and II, based on the given process directions, the net heat interaction is positive. From the first law for a cycle, the net heat interaction is also equal to the net work interaction during the cycle. Therefore, both the cycles are power producing cycles.

Heat is supplied during process *1-2* in cycle I and during processes *1-2* and *2-3* in cycle II. It is easy to establish graphically that the area under the process curve *1-2* in cycle I is more than the combined area under the process curves *1-2* and *2-3* in cycle II. Consequently, heat supplied in cycle I is more. Since the net work produced in

$$Q_C = 2\left(T_C \frac{\Delta S}{2} + \frac{1}{2}\frac{\Delta S}{2}(T_H - T_C)\right) = \frac{T_H + T_C}{2}\Delta S$$

Therefore, the temperature limits of a Carnot engine operating between the same entropy limits are T_H and $(T_H + T_C)/2$.

Similarly, for cycle II, heat is rejected by the engine at temperature T_C. A Carnot engine operating between the same entropy limits will operate between $(T_H + T_C)/2$ and T_C.

9.5 Calculation of entropy change

Although the change in entropy between two states (the left hand side in Eqn. 9.5) is independent of the process, information about the process is required in order to calculate it *i.e.,* to evaluate the right hand side of Eqn. 9.5. In this context, it is important to realize that, even if it is possible to evaluate the integral on the right hand side of Eqn. 9.5 for an irreversible process, it is not possible to evaluate σ_{int}. This is a theoretical construct that is intended to make it easy to draw qualitative inferences but cannot be evaluated directly. Hence, entropy change is always calculated using Eqn. 9.4.

Solids and liquids : Since solids and liquids are incompressible, the $T\,dS$ relation may be written as

$$T\,dS = mc\,dT$$

where c is the specific heat capacity. Therefore, entropy change is given as

$$\Delta S = S_2 - S_1 = mc\ln\frac{T_2}{T_1} \tag{9.12}$$

Perfect gases and mixtures : Integration of Eqns. 9.9 - 9.10 gives

$$\Delta S = m\left(C_v \ln\frac{T_2}{T_1} + R\ln\frac{V_2}{V_1}\right) \tag{9.13}$$

$$= m\left(C_p \ln\frac{T_2}{T_1} - R\ln\frac{P_2}{P_1}\right) \tag{9.14}$$

$$= m\left(C_v \ln\frac{P_2}{P_1} + C_p \ln\frac{V_2}{V_1}\right) \tag{9.15}$$

In case of a mixture of perfect gases, the entropy change of each component i in the mixture may be evaluated using the above equations, with m replaced by m_i and

$$\Delta s = \sum_i x_i \left(C_{p,i} \ln \frac{T_2}{T_1} - R_i \ln \frac{P_{i,2}}{P_{i,1}} \right) \qquad (9.17)$$

$$\Delta \bar{s} = \sum_i y_i \left(\bar{C}_{p,i} \ln \frac{T_2}{T_1} - \mathcal{R} \ln \frac{P_{i,2}}{P_{i,1}} \right) \qquad (9.18)$$

Water and R134a : In this case, the entropy values may be retrieved from Tables A - F. Specific entropy of compressed/subcooled liquids may be approximated as follows: $s(T, P) \approx s_f(T)$.

■ EXAMPLE 9.2

Two kg of ice at -15°C is dropped into 10 kg of liquid water initially at 27°C in an insulated container. Simultaneously, 3500 kJ of work is transferred to the container by means of a paddle wheel. Determine (a) the final temperature and (b) entropy change. The latent heat of melting of ice may be taken to be 334 kJ/kg, specific heat capacity of ice to be 2.03 kJ/kg.K and specific heat of water to be 4.2 kJ/kg.K.

Solution : The ice and the water together may be taken as the system. Application of first law to this system gives

$$\Delta E = \Delta U = \cancel{Q}^{\,0} - W \Rightarrow \Delta U_{ice} + \Delta U_{water} = 3500$$

Assuming that all the ice melts, the above expression may be written as

$$m_{ice} c_{ice} (273 - 258) \quad + \quad m_{ice} L_{ice} + m_{ice} c_{water} (T_2 - 273)$$
$$+ \quad m_{water} c_{water} (T_2 - 300) = 3500$$

Here, L_{ice} is the latent heat of melting of ice. Upon substituting the numerical values and solving, we get $T_2 = 77°C$.

Change in entropy of the ice and water may be evaluated using the expression given above. Accordingly,

$$\begin{aligned} \Delta S_{ice} &= m_{ice} c_{ice} \ln \frac{273}{273 - 15} + \frac{m_{ice} L_{ice}}{273} + m_{ice} c_{water} \ln \frac{273 + 77}{273} \\ &= 4.7634 \, \text{kJ/K} \end{aligned}$$

[†]It must be noted that this is possible only because the composition of the mixture remains the same during the process. If the composition changes, as it does in the presence of chemical reaction(s), the procedure has to be modified. This is discussed in Chapter 13.

Therefore,

$$\Delta S_{system} = \Delta S_{ice} + \Delta S_{water} = 11.2377 \, \text{kJ/K}$$

Since the vessel is insulated, this increase in entropy of the system is entirely due to the internal irreversibility arising from mixing and stirring.

■ EXAMPLE 9.3

An insulated box is divided into two compartments A and B by a negligibly thin partition. Compartment A initially contains 10 kg of N_2 at 500 K and 500 kPa and compartment B initially contains 2 kg of H_2 at 300 K and 100 kPa. The partition is now removed and the gases are allowed to mix and reach an equilibrium state. Taking the box as the system, determine its entropy change.

Solution : Application of first law to this system gives

$$\Delta E = \Delta U = \cancel{Q}^0 - \cancel{W}^0$$
$$m_{N_2} C_{v,N_2} (T_2 - T_{1,N_2}) + m_{H_2} C_{v,H_2} (T_2 - T_{1,H_2}) = 0$$

For N_2, $M = 28$ kg/kmol and $\gamma = 7/5$. Therefore, $R = 8314/28 = 296.93$ J/kg.K and $C_v = R/(\gamma - 1) = 742.32$ J/kg.K. For H_2, $M = 2$ kg/kmol and $\gamma = 7/5$. Therefore, $R = 8314/2 = 4157$ J/kg.K and $C_v = R/(\gamma - 1) = 10392.5$ J/kg.K.

Upon substituting the values into the above expression and solving, we get the final temperature $T_2 = 353$ K.

Initial volume occupied by both the gases may be evaluated as

$$V_{1,H_2} = \left. \frac{mRT_1}{P_1} \right|_{H_2} = \frac{2 \times 4157 \times 300}{100 \times 10^3} = 24.942 \, \text{m}^3$$

and

$$V_{1,N_2} = \left. \frac{mRT_1}{P_1} \right|_{N_2} = \frac{10 \times 296.93 \times 500}{500 \times 10^3} = 2.9696 \, \text{m}^3$$

The volume occupied by the gases after mixing, V_2, is therefore $24.942 + 2.9696 = 27.9116 \, \text{m}^3$.

The partial pressure of H_2 in the final state is given by

$$P_{2,H_2} = \left. \frac{mRT_2}{V_2} \right|_{H_2} = 105.15 \, \text{kPa}$$

Entropy of the individual gases may be evaluated using Eqn. 9.14 as follows.

$$\Delta S_{H_2} = m_{H_2} \left(C_p \ln \frac{T_2}{T_1} - R \ln \frac{P_2}{P_1} \right)_{H_2} = 4.3165 \, \text{kJ/K}$$

$$\Delta S_{N_2} = m_{N_2} \left(C_p \ln \frac{T_2}{T_1} - R \ln \frac{P_2}{P_1} \right)_{N_2} = 4.0693 \, \text{kJ/K}$$

Therefore, entropy change of the system, $\Delta S = 4.3165 + 4.0693 = 8.3858$ kJ/K.

Just as in the previous example, here too, the increase in entropy is entirely due to the internal reversibility arising from mixing.

EXAMPLE 9.4

A rigid insulated vessel of volume 1.5 m³ initially contains water at 20 bar, 50 percent dryness fraction. A valve on the top of the vessel is now opened and the content of the vessel is allowed to escape slowly until the pressure reduces to 200 kPa at which point, the valve is closed. Determine (a) the final state of the contents of the vessel and (b) the mass that escapes.

Solution : At the initial state, from Table B, we have at 20 bar, $v_f = 1.1766 \times 10^{-3} \, \text{m}^3/\text{kg}$, $v_g = 0.0996 \, \text{m}^3/\text{kg}$, $s_f = 2.4470$ kJ/kg.K and $s_g = 6.3397$ kJ/kg.K. The specific volume and specific entropy at the initial state may be evaluated:

$$v_1 = v_f + x_1 (v_g - v_f) = 0.0504 \, \text{m}^3/\text{kg}$$

$$s_1 = s_f + x_1 (s_g - s_f) = 4.39335 \, \text{kJ/kg.K}$$

Mass initially contained in the vessel is

$$m_1 = \frac{V}{v_1} = \frac{1.5}{0.0504} = 29.762 \, \text{kg}$$

(a) Since the vessel is insulated and the vapor escapes slowly, we may assume the content of the vessel to have undergone an isentropic process. Hence, $s_2 = s_1 = 4.39335$ kJ/kg.K[†]. Since $P_2 = 200$ kPa, from Table B, we can retrieve $s_f = 1.5301$ kJ/kg.K and $s_g = 7.1275$ kJ/kg.K. Therefore,

$$x_2 = \frac{s_2 - s_f}{s_g - s_2} = 0.5115$$

[†]Note that $S_2 (= m_2 s_2) \neq S_1 (= m_1 s_1)$, since $m_2 \neq m_1$

$$m_2 = \frac{V}{v_2} = \frac{1.5}{0.454} = 3.304\,\text{kg}$$

Mass that escapes is thus 29.762 - 3.304 = 26.458 kg.

▮ EXAMPLE 9.5

Steam is contained in an insulated rigid tank of volume 2 m³ initially at a pressure and temperature of 15 bar and 420°C respectively. The steam is discharged slowly through an insulated turbine into the atmosphere at 100 kPa. It may be assumed that the steam is always expanded to the atmospheric pressure. The mass of steam in the turbine at any instant and its energy may be neglected. Determine the work developed by the turbine.

Solution : For the steam initially in the tank, the following properties may be retrieved from Table C: $v_1 = 0.2095$ m³/kg, $u_1 = 2984.8$ kJ/kg and $s_1 = 7.3322$ kJ/kg.K. Therefore, the mass of steam initially in the tank

$$m_1 = \frac{V}{v_1} = \frac{2}{0.2095} = 9.5465\,\text{kg}$$

Considering the tank and turbine together as the control volume, the unsteady energy and mass balance equation reduce to (with $\dot{Q} = 0$, $\dot{m}_i = 0$ and after neglecting KE and PE terms)

$$\frac{dU_{CV}}{dt} = -\dot{W}_x - \dot{m}_e\,h_e$$

$$\frac{dm_{CV}}{dt} = -\dot{m}_e$$

Upon combining these two expressions, we get,

$$\frac{d(mu)_{tank}}{dt} = -\dot{W}_x + \frac{dm_{tank}}{dt}\,h_e$$

where the mass inside the turbine has been neglected.

The steam may be assumed to undergo an isentropic expansion and hence the specific entropy of the steam in the tank at any instant, s and the specific entropy of the steam at the turbine exit, s_e, are both equal to s_1. Corresponding to 100 kPa, we can get from Table B, $s_f = 1.3024$ kJ/kg.K and $s_g = 7.3592$ kJ/kg.K. Therefore,

$$x_e = \frac{s_e - s_f}{s_g - s_f} = 0.9955$$

and $h_e = 2664.94$ kJ/kg.

We may now integrate the above expression for work to get

$$W_x = h_e \left(m_2 - m_1 \right) - \left(m_2 \, u_2 - m_1 \, u_1 \right) = 3253.57 \, \text{kJ}$$

9.6 Entropy change of a control volume

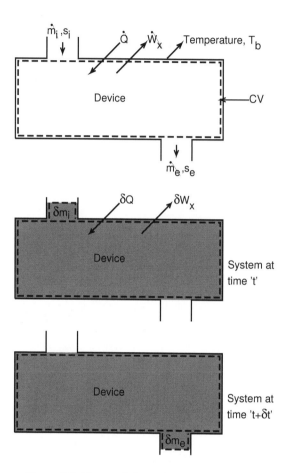

Figure 9.7: Entropy balance for a control volume

Consider the device shown in Fig. 4.2 again, which is reproduced here in Fig. 9.7 for convenience. The change in entropy of the system shown in this figure, between

$$dS = S_{CV,t+\delta t} + s_e\,\delta m_e - S_{CV,t} - s_i\,\delta m_i \qquad (9.19)$$

The left hand side may be expanded using the differential form of Eqn. 9.5 to give

$$dS = \int_{CS} \frac{\delta Q}{T_b} + \delta \sigma_{int}$$

where CS denotes the surface of the control volume or control surface and T_b denotes the temperature at a location on the control surface. The surface integral on the right hand side is required since the temperature is not the same across the entire control volume, unlike in a system. If we combine the above two expressions and rearrange, we get

$$S_{CV,t+\delta t} - S_{CV,t} = s_i\,\delta m_i - s_e\,\delta m_e + \int_{CS} \frac{\delta Q}{T_b} + \delta \sigma_{int}$$

Upon dividing both sides by δt, and taking the limit as $\delta t \to 0$, we finally get

$$\frac{dS_{CV}}{dt} = \dot{m}_i\,s_i - \dot{m}_e\,s_e + \int_{CS} \frac{\delta \dot{Q}}{T_b} + \dot{\sigma}_{int} \qquad (9.20)$$

If the control volume is at a steady state, then the time derivative in the left hand side becomes zero and we get

$$0 = \dot{m}_i\,s_i - \dot{m}_e\,s_e + \int_{CS} \frac{\delta \dot{Q}}{T_b} + \dot{\sigma}_{int}$$

For a device with a single inlet and single outlet, we may write $\dot{m}_i = \dot{m}_e = \dot{m}$, and so the above expression becomes

$$0 = \dot{m}\,(s_i - s_e) + \int_{CS} \frac{\delta \dot{Q}}{T_b} + \dot{\sigma}_{int} \qquad (9.21)$$

▪ EXAMPLE 9.6

Calculate the rate of entropy generation in the mixing chamber of Example 7.5 per unit mass flow rate of the exiting stream.

Solution : Since the mixing chamber operates at steady state, Eqn. 9.21 is applicable. For the present case, since the mixing chamber is insulated, we have

$$0 = \dot{m}_1\,s_1 + \dot{m}_2\,s_2 - \dot{m}_3\,s_3 + \dot{\sigma}_{int}$$

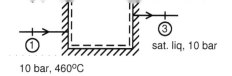

10 bar, 460°C

for the CV shown in the figure.

From Table C, we can retrieve, $s_1 = 7.6475$ kJ/kg.K. State 2 is a compressed liquid state. Hence, $s_2(10\,\text{bar}, 45°C) \approx s_f(45°C) = 0.6385$ kJ/kg.K, from Table A. State 3 is a saturated liquid state at 10 bar and so $s_3 = s_f(10\,\text{bar}) = 2.1387$ kJ/kg.K, from Table B. Therefore,

$$\frac{\dot{\sigma}_{int}}{\dot{m}} = -\frac{0.2181}{1.2181} s_1 - \frac{1}{1.2181} s_2 + s_3 = 0.2452\,\text{kJ/kg.K}$$

The entropy generation in this case is due to the internal irreversibility arising from mixing.

■ EXAMPLE 9.7

Saturated liquid R134a at 30°C is throttled to 200 kPa inside a domestic refrigerator. Determine the rate of entropy generation neglecting any heat loss. KE and PE changes may also be neglected.

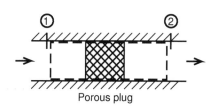

Porous plug

Solution : For the steady state throttling process, application of SFEE to the CV shown in the figure gives (after setting $\dot{Q} = 0$ and neglecting KE and PE changes), $h_2 = h_1$. The specific enthalpy at the inlet may be retrieved from Table D as 93.58 kJ/kg. Hence, $h_2 = 93.58$ kJ/kg.

From Table E, corresponding to 200 kPa, $h_f = 38.41$ kJ/kg and $h_g = 244.5$ kJ/kg. Since $h_f < h_2 < h_g$, the exit state after throttling is a saturated mixture. Hence, the final temperature is $T_2 = T_{sat}(200\,\text{kPa}) = -10°C$. The dryness fraction may be

Application of Eqn. 9.21 to the CV shown in the figures gives

$$0 = \dot{m}_1 s_1 - \dot{m}_2 s_2 + \dot{\sigma}_{int}$$

From Table D, we can retrieve $s_1 = s_f(30°C) = 0.3479$ kJ/kg.K. From Table E, we can retrieve $s_f = 0.1545$ kJ/kg.K and $s_g = 0.9379$ kJ/kg.K. The specific entropy at the exit may now be calculated as $s_2 = 0.3642$ kJ/kg.K. Hence,

$$\frac{\dot{\sigma}_{int}}{\dot{m}} = s_2 - s_1 = 0.0163 \text{ kJ/kg.K}$$

The entropy generation in this case is due the irreversibility associated with throttling.

■ **EXAMPLE 9.8**

Calculate the rate of entropy generation inside the heat exchanger in Example 7.13.

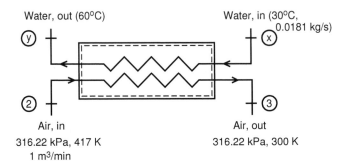

Water, out (60°C) Water, in (30°C, 0.0181 kg/s)

Air, in
316.22 kPa, 417 K
1 m³/min

Air, out
316.22 kPa, 300 K

Solution : For operation at steady state, we have

$$0 = \dot{m}_{water}(s_x - s_y) + \dot{m}_{air}(s_2 - s_3) + \dot{\sigma}_{int}$$

for the CV shown in the figure.

From Table A, $s_x = s_f(30°C) = 0.4365$ kJ/kg.K and $s_y = s_f(60°C) = 0.8312$ kJ/kg.K. From Eqn. 9.14, we have

$$s_3 - s_2 = C_p \ln\frac{T_3}{T_2} = \frac{1.4}{0.4}\frac{8.314}{28.8} \ln\frac{300}{417} = -0.3327 \text{ kJ/kg.K}$$

where we have assumed that there is no pressure loss. Therefore,

$$\dot{\sigma}_{int} = \dot{m}_{water}(s_y - s_x) + \dot{m}_{air}(s_3 - s_2) = 0.743 \text{ W/K}$$

Steam enters an insulated turbine operating at steady state at 60 bar, 400°C and exits at 10 bar, 190°C. Determine the power developed and rate of entropy generation if the mass flow rate is 10 kg/s. Neglect KE and PE changes.

Solution : SFEE applied to this case reduces to (with $\dot{Q} = 0$ and KE, PE changes neglected),

$$0 = -\dot{W}_x + \dot{m}\,(h_1 - h_2),$$

where 1 and 2 denote the inlet and exit states respectively.

From Table C, we can get $h_1 = 3177$ kJ/kg, $s_1 = 6.5404$ kJ/kg.K, $h_2 = 2801.85$ kJ/kg and $s_2 = 6.6372$ kJ/kg.K. Therefore, the power developed by the turbine is $\dot{W}_x = 3751.5$ kW.

Equation 9.21 becomes, for this case

$$0 = \dot{m}\,(s_1 - s_2) + \dot{\sigma}_{int}$$

The rate of entropy production may be evaluated as $\dot{\sigma}_{int} = 968$ W/K. This entropy production is due to the internal irreversibility arising from friction (both mechanical as well as fluid).

■ EXAMPLE 9.10

R134a enters an adiabatic compressor as saturated vapor at -5°C and leaves at 700 kPa, 40°C. If the mass flow rate is 0.5 kg/min, determine the compressor power and the rate of entropy generation. Assume steady state operation and neglect KE and PE changes.

Solution : SFEE for this case reduces to (with $\dot{Q} = 0$, KE, PE changes neglected),

$$0 = -\dot{W}_x + \dot{m}\,(h_1 - h_2)$$

where 1 and 2 denote the inlet and exit states respectively.

From Table D, we can get $h_1 = 247.55$ kJ/kg, $s_1 = 0.9345$ kJ/kg.K and from Table F, $h_2 = 278.6$ kJ/kg and $s_2 = 0.9642$ kJ/kg.K. Therefore, $\dot{W}_x = -15.525$ kJ/min. Required compressor power is 15.525 kJ/min.

Equation 9.21 becomes, for this case

$$0 = \dot{m}\,(s_1 - s_2) + \dot{\sigma}_{int}$$

Steam at 10 bar, 420°C enters an adiabatic nozzle with negligible velocity. It is expanded in the nozzle to 1 bar, 160°C. Assuming steady state operation and neglecting PE changes, determine the exit velocity and rate of entropy generation.

Solution : SFEE for this case reduces to ($\dot{Q} = 0$, $V_1 = 0$, PE change neglected),

$$V_2 = \sqrt{2\,(h_1 - h_2)}$$

where 1 and 2 denote the inlet and exit states respectively.

From Table C, we have h_1 = 3306.5 kJ/kg, s_1 = 7.5273 kJ/kg.K, h_2 = 2795.8 kJ/kg and s_2 = 7.6591 kJ/kg.K. Therefore, the exit velocity, V_2 = 1010 m/s. From Eqn. 9.21, we get

$$\frac{\dot{\sigma}_{int}}{\dot{m}} = s_2 - s_1 = 0.1318\,\mathrm{kJ/kg.K}$$

In this case also, the entropy production is due to the internal irreversibility arising from friction.

9.6.1 Work interaction of internally reversible steady flow processes

It was shown in section 3.2.1 that the displacement work for a system is $\int_1^2 P\,dV$ and that this is the area under the process curve drawn in $P-V$ coordinates. Equivalently, the specific displacement work, W/m, where m is the mass of the system, is equal to $\int_1^2 P\,dv$ and this is the area under the process curve in $P - v$ coordinates. A similar representation is possible in the case of control volumes that operate at steady state while executing, not surprisingly, an adiabatic reversible or a reversible isothermal process.

In the absence of any internal irreversibilities, Eqn. 9.21 may be written as

$$\begin{aligned}
0 &= \dot{m}\,(s_1 - s_2) + \int_{CS} \frac{\delta\dot{Q}}{T_b} \\
&= -\dot{m}\int_1^2 ds + \int_{CS} \frac{\delta\dot{Q}}{T_b}
\end{aligned} \qquad (9.22)$$

where *1* and *2* denote the inlet and outlet states respectively. SFEE (Eqn. 4.7) applied to this control volume gives (after neglecting KE and PE changes),

$$\dot{W}_x = \dot{Q} + \dot{m}\,(h_1 - h_2)$$

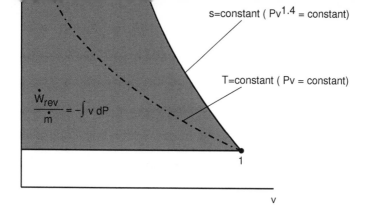

Figure 9.8: Work interaction for a steady state control volume executing an isentropic and an internally reversible isothermal compression process

$$= \dot{Q} - \dot{m} \int_1^2 dh$$

$$= \dot{Q} - \dot{m} \int_1^2 (T\,ds + v\,dP) \qquad (9.23)$$

Here, we have used Eqn. 9.8 to arrive at Eqn. 9.23.

For a control volume executing an isentropic process, Eqn. 9.23 becomes, with $\dot{Q} = 0$ and $ds = 0$,

$$\frac{\dot{W}_x}{\dot{m}} = -\int_1^2 v\,dP$$

The similarity between this expression and Eqn. 3.1 is quite obvious. The area enclosed by the integral in the above expression is illustrated graphically in Fig. 9.8 for a compression process undergone by an ideal gas. It is instructive to compare this with the shaded area shown in Fig. 3.2.

If the control volume executes an isothermal process, the second integral in the right hand side of Eqn. 9.22 may be simplified since, in this case, $T_b = T_1 = T_2 = T$ and so

$$0 = -\dot{m} \int_1^2 ds + \frac{1}{T} \int_{CS} \delta\dot{Q}$$

where the temperature has been taken inside the integral since it is a constant. Upon substituting this expression into Eqn. 9.23, we get

$$\frac{\dot{W}_x}{\dot{m}} = -\int_1^2 v\, dP$$

for a control volume executing an internally reversible, isothermal process. The process curve is shown in Fig. 9.8. It is evident that the work required in a reversible, isothermal compression is less than that required in an isentropic process in a steady flow device. This is physically understandable, since the work required to compress a fluid is less when its temperature is not allowed to increase during the compression process.

9.7 Isentropic efficiency

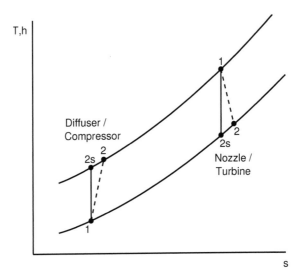

Figure 9.9: Adiabatic, irreversible and the corresponding isentropic process for various devices

When internal irreversibilities such as mixing, stirring and throttling are present, the attendant increase in entropy remains what it is without any scope to further reduce it. In contrast, when the internal irreversibility is friction as in the case of turbines, compressors, nozzles and diffusers (assumed to be adiabatic), it is possible,

same exit pressure as in the actual process. This is illustrated in Fig. 9.9 and in Fig. 9.6. Note that, state 2 always lies to the right of state 1, since process 1-2 is an irreversible adiabatic process and hence $s_2 > s_1$.

The discussion above suggests that a performance metric, denoted η, may be defined for such devices as follows:

Turbine, Nozzle :

$$\eta = \frac{h_1 - h_2}{h_1 - h_{2s}} \qquad (9.24)$$

Compressor, Diffuser :

$$\eta = \frac{h_{2s} - h_1}{h_2 - h_1} \qquad (9.25)$$

The quantity η is called the isentropic efficiency. It compares the enthalpy change in the actual and the ideal (isentropic) process. The enthalpy change is the work output and work input in the case of a turbine and compressor respectively. It is also the increase and decrease in the kinetic energy in a nozzle and diffuser respectively. The interpretation of isentropic efficiency is straight forward except in the case of the diffuser. In this case, since $\eta < 1$ and $P_{2s} = P_2$, the reduction in kinetic energy required to diffuse the flow to the same exit pressure is less in an isentropic process. In other words, for a given decrease in the kinetic energy, the pressure rise is more in an isentropic process when compared to an adiabatic irreversible process.

■ **EXAMPLE 9.12**

Calculate the isentropic efficiency of the turbine in Example 9.9.

Solution : At the end of an isentropic expansion to the same exit pressure, $P_{2s} = 10$ bar and $s_{2s} = s_1 = 6.5404$ kJ/kg.K. From Table B, with $s_f = 2.1387$ kJ/kg.K and $s_g = 6.5861$ kJ/kg.K, it is clear that, state $2s$ is in the mixture region. The dryness fraction may be evaluated as

$$x_{2s} = \frac{s_{2s} - s_f}{s_g - s_f} = 0.9897$$

From Table B, with $h_f = 762.8$ kJ/kg and $h_g = 2777.7$ kJ/kg, the specific enthalpy at state $2s$, is given as

$$h_{2s} = h_f + x_{2s}(h_g - h_f) = 2756.95 \text{ kJ/kg}$$

Therefore, the isentropic efficiency

$$\eta_t = \frac{h_1 - h_2}{h_1 - h_{2s}} = 0.89$$

we have $P_{2s} = 700$ kPa, $s_{2s} = s_1 = 0.9345$ kJ/kg.K. This is a superheated state. From Table F, we can retrieve $h_{2s} = 269.45$ kJ/kg after interpolation. Therefore, the isentropic efficiency

$$\eta_c = \frac{h_{2s} - h_1}{h_2 - h_1} = 0.71$$

EXAMPLE 9.14

Calculate the isentropic efficiency of the nozzle in Example 9.11.

Solution : At the end of an isentropic expansion process to the same pressure, we have $P_{2s} = 1$ bar, $s_{2s} = s_1 = 7.5273$ kJ/kg.K. This is a superheated state. From Table C, we can retrieve $h_{2s} = 2741.4$ kJ/kg after interpolation. Therefore, the isentropic efficiency

$$\eta_n = \frac{h_1 - h_2}{h_1 - h_{2s}} = 0.9$$

EXAMPLE 9.15

If the diffuser in Example 7.4 has an isentropic efficiency of 94 percent, determine the final speed to which the air is decelerated for the same pressure rise.

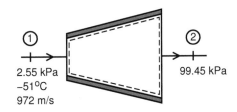

①
2.55 kPa
−51°C
972 m/s

②
99.45 kPa

Solution : For an isentropic process, we have, from Example 7.4, $T_{2s} = 632$ K and $V_{2s} = 341$ m/s.

Equation 9.25 may be written for an ideal gas as

$$\eta = \frac{h_{2s} - h_1}{h_2 - h_1} = \frac{T_{2s} - T_1}{T_2 - T_1}$$

Upon substituting the known values into this expression, we get $T_2 = 658$ K. The final speed in this case may be calculated in the same manner as in Example 7.4 as

Superheated steam at 3 MPa enters an adiabatic turbine at the rate of 10 kg/s where it is expanded to 100 kPa. The turbine is expected to produce 5556 kW of power while ensuring that there is no liquid at the turbine exit. Calculate (a) the minimum permissible entry temperature and (b) the isentropic efficiency of the turbine. Assume steady flow operation and neglect KE and PE changes.

Solution :
(a) Since there should be no liquid at the turbine exit, the minimum permissible turbine entry temperature is such that the exit state is a saturated vapor state. Hence, $h_2 = 2675.1$ kJ/kg, from Table B.

SFEE applied to this case reduces to (with $\dot{Q} = 0$ and KE, PE changes neglected),

$$0 = -\dot{W}_x + \dot{m}\,(h_1 - h_2)$$

where 1 and 2 denote the inlet and exit states respectively.

Upon substituting the known values into this expression, we get $h_1 = 3230.7$ kJ/kg. From Table C, we can determine that this corresponds to $400°C$.

(b) From Table C, we can retrieve $s_1 = 6.9210$ kJ/kg.K. Therefore, at state $2s$, $s_{2s} = s_1 = 6.9210$ kJ/kg.K and $P_{2s} = 100$ kPa. At 100 kPa, from Table B, we have $s_f = 1.3024$ kJ/kg.K and $s_g = 7.3592$ kJ/kg.K. Since $s_f < s_{2s} < s_g$, state $2s$ lies in the mixture region. Accordingly,

$$x_{2s} = \frac{s_{2s} - s_f}{s_g - s_f} = 0.9277$$

With $h_f = 417.4$ kJ/kg and $h_g = 2675.1$ kJ/kg (from Table B), we can get $h_{2s} = 2511.87$ kJ/kg. The isentropic efficiency of the turbine

$$\eta_t = \frac{h_1 - h_2}{h_1 - h_{2s}} = 0.77$$

9.8 Principle of increase of entropy

Equations 9.5 and 9.20 respectively allow the entropy change of a system or control volume to be evaluated in the most general case. However, all the examples worked out in the earlier sections have been for adiabatic devices. Consequently, the entropy

Let a thermodynamic system at temperature T have a heat interaction δQ with the surroundings at temperature T_0. The change in entropy of the system, denoted dS_{sys} is given as (Eqn. 9.5)

$$dS_{sys} = \frac{\delta Q_{sys}}{T} + \delta\sigma_{int}$$

where $\delta Q_{sys} = \delta Q$. The change in entropy of the surroundings, denoted dS_{surr}, is given simply as

$$dS_{surr} = \frac{\delta Q_{surr}}{T_0}$$

Note that $\delta Q_{surr} = -\delta Q_{sys}$. Furthermore, there is no internal irreversibility in the surroundings since it is the system that executes a process - not the surroundings. Since the system and the surroundings together comprise the universe, the entropy change of the universe may be evaluated by combining the two expressions above. Thus

$$dS_{universe} = dS_{sys} + dS_{surr} = \delta Q \left(\frac{1}{T} - \frac{1}{T_0} \right) + \delta\sigma_{int} \qquad (9.26)$$

Note that δQ in this expression is positive if the system receives heat and negative if heat is rejected by it. The quantity within the parentheses in the right hand side of this equation arises from the external irreversibility associated with heat transfer across a finite temperature difference (T and T_0). Let us now determine the sign for $dS_{universe}$ by considering the following possibilities:

No internal or external irreversibility : In this case, $\delta\sigma_{int} = 0$ and $T = T_0$. Consequently, both terms in the right hand side are zero. Hence, the entropy of the universe remains the same.

Internal irreversibility; no external irreversibility : In this case, $T = T_0$. Since $\delta\sigma_{int}$ is always greater than zero, the entropy of the universe increases.

External irreversibility; no internal irreversibility: In this case, $\delta\sigma_{int} = 0$. However, now, the following possibilities must be considered.

> **$T > T_0$:** In this case, heat is rejected by the system and so δQ is negative. The quantity within the parentheses is also negative. Hence, the product term in the right hand side is positive. Thus, the entropy of the universe increases.
>
> **$T < T_0$:** In this case, heat is supplied to the system and so δQ is positive. The quantity within the parentheses is also positive. Hence, the product term in the right hand side is positive. The entropy of the universe increases in this case as well.
>
> **Internal irreversibility; External irreversibility :** In this case, both the terms in the right hand side are non-zero and positive. Hence the entropy of the universe increases.

For something to become cleaner, something else must become dirtier and *Left to themselves, things tend to go from bad to worse.*

In view of the universality of the Principle of increase of entropy, it is important to establish that (a) any device that violates the Kelvin-Planck and Clausius statements also violates this principle and (b) the thermodynamic temperature scale is consistent with this principle. To this end, we first integrate Eqn. 9.26 to get

$$\Delta S_{universe} = \Delta S_{sys} + \Delta S_{surr} \tag{9.27}$$

for a system executing a process.

Kelvin-Planck statement : Consider the direct engine in Fig. 8.5 that violates the Kelvin-Planck statement. If we take this engine to be the system, then the entropy change is zero since it is supposed to operate in a cycle. Hence, $\Delta S_{sys} = 0$. The entropy change of the surroundings during each cycle is given as

$$\Delta S_{surr} = -\frac{Q_H}{T_H}.$$

Therefore, the entropy change of the universe during every cycle is given as

$$\Delta S_{universe} = \Delta S_{sys} + \Delta S_{surr} = -\frac{Q_H}{T_H}$$

Since this is negative, it is clear that the engine violates the entropy increase principle also.

Clausius statement : Consider the reverse engine in Fig. 8.6 that violates the Clausius statement. Once again, if we take the engine as the system, its entropy change is zero during each cycle. The entropy change of the surroundings during each cycle is given as

$$\Delta S_{surr} = -\frac{Q_C}{T_C} + \frac{Q_C}{T_H}$$

Therefore, the entropy change of the universe during every cycle is given as

$$\Delta S_{universe} = \Delta S_{sys} + \Delta S_{surr} = Q_C \left(\frac{1}{T_H} - \frac{1}{T_C}\right)$$

[†]A special case relating to Eqn. 9.26 is the isolated system. By definition, this is a system that has no interaction with the surroundings. It is easy to establish from Eqn. 9.26 that the entropy of an isolated system must increase or remain the same. In this context, an isolated system containing a demon (popularly called Maxwell's demon) proposed by James Clerk Maxwell in 1871 was, for long, thought to violate this principle and hence the second law. It has now been proven convincingly using the theory of computing that this is not so. Interested readers may read the article "Demons, Engines and the Second Law" by Charles Bennett that appeared in Scientific American in 1987 (pp. 108-116).

Thermodynamic temperature scale : Consider the Carnot engine in Fig. 8.7 and the Carnot cycle shown in Fig. 8.8 that it executes. For the Carnot engine as the system, the entropy change during each cycle is zero. The entropy change of the surroundings is given as

$$\Delta S_{surr} = -\frac{Q_H}{T_H} + \frac{Q_C}{T_C}$$

The entropy change of the universe during each cycle is thus

$$\Delta S_{universe} = \Delta S_{sys} + \Delta S_{surr} = -\frac{Q_H}{T_H} + \frac{Q_C}{T_C}$$

Since there are no internal or external irreversibilities, the entropy change of the universe should be zero. It then follows that

$$\frac{Q_H}{T_H} = \frac{Q_C}{T_C}$$

Hence, the absolute or thermodynamic temperature scale is consistent with the entropy increase principle.

9.8.1 Entropy generation

We have established that the entropy change of the universe, if non-zero, is always positive. From Eqn. 9.26, it is evident that the magnitude of the entropy change is solely due to contributions from internal and external irreversibilities. Hence, entropy generated due to irreversibilities is equal to the entropy change of the universe. The entropy generated when a system executes a process can be used as a metric for a comparative assessment of the process against other processes or an appropriately defined ideal process. Consequently, calculation of the entropy generated during a process is very important and this is discussed next.

Entropy generation during a non-flow process Entropy generated during a non-flow process, σ, may simply be evaluated from Eqn. 9.27 as

$$\sigma = \Delta S_{universe} = \Delta S_{sys} + \Delta S_{surr} = \Delta S_{sys} + \frac{Q_{surr}}{T_0} \tag{9.28}$$

The subscript $surr$ is used for Q in the right hand side to explicitly indicate the sign convention to be used for the heat interaction.

at 140°C and then moved to a water bath that is initially at room temperature. It is finally allowed to cool in the ambient air. Determine the entropy generated during this process. The water bath may be assumed to contain 80 kg of water (specific heat capacity 4200 J/kg.K) and heat loss from the bath to the ambient may be neglected.

Solution : Let us take the casting as the system and calculate its entropy change using Eqn. 9.12 and that of the universe during the process step by step.

Step 1 :

$$\Delta S_{sys} = 20 \times 0.5 \times \ln \frac{273 + 140}{273 + 200} = -1.3565 \, \text{kJ/kg.K}$$

The heat transferred to the surroundings during this process may be obtained using first law as $Q = 20 \times 0.5 \times (200 - 140) = 600 \, \text{kJ}$. Therefore, the entropy change of the surroundings during this step is

$$\Delta S_{surr} = \frac{600}{273 + 140} = 1.4528 \, \text{kJ/K}$$

Step 2 : The final temperature of the water bath may be obtained by applying first law to the casting and the water bath. Accordingly, with $Q = 0$ and $W = 0$, the final temperature of the bath may be evaluated as

$$T = \frac{20 \times 500 \times 140 + 80 \times 4200 \times 27}{20 \times 500 + 80 \times 4200} = 30.266°C$$

Therefore,

$$\Delta S_{sys} = 20 \times 0.5 \times \ln \frac{273 + 30.266}{273 + 140} = -3.088 \, \text{kJ/kg.K}$$

and

$$\Delta S_{surr} = 80 \times 4.2 \times \ln \frac{273 + 30.266}{273 + 27} = 3.638 \, \text{kJ/kg.K}$$

Step 3 :

$$\Delta S_{sys} = 20 \times 0.5 \times \ln \frac{273 + 27}{273 + 30.266} = -0.1083 \, \text{kJ/kg.K}$$

The heat transferred to the surroundings during this process may be obtained as $Q = 20 \times 500 \times (30.266 - 27) = 32.66 \, \text{kJ}$. Therefore, the entropy change of the surroundings during this step is

$$\Delta S_{surr} = \frac{32.66}{273 + 27} = 0.1089 \, \text{kJ/K}$$

EXAMPLE 9.18

Calculate the entropy generated during the polytropic process given in Example 6.2, wherein two kg of air contained in a piston cylinder assembly is compressed from 100 kPa, 300 K to 800 kPa. The ambient may be assumed to be at 300 K.

Solution : The final temperature of the air was evaluated in Example 6.2 to be 467 K and the heat transfer, -116.06 kJ.

Taking the air as the system, the entropy change of the system may be evaluated using Eqn. 9.14 as

$$\Delta S_{sys} = m \left(C_p \ln \frac{T_2}{T_1} - R \ln \frac{P_2}{P_1} \right)$$

$$= 2 \times \frac{8314}{28.8} \left(\frac{1.4}{0.4} \ln \frac{467}{300} - \ln \frac{800}{100} \right)$$

$$= -306.31 \text{ J/K}$$

The entropy change of the surroundings may be evaluated as

$$\Delta S_{surr} = \frac{Q_{surr}}{T_0} = \frac{116.06 \times 1000}{300} = 386.87 \text{ J/K}$$

Therefore, entropy generated may be evaluated from Eqn. 9.28 as

$$\sigma = \Delta S_{sys} + \Delta S_{surr} = 80.56 \text{ J/K}$$

EXAMPLE 9.19

Five kg of saturated R134a vapor at -15°C is contained in a rigid vessel. The R134a is stirred by transferring an amount of work equal to 500 kJ. Simultaneously, heat transfer to the ambient at 30°C also takes place until the content of the vessel reaches the ambient temperature. Determine the entropy generated during the process.

Solution : At the initial state, the required property values for the R134a in the vessel may be retrieved from Table D as: $v_1 = 0.12066$ m³/kg, $u_1 = 221.72$ kJ/kg, $s_1 = 0.9415$ kJ/kg.K. At the final state, $v_2 = v_1 = 0.12066$ m³/kg and $T_2 = 30°C$. From Table D, we can establish that, this is a superheated state, since $v_2 > v_g(30°C)$. Thus, we can retrieve from Table F, $u_2 = 255.23$ kJ/kg and $s_2 = 1.0613$ kJ/kg.K.

$$= 5 \times (255.23 - 221.72) - 500 = -332.45\,\text{kJ}$$

Heat is thus lost to the ambient. Entropy change of the system amounts to

$$\Delta S_{sys} = m\,(s_2 - s_1) = 0.599\,\text{kJ/K}$$

It is important to note that the internal irreversibility in this process is so high that the entropy increases despite the heat loss to the ambient. From Eqn. 9.28, the entropy generated during the process is given as

$$\sigma = \Delta S_{sys} + \frac{Q_{surr}}{T_0} = 0.6 + \frac{332.45}{273 + 30} = 1.6962\,\text{kJ/K}$$

■ EXAMPLE 9.20

A rigid container of volume of 400 L is divided into two equal compartments A and B by a thin partition. Both compartments contain air; compartment A is at 2 MPa, 200°C, and compartment B is at 200 kPa, 100°C. The partition ruptures and the contents eventually reach an equilibrium state while losing heat to the ambient at 27°C. Determine (a) the final pressure, (b) heat lost to the surroundings and (c) the entropy generated.

Solution : Mass of air in compartment A may be evaluated as

$$m_A = \left.\frac{PV}{RT}\right|_{1,A} = \frac{2000 \times 10^3 \times 0.2}{288.68 \times 473} = 2.93\,\text{kg}$$

Similarly, mass of air in compartment B is given as

$$m_B = \left.\frac{PV}{RT}\right|_{1,B} = \frac{200 \times 10^3 \times 0.2}{288.68 \times 373} = 0.3715\,\text{kg}$$

(a) Since the final temperature T_2 is equal to the ambient temperature, the final pressure may be calculated as

$$P_2 = \frac{(m_A + m_B)\,R\,T_2}{V_A + V_B} = 714.81\,\text{kPa}$$

(b) Taking the air in both the compartments as the system, application of first law gives (with $W = 0$)

$$\Delta E = \Delta U = Q$$
$$\Rightarrow \quad Q = m_A\,C_v\,(T_2 - T_{1,A}) + m_B\,C_v\,(T_2 - T_{1,B})$$

$$\Delta S_{sys} = m_A \left(C_p \ln \frac{T_2}{T_{1,A}} - R \ln \frac{P_2}{P_{1,A}} \right) + m_B \left(C_p \ln \frac{T_2}{T_{1,B}} - R \ln \frac{P_2}{P_{1,B}} \right)$$

$$= -477.65 \, \text{J/K} - 218.35 \, \text{J/K} = -696 \, \text{J/K}$$

From Eqn. 9.28, the entropy generated during the process is given as

$$\sigma = \Delta S_{sys} + \frac{Q_{surr}}{T_0} = -696 + \frac{385.39 \times 10^3}{273 + 27} = 588.63 \, \text{J/K}$$

Entropy generation during a flow process Consider the device shown in Fig. 9.7, which executes a flow process. The change in entropy of the system shown in this figure, between time $t + \delta t$ and t, has already been shown to be (Eqn. 9.19)

$$dS_{sys} = S_{CV, t+\delta t} + s_e \, \delta m_e - S_{CV, t} - s_i \, \delta m_i$$

The entropy change of the surroundings during this time interval is given as

$$dS_{surr} = \frac{\delta Q_{surr}}{T_0}.$$

Therefore the entropy generated during this time interval is

$$\delta \sigma = dS_{sys} + dS_{surr} = S_{CV, t+\delta t} + s_e \, \delta m_e - S_{CV, t} - s_i \, \delta m_i + \frac{\delta Q_{surr}}{T_0}$$

The rate at which entropy is generated during the flow process executed by the device is thus

$$\dot{\sigma} = \frac{dS_{CV}}{dt} + \dot{m}_e \, s_e - \dot{m}_i \, s_i + \frac{\dot{Q}_{surr}}{T_0} \tag{9.29}$$

For steady state operation, the time derivative in the right hand side vanishes.

■ **EXAMPLE 9.21**

Calculate the rate of entropy generation during the throttling process in Example 7.6. The ambient temperature may be taken as 27°C.

Solution : Equation 9.29 may be simplified for this process to give

$$\dot{\sigma} = \dot{m} \, (s_2 - s_1) + \frac{\dot{Q}_{surr}}{T_0}$$

Porous plug

From Table D, we can get $s_1 = s_f = 0.3479$ kJ/kg.K. Corresponding to 200 kPa, we can get from Table E, $s_f = 0.1545$ kJ/kg.K and $s_g = 0.9379$ kJ/kg.K. With $x_2 = 0.258$ from Example 7.6, we can calculate $s_2 = 0.3566$ kJ/kg.K.

Upon substituting the known values into the above expression, we get

$$\frac{\dot{\sigma}}{\dot{m}} = (0.3566 - 0.3479) + \frac{2}{273 + 27} = 15.37 \, \text{J/kg.K}$$

EXAMPLE 9.22

Refrigerant 134-a is compressed steadily in a compressor from 140 kPa, -10°C to 700 kPa, 120°C. The volumetric flow rate at the inlet is 0.001 m³/s and the input power is 1.0 kW. Neglecting KE and PE changes, determine the rate of entropy generation. Assume the ambient temperature to be 27°C.

Solution : Application of SFEE in this case leads to (with KE, PE changes neglected),

$$0 = \dot{Q} - \dot{W}_x + \dot{m}(h_1 - h_2)$$

From Table F, we can retrieve, $v_1 = 0.1461$ m³/kg, $h_1 = 246.4$ kJ/kg, $s_1 = 0.9724$ kJ/kg.K, $h_2 = 358.9$ kJ/kg and $s_2 = 1.192$ kJ/kg.K.

The mass flow rate may be evaluated as

$$\dot{m} = \frac{\dot{V}_1}{v_1} = \frac{0.001}{0.1461} = 0.006845 \, \text{kg/s}$$

Therefore

$$\dot{Q} = \dot{W}_x - \dot{m}(h_1 - h_2) = -0.23 \, \text{kW}$$

The rate of entropy generation may be calculated from Eqn. 9.29 as

$$\dot{\sigma} = \dot{m}(s_2 - s_1) + \frac{\dot{Q}_{surr}}{T_0}$$

EXAMPLE 9.23

Steam enters a turbine steadily at 30 bar and 440°C at a rate of 8 kg/s and exits at 3 bar and 160°C. Heat is lost to the surroundings at 27°C at a rate of 300 kW. KE and PE changes are negligible. Determine (a) the actual power output and (b) the rate at which entropy is generated.

Solution :

(a) Application of SFEE in this case leads to (with KE, PE changes neglected),

$$0 = \dot{Q} - \dot{W}_x + \dot{m}(h_1 - h_2)$$

From Table C, we can retrieve $h_1 = 3321.5$ kJ/kg, $s_1 = 7.0521$ kJ/kg.K, $h_2 = 2782.1$ kJ/kg and $s_2 = 7.1274$ kJ/kg.K. If we substitute the numerical values into the above expression, we get the actual power output to be

$$\dot{W}_x = -300 + 8 \times (3321.5 - 2782.1) = 4015.2\,\text{kW}$$

(b) The rate of entropy generation may be calculated from Eqn. 9.29 as

$$
\begin{aligned}
\dot{\sigma} &= \dot{m}(s_2 - s_1) + \frac{\dot{Q}_{surr}}{T_0} \\
&= 8 \times (7.1274 - 7.0521) + \frac{300}{273 + 27} \\
&= 1.6024\,\text{kW/K}
\end{aligned}
$$

EXAMPLE 9.24

Water at 200 kPa and 30°C enters the mixing chamber of a waste heat recovery unit at a rate of 5 kg/s where it is mixed with steam at 200 kPa and 200°C. Heat loss to the surrounding air at 27°C occurs at a rate of 10 kW. If the mixture is required to leave at 200 kPa and 80°C, determine (a) the required mass flow rate of steam and (b) the rate of entropy generation during this mixing process.

Solution : From Table B, it may be ascertained that water at 200 kPa and 30°C as well as 200 kPa and 80°C are both compressed liquid states, since the saturation temperature corresponding to 200 kPa is 120.2°C. Let these states be labelled 1 and 3 respectively. Accordingly, $h_1 \approx u_f(30°C) + 200\,v_f(30°C) = 125.871$ kJ/kg, $s_1 \approx s_f(30°C) = 0.4365$ kJ/kg.K, $h_3 \approx u_f(80°C) + 200\,v_f(80°C) = 335.09$ kJ/kg and $s_3 \approx s_f(80°C) = 1.0753$ kJ/kg.K. From Table C, corresponding to the superheated state at 200 kPa and 200°C (labelled 2), $h_2 = 2870$ kJ/kg and $s_2 = 7.5319$ kJ/kg.K.

$$Q - \dot{W}_x + \dot{m}_1\,h_1 + \dot{m}_2\,h_2 - (\dot{m}_1 + \dot{m}_2)\,h_3 = 0$$

If we substitute the known numerical values into this expression, we get the required mass flow rate of steam, $\dot{m}_2$ to be 0.4166 kg/s.

(b) Application of Eqn. 9.29 leads to

$$
\begin{aligned}
\dot{\sigma} &= \dot{m}_3\,s_3 - \dot{m}_1\,s_1 - \dot{m}_2\,s_2 + \frac{\dot{Q}_{surr}}{T_0} \\
&= 5.4166 \times 1.0753 - 5 \times 0.4365 - 0.4166 \times 7.5319 + \frac{10}{273 + 27} \\
&= 537.51 \ \mathrm{W/K}
\end{aligned}
$$

■ EXAMPLE 9.25

Determine the entropy generated during the transient process described in Example 7.18, wherein the pressure of the air in the tank decreased from 1 MPa to 100 kPa with the temperature remaining constant at 300 K. What would be the entropy generated if the temperature of the air in the tank remains constant as a result of heat transfer with the ambient at 27°C, instead of electrical work.

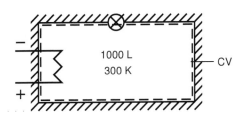

1000 L
300 K
— CV

Solution : For the CV shown in the figure, Eqn. 9.29 simplifies, with $\dot{m}_i = 0$, as

$$\dot{\sigma} = \frac{dS_{CV}}{dt} + \dot{m}_e\,s_e + \frac{\dot{Q}_{surr}}{T_0}$$

The unsteady mass balance equation, Eqn. 4.5, reduces to

$$\frac{dm_{CV}}{dt} = -\dot{m}_e$$

Upon combining the two expressions above, we get

$$\dot{\sigma} = \frac{d}{dt}\left(m_{CV}\,s_{CV}\right) - s_{CV}\,\frac{dm_{CV}}{dt} + \frac{\dot{Q}_{surr}}{T_0}$$

If we multiply both sides by dt and use Eqn. 9.10 to replace ds (while noting that $dT = 0$), we get

$$\dot{\sigma}\, dt = \left(\frac{PV}{RT}\right)\left(-R\,\frac{dP}{P}\right) + \frac{\dot{Q}_{surr}}{T_0}\, dt$$

The subscript CV has been dropped as there is no danger of ambiguity. Upon integrating this equation from the initial to the final state, we get

$$
\begin{aligned}
\sigma &= -\frac{V}{T}(P_2 - P_1) + \overset{0}{\cancel{\frac{Q}{T_0}}} \\
&= -\frac{1}{300}(100 - 1000) \\
&= 3\,\text{kJ/K}
\end{aligned}
$$

The entropy generation is due to internal irreversibility associated with the work transfer.

If the temperature of the air inside the tank is maintained constant as a result of heat transfer with the ambient, then $\dot{W}_x = 0$ and $Q = 900$ kJ. Hence $Q_{surr} = -900$ kJ. Therefore,

$$
\begin{aligned}
\sigma &= -\frac{V}{T}(P_2 - P_1) + \frac{Q_{surr}}{T_0} \\
&= -\frac{1}{300}(100 - 1000) - \frac{900}{273 + 27} \\
&= 0\,\text{kJ/K}
\end{aligned}
$$

There is no entropy generation on account of the fact that there is no internal or external irreversibility (the system and the surrounding are at the same temperature).

■ EXAMPLE 9.26

Determine the entropy generated during the transient process described in Example 7.19. Assume that the heat is supplied from a reservoir at 50°C.

Solution : The specific entropy of the content of the vessel at the initial and final state may be calculated as, $s_1 = 0.4078$ kJ/kg.K and $s_2 = s_g = 0.9171$ kJ/kg.K.

For the control volume used in Example 7.19, Eqn. 9.29 simplifies, with $\dot{m}_i = 0$, as

$$\dot{\sigma} = \frac{dS_{CV}}{dt} + \dot{m}_e\, s_e + \frac{\dot{Q}_{surr}}{T_0}$$

Upon combining the two expressions above, we get

$$\dot{\sigma} = \frac{dS_{CV}}{dt} - s_g \frac{dm_{CV}}{dt} + \frac{\dot{Q}_{surr}}{T_0}$$

where we have used the fact the specific entropy of the exiting stream is equal to the specific entropy of saturated vapor at 900 kPa, since only saturated vapor is allowed to escape. If we multiply both sides by dt and integrate from the initial to the final state, we get

$$
\begin{aligned}
\sigma &= (m_2 s_2 - m_1 s_1) - s_g (m_2 - m_1) + \frac{Q_{surr}}{T_0} \\
&= (1.57 \times 0.9171 - 16 \times 0.4078) - 0.9171 \times (1.57 - 16) - \frac{2515.285}{273 + 50} \\
&= 361.54 \, \text{J/K}
\end{aligned}
$$

■ EXAMPLE 9.27

Determine the entropy generated during the transient process described in Example 7.20. Assume that the heat is supplied from a reservoir at 250°C.

Solution : The specific entropy of the content of the vessel at the initial and final state may be calculated as, $s_1 = 5.3665$ kJ/kg.K and $s_2 = 6.3397$ kJ/kg.K.

Proceeding in the same manner as in the previous example, it is easy to show that

$$\dot{\sigma} = \frac{dS_{CV}}{dt} - s_f \frac{dm_{CV}}{dt} + \frac{\dot{Q}_{surr}}{T_0}$$

If we multiply both sides by dt and integrate from the initial to the final state, we get

$$
\begin{aligned}
\sigma &= (m_2 s_2 - m_1 s_1) - s_f (m_2 - m_1) + \frac{Q_{surr}}{T_0} \\
&= (6.024 \times 6.3397 - 8 \times 5.3665) - 2.447 \times (6.024 - 8) - \frac{45.22}{273 + 250} \\
&= 6.3621 \, \text{J/K}
\end{aligned}
$$

EXERGY

It was demonstrated in Chapter 8 that, among all engines that operate in a cycle between two thermal reservoirs, the Carnot engine produces the maximum work and its efficiency is the highest possible. Although the Carnot efficiency of simple cycles may be calculated in a straightforward manner, it is not easy to evaluate for complicated and realistic cycles and so it is essential to be able to determine, in some manner, the maximum possible efficiency in such cases. In addition, for devices that execute non-cyclic processes, it is desirable to be able to define an appropriate ideal process between the same initial and final state and evaluate its performance. Although the isentropic efficiency is such a process, its applicability is restricted to adiabatic devices for which the ideal process is an isentropic process, for instance, turbines, compressors, nozzles and diffusers. It would be extremely useful to be able to quantitatively assess the performance of devices such as mixing chambers and heat exchangers, to name a few, or turbines, compressors, nozzles and diffusers when they do not operate adiabatically. With these issues in mind, in the present chapter, a new quantity termed exergy, is defined. It is quite general and allows us to calculate a limiting value against which the actual performance of any device or cycle may be compared.

is such that it is not possible to develop any work from a system that exists at the dead state. The ambient state at 25°C and 100 kPa, is taken to be the dead state.

The exergy of a system at a given state, denoted X, is defined as the maximum theoretical work that can be developed as the system goes from the given state to the dead state. By definition, the exergy of a system that is at the dead state is zero. It follows then, that the exergy of a system that is not at the dead state is non-zero, but not whether it is positive or negative. The latter aspect is best established through the examples illustrated in Fig. 10.1.

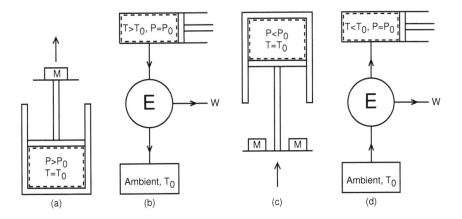

Figure 10.1: Demonstrations of systems (shown shaded) developing work while approaching the dead state

The systems shown in Fig. 10.1(a) and (b) are initially at a state where the pressure or the temperature is *above* that of the ambient. It is easy to see how positive work can be developed as these systems attain the dead state. The exergy of these systems at their initial state is thus positive. The systems in Fig. 10.1(c) and (d) are initially at a pressure or temperature that is *below* the corresponding ambient value but it is evident that these systems can also develop positive work as they attain the ambient state. Hence, the exergy of the initial state of these systems is also positive. In general, we can thus state that the exergy of a system is *always* positive and non-zero, as long as it is not at the dead state. Let us now turn to the evaluation of the exergy of a system.

If we apply first law to any of the systems shown in Fig. 10.1, we get

$$\Delta E_{sys} = Q_{sys} - W_{sys} \qquad (10.1)$$

$$\Delta E_{sys} \;=\; E_0 - E_1 = U_0 + \frac{m\cancel{V_0^2}}{2} + mg\cancel{z_0}^{\,0} - U_1 - \frac{mV_1^2}{2} - mgz_1$$

$$= \; U_0 - U_1 - \frac{mV_1^2}{2} - mgz_1$$

where m is the mass contained in the system.

Next, we write the work term in Eqn. 10.1 as

$$W_{sys} = W_u + P_0\,\Delta V_{sys} = W_u + P_0\,(V_0 - V_1)$$

Here, the total work developed by the system has been decomposed into a useful part *i.e.*, one which can be used for lifting a weight and a part which is used for just displacing the atmosphere. The exergy of the system at its initial state is thus the maximum value for W_u that can be realized as it goes to the dead state.

We now turn to the remaining term in Eqn. 10.1, namely, Q_{sys}. For any of the systems shown in Fig. 10.1, we may write

$$\Delta S_{sys} + \Delta S_{surr} \;=\; \sigma$$

$$\Delta S_{sys} + \frac{Q_{surr}}{T_{surr}} \;=\; \sigma$$

$$\Delta S_{sys} - \frac{Q_{sys}}{T_0} \;=\; \sigma$$

where we have used the fact that $Q_{surr} = -Q_{sys}$. Therefore,

$$Q_{sys} = T_0\,\Delta S_{sys} - T_0\sigma = T_0\,(S_0 - S_1) - T_0\sigma$$

Equation 10.1 may now be written as

$$W_u \;=\; T_0\,(S_0 - S_1) - \left(U_0 - U_1 - \frac{mV_1^2}{2} - mgz_1\right) - P_0\,(V_0 - V_1) - T_0\sigma$$

$$= \; (U_1 - U_0) + P_0\,(V_1 - V_0) - T_0\,(S_1 - S_0) + \frac{mV_1^2}{2} + mgz_1 - T_0\sigma$$

$$\text{(10.2)}$$

It is easy to see that, since σ is always greater than or equal to zero, W_u is a maximum when $\sigma = 0$, *i.e.*, when there are no internal or external irreversibilities and the process is fully reversible. Thus,

$$X = W_{u,\,max} = W_{u,\,rev} = (U - U_0) + P_0\,(V - V_0) - T_0\,(S - S_0) + \frac{mV^2}{2} + mgz$$

$$\text{(10.3)}$$

As already mentioned, $X > 0$ and so $\phi > 0$. We can combine Eqns. 10.2 and 10.3 and write

$$\underbrace{W_u}_{\text{Actual work}} = \underbrace{W_{u,rev}}_{\text{Maximum possible work}} - \underbrace{T_0\sigma}_{\text{Lost work}}$$

It can be seen that the irreversibilities – both internal and external, contribute to the lost work.

■ EXAMPLE 10.1

A storage tank of volume 10 m^3 contains air at 1 MPa and 298 K. Determine the exergy available.

Solution : The mass of the air in the tank may be calculated as

$$m = \frac{PV}{RT} = \frac{10^6 \times 10}{\frac{8314}{28.8} \times 298} = 116.243\,\text{kg}$$

The exergy of the air in the tank is given as

$$\begin{aligned}
X &= m\left[(u - u_0) + P_0(v - v_0) - T_0(s - s_0)\right] \\
&= m\left[C_v\underbrace{(T - T_0)}_{=0} + R\left(T\frac{P_0}{P} - T_0\right) - T_0\left(C_p\ln\cancel{\frac{T}{T_0}}^{0} - R\ln\frac{P}{P_0}\right)\right] \\
&= mRT_0\left[\left(\frac{T}{T_0}\frac{P_0}{P} - 1\right) + \ln\frac{P}{P_0}\right] \\
&= 14.026\,\text{MJ}
\end{aligned}$$

This example numerically illustrates the situation in Fig. 10.1(a). Note that the value for exergy obtained here is less than the work that can be produced if the air is allowed to expand isentropically through a turbine as in Example 7.22(b). This is on account of the facts that, in the latter case (a) the air is expanded to a temperature less than the ambient temperature and (b) the process is a flow process in which the *enthalpy* is converted to work.

■ EXAMPLE 10.2

Calculate the exergy of a steel casting (specific heat 500 J/kg.K) of mass 20 kg initially at 300°C.

$$\begin{aligned} &= m \left[c(T - T_0) - T_0 c \ln \frac{T}{T_0} \right] \\ &= mcT_0 \left[\frac{T}{T_0} - 1 - \ln \frac{T}{T_0} \right] \end{aligned}$$

It may be noted that the final expression above is identical to the one derived in Example 8.10. Upon substituting the known values, we get the exergy of the casting to be 801.7 kJ.

10.2 Flow exergy

The specific exergy of a stream (flow exergy), ψ, that flows steadily with a velocity V and an elevation z, may be written in a similar manner as

$$\psi = (h - h_0) - T_0 (s - s_0) + \frac{V^2}{2} + gz \tag{10.5}$$

Note that, in this case, no work is spent in displacing the atmosphere and hence the term $P_0 (v - v_0)$ is absent. The rate at which work is developed by the stream is simply $\dot{W}_{rev} = \dot{m}\,\psi$.

10.3 Exergy destruction principle

Application of first law to an isolated system that undergoes a process from state 1 to 2 gives

$$\Delta E = \cancelto{0}{Q} - \cancelto{0}{W} \quad \Rightarrow E_2 - E_1 = 0$$

which may be stated as "Total energy of an isolated system remains constant". Entropy change of this system may be written as

$$\Delta S = \int_1^2 \frac{\cancelto{0}{\delta Q}}{T} + \sigma_{int} \quad \Rightarrow S_2 - S_1 = \sigma_{int}$$

which may be stated as "Entropy of an isolated system increases or remains the same", since $\sigma_{int} \geq 0$. If we subtract the second expression from the first one above, after multiplying the second one by T_0, we get

$$E_2 - E_1 - T_0 (S_2 - S_1) = -T_0\, \sigma_{int}$$

Since $\sigma_{int} \geq 0$, it is clear that $X_2 \leq X_1$. This may be stated as "Exergy of an isolated system decreases or remains the same". This is known as the exergy destruction principle.

10.4 Exergy transfer and exergy change

Exergy can be transferred – added to or removed from, a system as a result of its interaction with the surroundings. These interactions may be classified as, addition or removal of, work , heat and mass. Let us now examine these in turn.

Work : When a system does an amount of work, W, its exergy decreases by W. Conversely, when an amount of work W is done on the system, its exergy increases by W.

Mass : When a certain amount of mass m having a specific exergy ϕ is added/removed to/from a system, the exergy of the system increases/decreases by $m\phi$.

Heat : An important aspect of exergy change due to heat transfer is the dependence on the temperature of the system with reference to the dead state temperature, T_0. This is illustrated in Fig. 10.2. Contours of exergy (X = constant) are shown

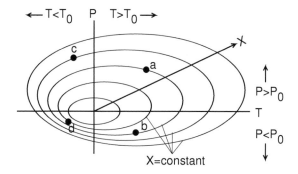

Figure 10.2: Contours of exergy and changes in exergy of systems at different states when heat is supplied/removed

in this figure (these are shown as ellipses for the sake of illustration only). The exergy is zero at the dead state and the exergy increases as one moves outwards away from the dead state. When heat is supplied to the systems labelled a and b, which are at temperatures higher than T_0, they move farther away from T_0 and

behaviour is owing to the fact that the exergy is always positive, as already mentioned.

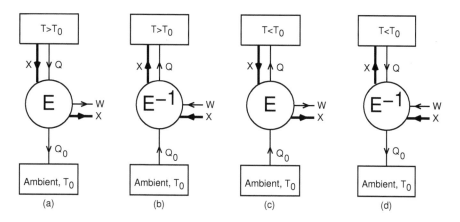

Figure 10.3: Direct and reverse heat engines with reservoirs at temperature greater and less than the ambient temperature

We will now quantify the magnitude of the exergy transferred as a result of heat supplied from/rejected to reservoirs that are at temperatures above and below the dead state temperature, T_0. Figure 10.3 depicts four *reversible* engines – two of them being direct engines and two being reverse engines. In all the cases, let us consider the engine as the system. In all the four cases, exergy recovered from the engine must be equal to the exergy supplied to the engine, since there is no destruction of exergy as the engines are reversible – both internally and externally. In addition, it should be noted that exchange of heat with the ambient by a system does not result in any transfer of exergy to/from the system.

T > T₀ : In this case (Fig. 10.3a), the transfer of heat Q to the engine (system) produces a work output (exergy recovered)

$$W = Q\,\eta = Q\left(1 - \frac{T_0}{T}\right)$$

It may be inferred that, when a reservoir at a temperature $T > T_0$ supplies an amount of heat Q to a system, the exergy *supplied* to the system is $Q\left(1 - T_0 \div T\right)$. The direction of heat transfer and the direction of exergy transfer are the same.

$$COP_{HP} \quad \frac{1}{1 - T_0 \div T} \quad \left(\quad T \right)$$

It may be inferred that, when a reservoir at a temperature $T > T_0$ *receives* an amount of heat Q from a system, the exergy *received* from the system is $Q(1 - T_0 \div T)$. In this case also, the direction of heat transfer and the direction of exergy transfer are the same.

$\mathbf{T < T_0}$: In this case (Fig. 10.3c), rejection of heat Q by the system to the reservoir produces a work output (exergy recovered)

$$W = Q_0 \eta = (Q + W) \left(1 - \frac{T}{T_0} \right) \quad \Rightarrow \quad W = Q \left(\frac{T_0}{T} - 1 \right)$$

It may be inferred that, when a reservoir at a temperature $T < T_0$ *receives* an amount of heat Q from a system, the exergy *supplied* to the system is $Q(T_0 \div T - 1)$. The direction of heat transfer and the direction of exergy transfer are *opposite* to each other in this case.

$\mathbf{T < T_0}$: In this case (Fig. 10.3d), transfer of heat Q to the system from the reservoir requires a work input (exergy supplied)

$$W = \frac{Q}{COP_{Ref}} = \frac{Q}{\frac{1}{T_0 \div T - 1}} = Q \left(\frac{T_0}{T} - 1 \right)$$

It follows that, when a reservoir at temperature $T < T_0$ *supplies* an amount of heat Q to a system, the exergy *received* from the system is $Q(T_0 \div T - 1)$. In this case too, the direction of heat transfer and the direction of exergy transfer are *opposite* to each other.

Consider the case of a non-isolated system with work interaction alone. For a system that goes from state 1 to 2, Eqn. 10.2 gives

$$\underbrace{W_{u, 1-2}}_{\text{Actual work}} = \underbrace{X_1 - X_2}_{\text{Exergy change}} - \underbrace{T_0 \sigma_{1-2}}_{\text{Exergy destroyed}}$$

$$= \underbrace{W_{u,rev}}_{\text{Ideal work}} - \underbrace{T_0 \sigma_{1-2}}_{\text{Lost work}}$$

The second expression shows that, as a result of lost work, (a) for a work producing process for which $W > 0$, the actual work produced is less than the ideal work and (b) for a work absorbing process for which $W < 0$, the actual work to be supplied

In the general case, the exergy of a system changes as a result of any and all of the interactions mentioned above. For such a case, we may write.

$$X_2 - X_1 = \text{Exergy change}$$
$$= \text{Exergy supplied} - \text{Exergy recovered} - \text{Exergy destroyed}$$

$$(10.8)$$

Equation 10.7, may now be interpreted as:

$$\underbrace{X_2 - X_1}_{\substack{\text{Exergy} \\ \text{change}}} = - \underbrace{W_u}_{\substack{\text{Exergy} \\ \text{recovered}}} - \underbrace{T_0\sigma}_{\substack{\text{Exergy} \\ \text{destroyed}}}$$

For an unsteady flow process with multiple inlet and outlet streams, we may write Eqn. 4.4 as

$$\frac{dE_{CV}}{dt} = \dot{Q} - \dot{W}_x + \sum_{in} \dot{m}\left(h + \frac{V^2}{2} + gz\right) - \sum_{out} \dot{m}\left(h + \frac{V^2}{2} + gz\right)$$

and Eqn. 9.29 as

$$\frac{dS_{CV}}{dt} = \sum_{in} \dot{m}\,s - \sum_{out} \dot{m}\,s - \frac{\dot{Q}_{res}}{T_{res}} + \dot{\sigma}$$

where the subscript *res* denotes the reservoir. If we multiply the above expression by T_0 and subtract from the previous expression, we get

$$\frac{dX_{CV}}{dt} = -\dot{W}_x + \sum_{in} \dot{m}\,\psi - \sum_{out} \dot{m}\,\psi - T_0\,\dot{\sigma} + \dot{Q} + \frac{T_0}{T_{res}}\dot{Q}_{res}$$
$$= -\dot{W}_x + \dot{Q}\left(1 - \frac{T_0}{T_{res}}\right) + \sum_{in} \dot{m}\,\psi - \sum_{out} \dot{m}\,\psi - T_0\,\dot{\sigma} \quad (10.9)$$

where we have used the fact that $\dot{Q}_{res} = -\dot{Q}$. Hence, the rate of exergy destruction in the universe,

$$\dot{X}_{destroyed} = T_0\,\dot{\sigma} \qquad (10.10)$$

Equation 10.9 may be interpreted as follows:

| Rate of change of exergy | Rate at which exergy is supplied | Rate at which exergy is recovered | which exergy is destroyed |

For a steady flow process

$$\dot{W}_x = \underbrace{\dot{Q}\left(1 - \frac{T_0}{T_{res}}\right) + \sum_{in}\dot{m}\psi - \sum_{out}\dot{m}\psi}_{\dot{W}_{x,ideal}} - T_0\,\dot{\sigma} \qquad (10.11)$$

◼ EXAMPLE 10.3

In Example 7.6, saturated liquid R134a at 30°C was throttled to 200 kPa. Heat loss to the ambient was 2 kW per unit mass flow rate of refrigerant. Neglecting KE and PE changes, determine the rate of exergy destruction during this process.

Solution : The following property values may be taken from Example 7.6 : $h_1 = 93.58$ kJ/kg, $s_1 = 0.3479$ kJ/kg.K, $h_2 = 91.58$ kJ/kg and $s_2 = 0.3566$ kJ/kg.K. The rate of entropy generation was calculated in Example 9.21 to be 15.37 W/K per unit mass flow rate.

Rate of exergy destruction may be evaluated as

$$\frac{\dot{X}_{destroyed}}{\dot{m}} = T_0\,\dot{\sigma} = 4580.26\,\text{J/kg}$$

The rate at which exergy enters is

$$\frac{\dot{X}_1}{\dot{m}} = \psi = (h_1 - h_0) - T_0(s_1 - s_0)$$

From Table F, corresponding to 25°C and 100 kPa, we can get $h_0 = 276.45$ kJ/kg and $s_0 = 1.106$ kJ/kg.K. Therefore, $\dot{X}_1 \div \dot{m} = 43.0438$ kJ/kg. Hence, 10.6 percent of the incoming exergy is destroyed during the process.

10.5 Second law efficiency

Equations 10.7 and 10.11 make clear how the issues raised in the beginning of this chapter can be addressed. These equations can be used to define an efficiency for any

$$\eta_{II} = \frac{W_u}{W_{u,rev}} = 1 - \frac{T_0\,\sigma}{W_{u,rev}} = 1 - \frac{T_0\,\sigma}{X_1 - X_2} \qquad (10.12)$$

and for a work consuming process (with work interactions being positive numbers)

$$\eta_{II} = \frac{W_{u,rev}}{W_u} = 1 - \frac{T_0\,\sigma}{X_2 - X_1 + T_0\,\sigma} \qquad (10.13)$$

Efficiency defined in this manner is called the second law efficiency (as the subscript II indicates) since it accounts for work lost due to irreversibilities. Similar expressions may be written for efficiencies of steady flow devices (whether adiabatic or not) with non-zero work interactions using Eqn. 10.11. In case of devices for which the work interaction is zero, we may write

$$\eta_{II} = \frac{\text{Exergy recovered}}{\text{Exergy supplied}} = \frac{\sum_{out} \dot{m}\psi}{\sum_{in} \dot{m}\psi} = 1 - \frac{T_0\,\dot{\sigma}}{\sum_{in} \dot{m}\psi} \qquad (10.14)$$

where we have used Eqn. 10.11 to derive the last equality.

◼ EXAMPLE 10.4

Consider Example 6.2 again: Two kg of air contained in a piston cylinder assembly undergoes a compression that obeys $PV^n = constant$, with $n = 1.27$. The air is initially at 100 kPa, 300 K and is compressed to a final pressure of 800 kPa. Determine the second law efficiency of this process.

Solution : Considering the air in the cylinder as the system, the work and heat interactions were determined to be -357.11 kJ and -116.06 kJ respectively. The final temperature of the air was calculated as 467 K.

Since the displacement work is non-zero in this case, the useful work may be evaluated from $W_{sys} = W_u + P_0 (V_2 - V_1)$:

$$W_u = -357.11 - mRT_0 \left(\frac{T_2}{T_0} \frac{P_0}{P_2} - \frac{T_1}{T_0} \frac{P_0}{P_1} \right) = -217.61\,\text{kJ}$$

The entropy generated during this process was determined to be 80.56 J/K in Example 9.18. Therefore, from Eqn. 10.7

$$W_{u,rev} = W_u + T_0\,\sigma = -217.61 + 300 \times \frac{80.56}{1000} = -193.442\,\text{kJ}$$

Hence

$$\eta_{II} = \frac{W_{u,rev}}{W_u} = \frac{193.442}{217.61} = 89\%$$

Simultaneously, heat transfer to the ambient at 30°C also takes place until the content of the vessel reaches the ambient temperature. Determine the second law efficiency of this process.

Solution : Since there is no displacement work in this case, W_u = -500 kJ. In Example 9.19, the entropy generated was evaluated to be $\sigma = 1.6972$ kJ/K. Therefore,

$$W_{u,rev} = W_u + T_0\,\sigma = -500 + 303 \times 1.6962 = 13.9486\,\text{kJ}$$

As already mentioned in Example 9.19, the process undergone by the R134a is highly irreversible. The extent of this irreversibility becomes even more clear now, since the state change would have *produced* 13.9486 kJ of work had it been carried out reversibly, instead of *consuming* 500 kJ of work! Consequently, the second law efficiency is not meaningful in this case.

As a check, the change in exergy of the system may be evaluated, using the property values retrieved in Example 9.19 as:

$$
\begin{aligned}
W_{u,rev} = X_1 - X_2 \;&=\; m\left[(u_1 - u_2) + P_0 \underbrace{(v_1 - v_2)}_{=0} - T_0\,(s_1 - s_2)\right] \\
&=\; 5 \times [221.72 - 255.23 - 303 \times (0.9415 - 1.0613)] \\
&=\; 13.947\,\text{kJ}
\end{aligned}
$$

■ **EXAMPLE 10.6**

A vertical piston-cylinder arrangement contains 2 kg of water initially at 1 bar. The piston initially rests on stops and will move when the pressure reaches 3 bar. The enclosed volume when the piston is on the stops is 250 L. Heat is now supplied from a source at 320°C until the temperature of the contents of the cylinder reach 320°C. Determine the exergy destroyed and the second law efficiency of this process.

Solution : The specific volume at the initial state, $v_1 = V_1 \div m = 0.125$ m³/kg. From Table B, corresponding to 1 bar, we can retrieve, $v_f = 1.0431 \times 10^{-3}$ m³/kg and $v_g = 1.6958$ m³/kg. Since $v_f < v_1 < v_g$, the water is initially a saturated mixture.

The dryness fraction at the initial state may be evaluated as

$$x_1 = \frac{v_1 - v_f}{v_g - v_f} = 0.07314$$

$$s_1 = s_f + x_1 (s_g - s_f) = 1.7454 \, \text{kJ/kg.K}$$

The final state (denoted 3) is superheated at 3 bar, 320°C. From Table C, we can get $v_3 = 0.9067 \, \text{m}^3/\text{kg}$, $u_3 = 2837.8 \, \text{kJ/kg}$ and $s_3 = 7.7716 \, \text{kJ/kg.K}$.

The water in the cylinder undergoes a constant volume process *1-2* followed by a constant pressure process *2-3*. Application of first law to the contents of the system gives

$$\Delta E = \Delta U = Q - W$$
$$m (u_3 - u_1) = Q - m P_2 (v_3 - v_2)$$
$$\Rightarrow Q = m (u_3 - u_1) + m P_2 (v_3 - v_2)$$

Upon substituting the known values, we get $Q = 5004.558 \, \text{kJ}$. The entropy generated during this process may be evaluated as

$$\sigma = m (s_3 - s_1) + \frac{Q_{source}}{T_{source}} = 3.613 \, \text{kJ/K}$$

where $Q_{source} = -Q$. Therefore, the exergy destroyed $T_0 \, \sigma = 1076.674 \, \text{kJ}$. The useful work developed during process *1-3* is

$$W_u = m P_2 (v_3 - v_2) - m P_0 (v_3 - v_2) = 312.68 \, \text{kJ}$$

The second law efficiency for this process is

$$\eta_{II} = \frac{W_u}{W_{u,rev}} = \frac{W_u}{W_u + T_0 \, \sigma} = 22.5\%$$

The second law efficiency is poor on account of the fact that the exergy destroyed during the process is more than three times the useful work produced. Most of this exergy destruction occurs during the heat transfer from the source to the system across a large temperature difference.

■ **EXAMPLE 10.7**

Air is compressed steadily in an 8 kW compressor from ambient condition to 650 kPa and 187°C. The mass flow rate is 2.1 kg/min. Neglecting the changes in kinetic and potential energies, determine the second law efficiency of the compressor.

Solution : Application of SFEE to this case results in

$$0 = \dot{Q} - \dot{W}_x + \dot{m} (h_1 - h_2)$$

$$= -2.2711 \, \text{kW}$$

The rates of entropy generation and hence exergy destruction are thus

$$\dot{\sigma} = \dot{m}\left(s_2 - s_1\right) + \frac{\dot{Q}_{surr}}{T_0}$$

$$= \dot{m}\left(C_p \ln\frac{T_2}{T_1} - R \ln\frac{P_2}{P_1}\right) - \frac{\dot{Q}}{T_0}$$

$$T_0\,\dot{\sigma} = \dot{m}RT_0\left(\frac{\gamma}{\gamma - 1}\ln\frac{T_2}{T_1} - \ln\frac{P_2}{P_1}\right) - \dot{Q}$$

$$= 1.2102 \, \text{kW}$$

The second law efficiency of the compressor is

$$\eta_{II} = \frac{|\dot{W}_{x,rev}|}{|\dot{W}_x|} = 1 - \frac{T_0\,\dot{\sigma}}{|\dot{W}_x|} = 85\%$$

■ EXAMPLE 10.8

Refrigerant 134-a is compressed steadily in a compressor from 140 kPa, -10°C to 700 kPa, 120°C. The volumetric flow rate at the inlet is 0.001 m³/s and the input power is 1.0 kW. Neglecting KE and PE changes, determine (a) the exergy destruction rate (b) reversible power input and (c) the second law efficiency.

Solution : From Table F, we can retrieve the following property values at the inlet and exit of the compressor:

State	v (m³/kg)	h (kJ/kg)	s (kJ/kg.K)
1	0.1461	246.4	0.9724
2	-	358.9	1.192

The mass flow rate may be calculated as

$$\dot{m} = \frac{\dot{V}_1}{v_1} = \frac{0.001}{0.1461} = 6.845 \, \text{g/s}$$

Application of SFEE to the compressor gives

$$\dot{Q} = \dot{W}_x - \dot{m}\left(h_1 - h_2\right)$$

$$= -1 - 6.845 \times 10^{-3}\left(246.4 - 358.9\right) = -0.23 \, \text{kW}$$

$$= \dot{m}\,(s_2 - s_1) - \frac{\dot{Q}}{T_0}$$

$$T_0\,\dot{\sigma} = T_0\,\dot{m}\,(s_2 - s_1) - \dot{Q}$$

$$= 0.6779\,\text{kW}$$

(b) The reversible power input may be evaluated as (as this is a work consuming process)

$$\dot{W}_{x,rev} = |\dot{W}_x| - T_0\,\dot{\sigma} = 0.3221\,\text{kW}$$

(c) The second law efficiency of the compressor is

$$\eta_{II} = \frac{|\dot{W}_{x,rev}|}{|\dot{W}_x|} = 32.2\%$$

■ **EXAMPLE 10.9**

Steam enters a turbine steadily at 3 MPa and 440°C at a rate of 8 kg/s and exits at 0.35 MPa and 160°C. Heat is lost to the surroundings at a rate of 500 kW. KE and PE changes are negligible. Determine (a) the actual power output, (b) the maximum possible power output and (c) the second law efficiency.

Solution : From Table C, we can retrieve the following property values at the inlet and exit of the turbine:

State	h (kJ/kg)	s (kJ/kg.K)
1	3321.5	7.0521
2	2782.1	7.1274

(a) The actual power output may be evaluated by applying SFEE to the turbine, as follows:

$$\dot{W}_x = \dot{Q} + \dot{m}\,(h_1 - h_2)$$

$$= -500 + 8 \times (3321.5 - 2782.1) = 3815.2\,\text{kW}$$

(b) The rate exergy destruction may be evaluated as

$$T_0\,\dot{\sigma} = T_0\,\dot{m}\,(s_2 - s_1) + T_0\,\frac{\dot{Q}_{surr}}{T_0}$$

$$= T_0\,\dot{m}\,(s_2 - s_1) - \dot{Q}$$

$$= 679.5152\,\text{kW}$$

$$\eta_{II} = \frac{\dot{W}_x}{\dot{W}_{x,max}} = 85\%$$

EXAMPLE 10.10

An open feedwater heater (mixing chamber) operates steadily at a pressure of 1 MPa. Steam at 1 MPa, 300°C (state 1) and saturated liquid water at 10 kPa whose pressure has been raised isentropically to 1 MPa (state 2) enter the heater. Heat loss to the ambient occurs at the rate of 100 kW. If saturated liquid at 1 MPa (state 3) is to exit the heater, determine the required mass flow rate of steam for every unit mass flow rate of the exiting stream. Determine the rate of exergy destruction and the second law efficiency of the feedwater heater. KE and PE changes may be neglected.

Solution : From Table C, we can retrieve $h_1 = 3050.6$ kJ/kg and $s_1 = 7.1219$ kJ/kg.K. From Table B, $h_3 = h_f(1\,\text{MPa}) = 762.8$ kJ/kg and $s_3 = s_f(1\,\text{MPa}) = 2.1387$ kJ/kg.K.

State 2 is a compressed liquid state. Since the temperature at state 2 is unknown, we cannot use our usual approximation for calculating the specific enthalpy of a compressed liquid state, namely, $h(T,P) = u_f(T) + P v_f(T)$. The $T\,ds$ relation may be used in this case to evaluate the specific enthalpy at state 2. From $T\,ds = dh - v\,dP$, we get $dh = v\,dP$, since $ds = 0$. Therefore, $h_2 = h_f(10\,\text{kPa}) + v_f(10\,\text{kPa})(1000 - 10) = 192.72$ kJ/kg and $s_2 = s_f(10\,\text{kPa}) = 0.6489$ kJ/kg.K, with $h_f(10\,\text{kPa}) = 191.72$ kJ/kg, $s_f(10\,\text{kPa}) = 0.6489$ kJ/kg.K and $v_f(10\,\text{kPa}) = 1.0103 \times 10^{-3}$ m^3/kg, from Table B.

From Table A, we can get $h_0 \approx u_f(25°C) + 100 \times v_f(25°C) = 104.85$ kJ/kg and $s_0 = s_f(25°C) = 0.367$ kJ/kg.K.

Let the mass flow rate of steam be x kg/s. Applying SFEE to the feedwater heater gives

$$0 = \dot{Q} - \cancel{\dot{W}_x}^{0} + x\,h_1 + (1-x)\,h_2 - h_3$$

$$\Rightarrow x = \frac{-\dot{Q} - h_2 + h_3}{h_1 - h_2} = 0.2345$$

The specific exergy of the various streams may be evaluated as follows:

$$\psi_3 = (h_3 - h_0) - T_0\,(s_3 - s_0) = 129.9834\,\text{kJ/kg}$$

The rate of exergy destruction is given as

$$\dot{X}_{destroyed} = x\,\psi_1 + (1-x)\,\psi_2 - \psi_3 = 91.7135\ \text{kW}$$

The rate at which exergy enters the mixing chamber is given as

$$\dot{X}_{in} = x\,\psi_1 + (1-x)\,\psi_2 = 221.7\ \text{kW}$$

Therefore, the second law efficiency of the feedwater heater may be evaluated using Eqn. 10.14 as

$$\eta_{II} = 1 - \frac{\dot{X}_{destroyed}}{\dot{X}_{in}} = 58.6\%$$

The second law efficiency is poor on account of the fact that 41 percent of the incoming exergy is destroyed during the mixing process, which is highly irreversible.

■ EXAMPLE 10.11

Air at 300 kPa and 420 K flows through a heat exchanger at a rate of 2 kg/s, where it is cooled to 300 K. Cooling water enters the heat exchanger as a separate stream at 30°C and leaves at 60°C. Neglecting heat losses to the ambient and KE, PE changes, determine the required mass flow rate of water and the second law efficiency of the heat exchanger.

Solution : Let the inlet and exit states of the air stream be denoted I and 2 and the corresponding states for the water stream be denoted x and y. SFEE applied to the heat exchanger reduces to (with $\dot{Q} = 0$, $\dot{W}_x = 0$, and KE, PE changes neglected),

$$0 = \dot{m}_{water}\,(h_x - h_y) + \dot{m}_{air}\,(h_1 - h_2)$$

From Table A, $h_x = h_f(30°C) = 125.67\ \text{kJ/kg}$ and $h_y = h_f(60°C) = 251.15\ \text{kJ/kg}$. Upon substituting the numerical values, we get $\dot{m}_{water} = 1.9325\ \text{kg/s}$.

The rate of entropy generation in the universe may be written as

$$\dot{\sigma} = \dot{m}_{air}\,(s_2 - s_1) + \dot{m}_{water}\,(s_y - s_x) + \frac{\overset{0}{\cancel{Q}_{surr}}}{T_0}$$

$$= \dot{m}_{air}\left(C_p \ln\frac{T_2}{T_1} - R\ln\frac{\overset{0}{\cancel{P_2}}}{\cancel{P_1}}\right) + \dot{m}_{water}\,(s_y - s_x)$$

From Table A, we can get $s_x = s_f(30°C) = 0.4365\ \text{kJ/kg.K}$ and $s_y = s_f(60°C) = 0.8312\ \text{kJ/kg.K}$. After substituting the numerical values, we get $\dot{\sigma} = 0.08283\ \text{kW/K}$.

$$X_1 = \dot{m}_{air}\left[(h_1 - h_0) - T_0(s_1 - s_0)\right]$$

$$= \dot{m}_{air}\left[C_p(T_1 - T_0) - RT_0\left(\frac{\gamma}{\gamma-1}\ln\frac{T_1}{T_0} - \ln\frac{P_1}{P_0}\right)\right]$$

$$= \dot{m}_{air}RT_0\left[\frac{\gamma}{\gamma-1}\left(\frac{T_1}{T_0} - 1\right) - \frac{\gamma}{\gamma-1}\ln\frac{T_1}{T_0} + \ln\frac{P_1}{P_0}\right]$$

$$= 228.906\ \text{kW}$$

The rate at which exergy is carried in by the water stream is

$$\dot{X}_x = \dot{m}_{water}\left[(h_x - h_0) - T_0(s_x - s_0)\right]$$

From Table A, we can retrieve $h_0 \approx u_f(25°C) + 100 \times v_f(25°C) = 104.85$ kJ/kg and $s_0 = s_f(25°C) = 0.367$ kJ/kg.K. Therefore, $\dot{X}_x = 0.2106$ kW.

The second law efficiency of the heat exchanger is

$$\eta_{II} = 1 - \frac{T_0\dot{\sigma}}{\sum_{in}\dot{m}\psi} = 1 - \frac{T_0\dot{\sigma}}{\dot{X}_1 + \dot{X}_x}$$

Upon substituting the numerical values from above, we get $\eta_{II} = 89.23$ percent.

■ EXAMPLE 10.12

A direct heat engine receives an amount of heat Q_H from a reservoir at temperature T_H and rejects heat Q_C to a reservoir at T_C during each cycle. Here, $T_H > T_C > T_0$. Derive an expression for the second law efficiency of the heat engine.

Solution : The exergy recovered during each cycle is the sum of the work output and the exergy associated with Q_C. This may be expressed as $W + Q_C(1 - T_0 \div T_C)$. The exergy supplied during each cycle is $Q_H(1 - T_0 \div T_H)$. The second law efficiency of the heat engine, from Eqn. 10.14, is

$$\eta_{II} = \frac{W + Q_C\left(1 - \frac{T_0}{T_C}\right)}{Q_H\left(1 - \frac{T_0}{T_H}\right)}$$

$$= \frac{Q_H - Q_C + Q_C\left(1 - \frac{T_0}{T_C}\right)}{Q_H\left(1 - \frac{T_0}{T_H}\right)}$$

$$T_H$$

$$= \frac{1 - \frac{T_0}{T_C}(1 - \eta)}{\eta_{Carnot}}$$

Here, we have written $W = Q_H - Q_C$ as the engine operates in a cycle and also used the definition of the thermal efficiency of a direct engine, namely, $\eta = 1 - Q_C \div Q_H$.

When heat is rejected to the ambient, $T_C = T_0$ and the above expression simplifies to

$$\eta_{II} = \frac{\eta}{\eta_{Carnot}}$$

Second law efficiencies for more complicated and realistic cycles will be evaluated in the next chapter.

THERMODYNAMIC CYCLES

In section 8.2, cycles executed in a steam and gas turbine power plant as well as in a vapor compression refrigeration system were introduced. In this chapter, the performance of these cycles are studied in detail. We start with a discussion of the basic cycle in each case and its performance and then study the effect of irreversibilities on the performance and also strategies adopted to improve the performance.

11.1 Rankine cycle

The block diagram of the basic Rankine cycle used in steam power plants is shown in Fig. 8.1. The $T - s$ diagram of the cycle is shown in Fig. 11.1.

The cycle operates between the isobars corresponding to the boiler pressure and the condenser pressure. Since heat is invariably rejected to the ambient at 30°C, the condenser pressure is equal to the corresponding saturation pressure, 0.0425 bar. It may be noted that this is less than the atmospheric pressure. The ideal cycle, one without internal irreversibilities, may be identified as *1-2s-3-4s-1* in Fig. 11.1. The

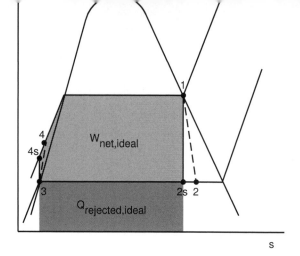

Figure 11.1: $T - s$ diagram of a basic Rankine cycle

actual cycle, *1-2-3-4-1*, accounts for internal irreversibilities in the turbine and the pump.

■ EXAMPLE 11.1

An ideal, basic Rankine cycle operates between a boiler pressure of 160 bar and a condenser temperature of 45°C. Determine the heat supplied, net power produced, thermal efficiency and second law efficiency of the cycle. Also calculate the rate of exergy destruction in each of the components. Assume steady operation and neglect KE and PE changes. Also assume that heat is supplied from a reservoir maintained at the highest temperature in the cycle and heat is rejected to a reservoir maintained at the condenser temperature.

Solution : From Table B, corresponding to 160 bar, we can get $h_1 = h_g = 2580.2$ kJ/kg and $s_1 = s_g = 5.245$ kJ/kg.K. From Table A , corresponding to 45°C, we can retrieve $h_f = 188.42$ kJ/kg, $h_g = 2582.3$ kJ/kg, $s_f = 0.6385$ kJ/kg.K and $s_g = 8.1629$ kJ/kg.K.

Since the expansion in the turbine is isentropic, $s_{2s} = s_1 = 5.245$ kJ/kg.K, from which we can evaluate

$$x_{2s} = \frac{s_{2s} - s_f}{s_g - s_f} = \frac{5.245 - 0.6385}{8.1629 - 0.6385} = 0.6122$$

remain constant. Therefore, $h_{4s} = 204.57$ kJ/kg.

State	h (kJ/kg)	s (kJ/kg.K)
1	2580.2	5.245
2s	1653.95	5.245
3	188.42	0.6385
4s	204.57	0.6385

Heat supplied may now be evaluated as

$$\frac{\dot{Q}_H}{\dot{m}} = h_1 - h_{4s} = 2375.63 \, \text{kJ/kg}$$

and heat rejected is

$$\frac{\dot{Q}_C}{\dot{m}} = h_{2s} - h_3 = 1465.53 \, \text{kJ/kg}$$

Work produced by the turbine is

$$\frac{\dot{W}_x}{\dot{m}} = h_1 - h_{2s} = 926.25 \, \text{kJ/kg}$$

and work supplied to the pump is

$$\frac{\dot{W}_x}{\dot{m}} = h_{4s} - h_3 = 16.15 \, \text{kJ/kg}$$

Net power generated is thus 910.1 kW per unit mass flow rate of steam. The thermal efficiency of the cycle may be evaluated as

$$\eta = \frac{\dot{W}_{x,net}}{\dot{Q}_H} = \frac{910.1}{2374.2} = 38.31\%$$

It should be noted that the pump work is only about 2 percent of the work produced by the turbine. Also, the average temperature at which heat is added may be evaluated as

$$T'_H = \frac{\dot{Q}_H \div \dot{m}}{s_1 - s_{4s}} = \frac{2375.63}{5.245 - 0.6385} = 516 \, \text{K} = 243°\text{C}$$

This is considerably less than $T_{sat}(160 \text{ bar}) = 347.4°\text{C}$.

Rate at which exergy is supplied is given as

$$\frac{\dot{X}_{supplied}}{\dot{m}} = \frac{\dot{W}_{x,pump}}{\dot{m}} + \frac{\dot{Q}_H}{\dot{m}} \left(1 - \frac{T_0}{T_H}\right) = 1250.68 \, \text{kJ/kg}$$

$$\quad m \qquad\qquad m \qquad\qquad m \quad \backslash \quad \iota_C \big/$$

where $T_C = 45°C$. The second law efficiency is thus

$$\eta_{II} = \frac{\text{Exergy recovered}}{\text{Exergy supplied}} = 81.43\%$$

The rate of exergy destruction in the turbine and pump are zero, since they are isentropic. The rate of exergy destruction in the boiler and condenser are given as

$$\frac{T_0 \,\dot{\sigma}_{boiler}}{\dot{m}} = T_0 \left[(s_1 - s_{4s}) - \frac{\dot{Q}_H}{\dot{m}} \frac{1}{T_H}\right] = 231.64\,\text{kJ/kg}$$

$$\frac{T_0 \,\dot{\sigma}_{condenser}}{\dot{m}} = T_0 \left[(s_3 - s_{2s}) + \frac{\dot{Q}_C}{\dot{m}} \frac{1}{T_C}\right] = 0\,\text{kJ/kg}$$

The rate of entropy generation in the condenser is zero since $\Delta s = (\dot{Q}_C \div \dot{m}) \div T_C$, on account of the fact this is a reversible isothermal process. It should be noted that exergy destruction occurs entirely in the boiler and so any scope for improvement must focus on reducing the exergy destruction rate associated with the heat addition process in the boiler. This insight demonstrates the usefulness of the the concept of exergy.

◼ EXAMPLE 11.2

Rework the previous example, assuming the isentropic efficiency of the turbine and pump to be respectively 95 percent and 90 percent.

Solution : The turbine work in this case is

$$\frac{\dot{W}_x}{\dot{m}} = \eta_t \frac{\dot{W}_{x,isen}}{\dot{m}} = 0.95 \times 926.25 = 879.94\,\text{kJ/kg}$$

and the work supplied to the pump is

$$\frac{\dot{W}_x}{\dot{m}} = \frac{1}{\eta_{pump}} \frac{\dot{W}_{x,isen}}{\dot{m}} = \frac{16.15}{0.9} = 17.94\,\text{kJ/kg}$$

The net power produced is thus 862 kW per unit mass flow rate of steam. Since

$$\eta_{pump} = \frac{h_{4s} - h_3}{h_4 - h_3}$$

the specific enthalpy at the exit of the pump is now 206.36 kJ/kg. Heat supplied is

$$\frac{\dot{Q}_H}{\dot{m}} = h_1 - h_4 = 2373.84\,\text{kJ/kg}$$

$$\frac{\dot{X}_{supplied}}{\dot{m}} = \frac{\dot{W}_{x,pump}}{\dot{m}} + \frac{\dot{Q}_H}{\dot{m}}\left(1 - \frac{T_0}{T_H}\right) = 1251.54\,\text{kJ/kg}$$

where we have assumed that the heat is supplied from a source at temperature $T_H = T_{sat}(160\text{ bar}) = 347.4°\text{C}$. Rate at which exergy is recovered may be evaluated as

$$\frac{\dot{X}_{recovered}}{\dot{m}} = \frac{\dot{W}_{x,turbine}}{\dot{m}} + \frac{\dot{Q}_C}{\dot{m}}\left(1 - \frac{T_0}{T_C}\right) = 975.02\,\text{kJ/kg}$$

where $T_C = 45°\text{C}$ and $\dot{Q}_C = \dot{Q}_H - \dot{W}_{x,net}$. The second law efficiency is thus

$$\eta_{II} = \frac{\text{Exergy recovered}}{\text{Exergy supplied}} = 77.9\%$$

The decrease in the second law efficiency is due to the additional exergy destruction in the turbine and the pump.

The basic Rankine cycle is not practicable since the expansion in the turbine takes place entirely in the two phase region and it is not possible to construct turbines that can handle two phase mixtures. Ideally, it is preferable to have the expansion occur in the superheated region such that the dryness fraction at the exit of the turbine is 0.9 or higher.

11.1.1   Superheat

The efficiency of the basic Rankine cycle can be improved by increasing the average temperature at which heat is added. This may be accomplished either by increasing the boiler pressure and/or superheating the vapor before it enters the turbine. Both the strategies are widely used and today, the boiler pressure in the so-called super-critical and ultra super-critical steam power units is well above the critical pressure of water (221 bar), reaching 300 bars or higher.

The addition of superheat to the basic Rankine cycle results in the cycle shown in Fig. 11.2. Since the isobars in the superheated region are steep, the increase in the heat addition for a given amount of superheat is much more than the increase in the heat rejected because the latter occurs in the two phase region where the isobar is horizontal. This results in an increase in the efficiency of the cycle. It can also be seen from this figure that superheating has the added advantage of increasing the dryness fraction at the exit of the turbine.

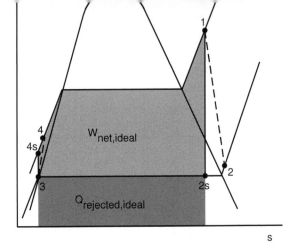

Figure 11.2: $T - s$ diagram of a Rankine cycle with superheat

▣ EXAMPLE 11.3

Redo Example 11.1, with the steam leaving the boiler with 212.6°C of superheat.

Solution :

State	h (kJ/kg)	s (kJ/kg.K)
1	3465.4	6.5133
2s	2057.56	6.5133
3	188.42	0.6385
4s	204.57	0.6385

From Table C, corresponding to 160 bar and 560 (=347.4 + 212.6)°C, we can get h_1 = 3465.4 kJ/kg and s_1 = 6.5133 kJ/kg.K. With $s_{2s} = s_1$, the dryness fraction at the end of the isentropic expansion in the turbine may be evaluated as

$$x_{2s} = \frac{s_{2s} - s_f}{s_g - s_f} = \frac{6.5133 - 0.6385}{8.1629 - 0.6385} = 0.7808$$

Thus, $h_{2s} = h_f + x_{2s}(h_g - h_f) = 2057.56$ kJ/kg. All other states remain the same and so $h_3 = 188.42$ kJ/kg and $h_{4s} = 204.57$ kJ/kg.

$$\frac{..}{\dot{m}} = h_1 - h_{4s} = 3260.83 \,\text{kJ/kg}$$

and heat rejected is

$$\frac{\dot{Q}_C}{\dot{m}} = h_{2s} - h_3 = 1869.14 \,\text{kJ/kg}$$

Work produced by the turbine is

$$\frac{\dot{W}_x}{\dot{m}} = h_1 - h_{2s} = 1407.84 \,\text{kJ/kg}$$

Since the power required by the pump is the same as before, net power produced in the cycle is 1391.69 kW per unit mass flow rate of steam, an increase of 53 percent. The thermal efficiency of the cycle may be evaluated as

$$\eta = \frac{\dot{W}_{x,net}}{\dot{Q}_H} = \frac{1391.69}{3260.83} = 42.68\%$$

This represents an increase of almost 4 percent. The dryness fraction at the exit of the turbine has also increased from 0.6122 to 0.78 now but is still less than the desirable value of 0.9. The average temperature at which heat is added may be evaluated as

$$T'_H = \frac{\dot{Q}_H \div \dot{m}}{s_1 - s_{4s}} = \frac{3260.83}{6.5133 - 0.6385} = 555 \,\text{K} = 282°\text{C}$$

As expected, this is higher by 25 percent in degree Celsius terms.

Rate at which exergy is supplied

$$\frac{\dot{X}_{supplied}}{\dot{m}} = \frac{\dot{W}_{x,pump}}{\dot{m}} + \frac{\dot{Q}_H}{\dot{m}} \left(1 - \frac{T_0}{T_H}\right) = 2110.44 \,\text{kJ/kg}$$

where we have assumed that the heat is supplied from a source at temperature $T_H = 560°\text{C}$. Rate at which exergy is recovered may be evaluated as

$$\frac{\dot{X}_{recovered}}{\dot{m}} = \frac{\dot{W}_{x,turbine}}{\dot{m}} + \frac{\dot{Q}_C}{\dot{m}} \left(1 - \frac{T_0}{T_C}\right) = 1525.4 \,\text{kJ/kg}$$

where $T_C = 45°\text{C}$. The second law efficiency is thus

$$\eta_{II} = \frac{\text{Exergy recovered}}{\text{Exergy supplied}} = 72.3\%$$

Although the thermal efficiency (first law efficiency) of the cycle has increased, the second law efficiency has decreased considerably. This can be attributed to

$$T_0 \frac{\dot{\sigma}_{boiler}}{\dot{m}} = T_0 \left[(s_1 - s_{4s}) - \frac{\dot{Q}_H}{\dot{m}} \frac{1}{T_H} \right] = 584.15 \, \text{kJ/kg}$$

$$T_0 \frac{\dot{\sigma}_{condenser}}{\dot{m}} = T_0 \left[(s_3 - s_{2s}) + \frac{\dot{Q}_C}{\dot{m}} \frac{1}{T_C} \right] = 0 \, \text{kJ/kg}$$

The total exergy destruction rate in the cycle is now more than twice the value calculated in Example 11.1.

It is left as an exercise to repeat this example taking into account internal irreversibilities in the pump and the turbine using the values for isentropic efficiency given in Example 11.2.

11.1.2   Regenerative feed water heating

Another strategy that is used for increasing the efficiency of the Rankine cycle is regenerative feed water heating. The basic idea here is to accomplish an increase in the enthalpy of the feedwater using steam that has been extracted from the turbine after undergoing partial expansion. This reduces the heat to be added in the boiler but has the disadvantage that the work output decreases since a part of the steam is extracted without expanding it to the condenser pressure.

The enthalpy transfer between the extracted steam and the feedwater is usually accomplished in two different ways, namely, open and closed feedwater heating. These are discussed in detail next.

Open feedwater heater   An open feedwater heater is a mixing chamber in which the extracted steam and the feedwater that has been pumped to the extraction pressure are physically mixed. The mass flow rate of the extracted steam is such that the exiting stream leaves as a saturated liquid at the extraction pressure (see Example 7.5) and it is then pumped to the boiler pressure. Since the streams are mixed in the feedwater heater, the incoming streams and the outgoing stream are at the same pressure. The block diagram of a Rankine cycle utilizing an open feedwater heater is shown in Fig. 11.3. The $T - s$ diagram of the resulting cycle is shown in Fig. 11.4.

■ EXAMPLE 11.4

Redo Example 11.3 with an open feedwater heater operating at 40 bar.

Solution : With reference to Fig. 11.4, state *1* is the same as before and hence $h_1 = 3465.4$ kJ/kg and $s_1 = 6.5133$ kJ/kg.K. State *2s* is now at the extraction

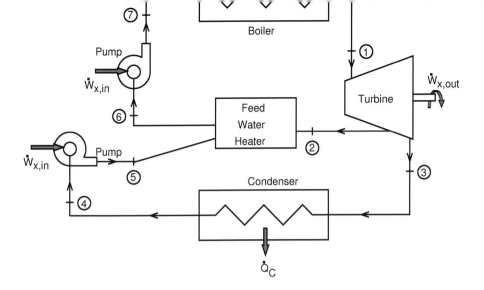

Figure 11.3: Block diagram of a Rankine cycle with regeneration using an open feedwater heater

State	h (kJ/kg)	s (kJ/kg.K)
1	3465.4	6.5133
2s	3050.93	6.5133
3s	2057.56	6.5133
4	188.42	0.6385
5s	192.45	0.6385
6	1087.2	2.7961
7s	1102.23	2.7961

pressure of 40 bar. From Table B, corresponding to this pressure, we have $s_g = 6.069$ kJ/kg.K. Since $s_{2s} = s_1 > s_g$, 2s is a superheated state. From Table C, we can get $h_{2s} = 3050.93$ kJ/kg, by interpolation. State 3s now is same as state 2s in Example 10.3 and so $h_{3s} = 2057.56$ kJ/kg. State 4 is saturated liquid at 45°C and hence $h_4 = 188.42$ kJ/kg. State 5s is a compressed liquid state at 40 bar and process 4-5s is an isentropic process. Hence, $h_{5s} - h_4 = v_4 (P_{5s} - P_4) = 4.03$ kJ/kg. State 6 is saturated liquid at 40 bar, and so $h_6 = 1087.2$ kJ/kg, from Table B. State 7s

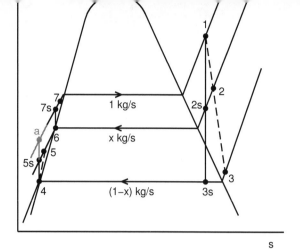

Figure 11.4: $T - s$ diagram of a Rankine cycle with regeneration using an open feedwater heater

is a compressed liquid state at 160 bar and process *6-7s* is an isentropic process. Therefore, $h_{7s} - h_6 = v_6 (P_{7s} - P_6) = 15.03$ kJ/kg.

SFEE applied to the feed water gives, after neglecting any heat loss,

$$0 = (x) h_{2s} + (1 - x) h_{5s} - (1) h_6$$

If we substitute the known values into this expression, we get $x = 0.313$.

Heat supplied may now be evaluated as

$$\frac{\dot{Q}_H}{\dot{m}} = h_1 - h_{7s} = 2363.17 \, \text{kJ/kg}$$

and heat rejected is

$$\frac{\dot{Q}_C}{\dot{m}} = (1 - x) (h_{3s} - h_4) = 1284.1 \, \text{kJ/kg}$$

Work produced by the turbine is

$$\frac{\dot{W}_x}{\dot{m}} = (x) (h_1 - h_{2s}) + (1 - x) (h_1 - h_{3s}) = 1096.92 \, \text{kJ/kg}$$

$$\frac{W_x}{\dot{m}} = (1-x)(h_{5s} - h_4) + (1)(h_{7s} - h_6) = 17.8 \, \text{kJ/kg}$$

Net power generated is thus 1079.12 kW per unit mass flow rate of steam. The thermal efficiency of the cycle may be evaluated as

$$\eta = \frac{\dot{W}_{x,net}}{\dot{Q}_H} = \frac{1079.12}{2363.17} = 45.66\%$$

In comparison with the previous example, both the net power and the heat added are less now, but the reduction in the heat supplied is more than the reduction in the power and hence the efficiency of the cycle increases.

Rate at which exergy is supplied

$$\frac{\dot{X}_{supplied}}{\dot{m}} = \frac{\dot{W}_{x,pump}}{\dot{m}} + \frac{\dot{Q}_H}{\dot{m}}\left(1 - \frac{T_0}{T_H}\right) = 1535.56 \, \text{kJ/kg}$$

where $T_H = 560°\text{C}$. Rate at which exergy is recovered may be evaluated as

$$\frac{\dot{X}_{recovered}}{\dot{m}} = \frac{\dot{W}_{x,turbine}}{\dot{m}} + \dot{Q}_C\left(1 - \frac{T_0}{T_C}\right) = 1177.68 \, \text{kJ/kg}$$

where $T_C = 45°\text{C}$. The second law efficiency is thus

$$\eta_{II} = \frac{\text{Exergy recovered}}{\text{Exergy supplied}} = 76.69\%$$

There is an increase in the second law efficiency of the cycle as well, almost to the value calculated in Example 11.1, suggesting that the exergy destruction rate in the cycle must be less now. The rate of exergy destruction in the boiler is

$$T_0\frac{\dot{\sigma}}{\dot{m}} = T_0\left[(s_1 - s_{7s}) - \frac{\dot{Q}_H}{\dot{m}}\frac{1}{T_H}\right] = 262.32 \, \text{kJ/kg}$$

where $s_{7s} = s_6 = s_f(40\,\text{bar}) = 2.7961$ kJ/kg.K. The rate of exergy destruction in the condenser is

$$T_0\frac{\dot{\sigma}}{\dot{m}} = T_0\left[(1-x)(s_4 - s_{3s}) + \frac{\dot{Q}_C}{\dot{m}}\frac{1}{T_C}\right] = 0 \, \text{kJ/kg}$$

In addition, the rate of exergy destruction in the feedwater heater is

$$T_0\frac{\dot{\sigma}}{\dot{m}} = T_0\left[s_6 - x\,s_{2s} - (1-x)\,s_{5s}\right] = 95 \, \text{kJ/kg}$$

higher temperature as a result of regeneration.

Closed feedwater heater In the case of a closed feedwater heater, there is no physical mixing of the feedwater and the extracted steam, but a heat exchanger is used instead. Consequently, the incoming and outgoing stream need not be at the same pressure. The block diagram of a Rankine cycle utilizing a closed feedwater heater is shown in Fig. 11.5.

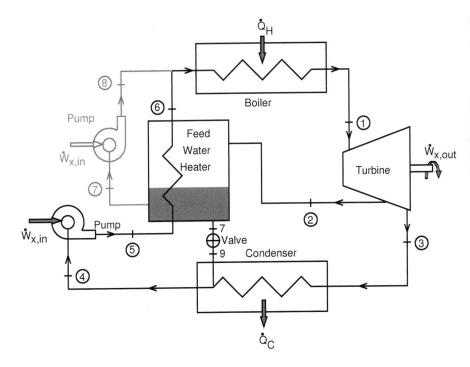

Figure 11.5: Block diagram of a Rankine cycle with regeneration using a closed feedwater heater

Depending upon how the saturated liquid condensate in the feedwater heater is handled, two designs are possible as shown in this figure. In one design, the condensate is throttled to the condenser pressure and mixed with the two phase mixture inside the condenser as shown in Fig. 11.5. Alternatively, the condensate may be pumped directly into the boiler. This is shown in gray in Fig. 11.5. The choice between these two designs is made based on whether the extraction pressure

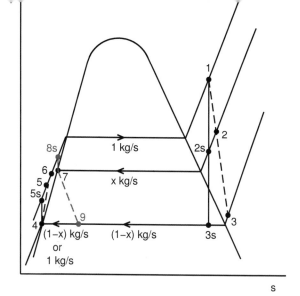

Figure 11.6: $T - s$ diagram of a Rankine cycle with regeneration using a closed feedwater heater

■ EXAMPLE 11.5

Redo Example 11.3 with a closed feedwater heater operating at 40 bar. The feedwater exits the heater at the saturation temperature corresponding to the extraction pressure. The liquid condensate from the feedwater heater is pumped into the boiler.

Solution : With reference to Fig. 11.6, state 6 is a compressed liquid state at 160 bar, 250.4°C. Hence, $h_6 = u_f(250.4°C) + v_f(250.4°C) \times 160 \times 100 = 1102.22$ kJ/kg and $s_6 = s_f(250.4°C) = 2.7961$ kJ/kg.K. Property values at the other states are known from the previous example.

SFEE applied to the feed water gives, after neglecting any heat loss,

$$0 = (x)(h_{2s} - h_7) + (1 - x)(h_{5s} - h_6).$$

If we substitute the known values into the above expression, we get $x = 0.3137$.

3s	2057.56	6.5133
4	188.42	0.6385
5s	204.57	0.6385
6	1102.22	2.7961
7	1087.2	2.7961
8s	1102.22	2.7961

Heat supplied may now be evaluated as

$$\frac{\dot{Q}_H}{\dot{m}} = (1)\, h_1 - (1-x)\, h_6 - (x)\, h_{8s} = 2363.18\,\text{kJ/kg}$$

and heat rejected is

$$\frac{\dot{Q}_C}{\dot{m}} = (1-x)\,(h_{3s} - h_4) = 1282.79\,\text{kJ/kg}$$

Work produced by the turbine is given as

$$\frac{\dot{W}_x}{\dot{m}} = (x)\,(h_1 - h_{2s}) + (1-x)\,(h_1 - h_{3s}) = 1096.22\,\text{kJ/kg}$$

Work supplied to the pump is

$$\frac{\dot{W}_x}{\dot{m}} = (1-x)\,(h_{5s} - h_4) + (x)\,(h_{8s} - h_7) = 15.8\,\text{kJ/kg}$$

Net power generated is thus 1080.42 kW per unit mass flow rate of steam. The thermal efficiency of the cycle may be evaluated as

$$\eta = \frac{\dot{W}_{x,net}}{\dot{Q}_H} = \frac{1080.42}{2363.18} = 45.72\%$$

As in the case of the open feedwater heater, there is an improvement in the efficiency and a reduction in the net work output as a result of adding the closed regenerative feedwater heater.

Rate at which exergy is supplied

$$\frac{\dot{X}_{supplied}}{\dot{m}} = \frac{\dot{W}_{x,pump}}{\dot{m}} + \frac{\dot{Q}_H}{\dot{m}}\left(1 - \frac{T_0}{T_H}\right) = 1533.57\,\text{kJ/kg}$$

where $T_C = 45°C$. The second law efficiency is thus

$$\eta_{II} = \frac{\text{Exergy recovered}}{\text{Exergy supplied}} = 76.74\%$$

This value is the same as the value calculated for the cycle with the open feedwater heater. The rate of exergy destruction in the boiler is

$$T_0 \frac{\dot{\sigma}}{\dot{m}} = T_0 \left[(x)(s_1 - s_{8s}) + (1 - x)(s_1 - s_6) - \frac{\dot{Q}_H}{\dot{m}} \frac{1}{T_H} \right] = 262.31 \text{ kJ/kg}$$

where $s_6 = s_f(250.4°C) = 2.7961$ kJ/kg.K. The rate of exergy destruction in the condenser is

$$T_0 \frac{\dot{\sigma}}{\dot{m}} = T_0 \left[(1 - x)(s_4 - s_{3s}) + \frac{\dot{Q}_C}{\dot{m}} \frac{1}{T_C} \right] = 0 \text{ kJ/kg}$$

In addition, the rate of exergy destruction in the feedwater heater is

$$\begin{aligned} T_0 \frac{\dot{\sigma}}{\dot{m}} &= T_0 \left[x(s_7 - s_{2s}) + (1 - x)(s_6 - s_{5s}) \right] \\ &= 93.77 \text{ kJ/kg} \end{aligned}$$

The total rate of exergy destruction in the cycle is now considerably less than the corresponding value calculated in Example 11.3 and almost the same as that with the open feedwater heater.

11.1.3 Reheat

As mentioned earlier, addition of regenerative feedwater heating results in an improvement in the thermal efficiency, albeit at the cost of net work output. This may be mitigated simply by increasing the mass flow rate of the steam. For instance, increasing the mass flow rate by a factor of $1391.69 \div 1079.12 = 1.29$ in the case of Example 11.4 or by $1391.69 \div 1081.61 = 1.287$ in the case of Example 11.5 will give the same net power output with regeneration as without (Example 11.3). The disadvantage of this strategy is the increase in physical size of all the components and the attendant implications. A better strategy to improve the power output is reheating as it does not require the mass flow rate of steam to be increased.

Here, the steam first undergoes expansion in a High Pressure Turbine (HPT) from the boiler pressure and temperature to an intermediate pressure. A fraction of the steam

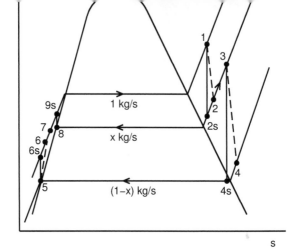

Figure 11.7: $T - s$ diagram of a Rankine cycle with reheat and regeneration using a closed feedwater heater

is extracted and sent to the feedwater heater while the remainder is sent to the boiler where it is re-heated at constant pressure. The temperature at the end of the reheat is usually the same as that at entry to the HPT, or even higher, in some cases. The low pressure, high temperature steam in then expanded in a Low Pressure Turbine (LPT) to the condenser pressure. The $T - s$ diagram of the resulting cycle is shown in Fig. 11.7. The reheating increases the enthalpy of the steam entering the LPT and hence produces more power, thereby offsetting the power output lost due to extraction for feedwater heating. However, the additional heat supplied during reheating has the potential to decrease the overall thermal efficiency. In general, since the steam is still superheated at the exit of the HPT, the increase in the power output is still more than the increase in the heat supplied. Therefore, the overall efficiency may be expected to increase slightly or remain the same due to the addition of reheat with regeneration.

■ EXAMPLE 11.6

Redo the previous example with one reheat stage where the steam is heated to the same temperature as in the HPT entry.

Solution : With reference to the feedwater heater in Fig. 11.7, it can be seen that the extraction fraction is the same as before, namely, $x = 0.3137$. State 3 is a superheated state 40 bar and 560°C. From Table C, we can get $h_3 = 3582.6$ kJ/kg and $s_3 = 7.2612$

$$x_{4s} = \frac{s_{4s} - s_f}{s_g - s_f} = \frac{7.2612 - 0.6385}{8.1629 - 0.6385} = 0.8802$$

Thus, $h_{4s} = h_f + x_{4s}(h_g - h_f) = 2295.51$ kJ/kg.

Values for the specific enthalpy at the other states are known from the previous example.

State	h (kJ/kg)	s kJ/kg.K
1	3465.4	6.5133
2s	3050.93	6.5133
3	3582.6	7.2612
4s	2295.51	7.2612
5	188.42	0.6385
6s	204.57	0.6385
7	1102.22	2.7961
8	1087.2	2.7961
9s	1102.22	2.7961

Heat supplied may now be evaluated as

$$\frac{\dot{Q}_H}{\dot{m}} = (1)\,h_1 - (1-x)\,h_7 - (x)\,h_{9s} + (1-x)\,(h_3 - h_{2s}) = 2728.07\,\text{kJ/kg}$$

and heat rejected is

$$\frac{\dot{Q}_C}{\dot{m}} = (1-x)\,(h_{4s} - h_5) = 1446.1\,\text{kJ/kg}$$

Work produced by the turbines is given as

$$\frac{\dot{W}_x}{\dot{m}} = (1)\,(h_1 - h_{2s}) + (1-x)\,(h_3 - h_{4s}) = 1297.8\,\text{kJ/kg}$$

Work supplied to the pump is the same as before,

$$\frac{\dot{W}_x}{\dot{m}} = 15.8\,\text{kJ/kg}$$

Net power generated is thus 1282 kW per unit mass flow rate of steam. The thermal efficiency of the cycle may be evaluated as

$$\eta = \frac{\dot{W}_{x,net}}{\dot{Q}_H} = \frac{1282}{2728.07} = 47\%$$

Rate at which exergy is supplied

$$\frac{\dot{X}_{supplied}}{\dot{m}} = \frac{\dot{W}_{x,pump}}{\dot{m}} + \frac{\dot{Q}_H}{\dot{m}}\left(1 - \frac{T_0}{T_H}\right) = 1767.92\,\text{kJ/kg}$$

where $T_H = 560°C$. Rate at which exergy is recovered may be evaluated as

$$\frac{\dot{X}_{recovered}}{\dot{m}} = \dot{W}_{x,turbine} + \dot{Q}_C\left(1 - \frac{T_0}{T_C}\right) = 1388.75\,\text{kJ/kg}$$

where $T_C = 45°C$. The second law efficiency is thus

$$\eta_{II} = \frac{\text{Exergy recovered}}{\text{Exergy supplied}} = 78.6\%$$

Both the first law and the second law efficiencies are the highest seen so far. This is due to the fact that the exergy destruction during the heat addition processes is less since the temperature at entry into the boiler is higher on account of regeneration and the partial expansion in the HPT. A summary of the performance of the Rankine cycles discussed so far is given below. It may be recalled that these cycles operate between a boiler pressure of 160 bar and a condenser pressure of 0.09593 bar.

Cycle	Peak temp (°C) T_H	Thermal efficiency η_{th}	Second law efficiency η_{II}	Specific work (kJ/kg) $\frac{\dot{W}_{x,net}}{\dot{m}}$
Basic	347.4	38.31	81.43	910.1
Superheat	560	42.68	72.3	1391.69
Regeneration (open)	560	45.66	76.69	1079.12
Regeneration (closed)	560	45.72	76.74	1080.42
Reheat with regeneration	560	47	78.6	1282

In actual steam power cycles, several feedwater heaters and two or more reheat stages are used in an optimal manner in order to maximize the power output as well as the thermal efficiency. With peak temperatures at values well above the

11.2 Air standard cycles

It may be recalled that in section 8.2, a heat engine was defined to be a thermody-namic system that executed a cyclic process. In the case of practical realizations such as steam power generation and vapor compression refrigeration equipment, the working substance does execute a cyclic process and hence the thermodynamic analysis of the Rankine and the vapor compression cycle presented here is quite useful and provides insights into the performance of practical cycles. In contrast, in the case of engines such as gas turbine engines and automotive engines (Fig. 2.10) that utilize air as the working substance, this is not true. In these engines, atmospheric air is taken inside the engine and the oxygen in the air is used for combustion. Hence, the working substance during processes that are executed before combustion is either air or air-fuel mixture. For the processes that are executed after combustion, the working substance is a mixture of combustion gases. Since the working fluid is no longer clean air, it cannot be sent back to the engine to complete the cycle and is hence discharged into the atmosphere and fresh air is taken in. Therefore, the processes executed in these engines do not form a cycle.

Notwithstanding this, thermodynamic analysis presented here for these engines assumes that air is the working substance and that it executes a cyclic process. This is an idealization and the cyclic process is usually called an air standard cycle. Although the practical realizations are quite different, the usefulness of the analysis lies in the fact that it allows us to identify the parameters that control the performance of the cycle and offers insights on the effect the various parameters have on the overall performance.

11.2.1 Brayton cycle

The operation of a gas turbine power plant that utilizes a basic Brayton cycle was discussed earlier in section 8.2. The block diagram of such a plant is illustrated in Fig. 8.2. The $T - s$ diagram of the cycle is shown in Fig. 11.8. Cyclic process 1-$2s$-3-$4s$ is referred to as the ideal basic cycle and process 1-2-3-4 is the actual basic cycle, which takes internal irreversibilities in the compressor and turbine into account. Assuming a calorically perfect gas to be the working substance and steady operation, SFEE applied to the individual components gives (after neglecting KE, PE changes and pressure losses),

---

[†]It may be recalled that the critical pressure of water is 221 bar. Steam power plants that operate at a boiler pressure above 221 bar but less than 300 bar are classified as super-critical; those that operate at a boiler pressure above 300 bar are classified as ultra super-critical.

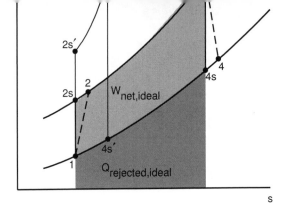

Figure 11.8: $T - s$ diagram of a basic Brayton cycle

$$\dot{W}_{x,comp} = \dot{m}\left(h_{2s} - h_1\right) = \dot{m}\,C_p\left(T_{2s} - T_1\right)$$

$$\dot{Q}_H = \dot{m}\left(h_3 - h_2\right) = \dot{m}\,C_p\left(T_3 - T_{2s}\right)$$

$$\dot{W}_{x,turbine} = \dot{m}\left(h_3 - h_{4s}\right) = \dot{m}\,C_p\left(T_3 - T_{4s}\right)$$

Here, $\dot{W}_{x,comp}$ is the required compressor power. The above expressions for work may be simplified further by using the fact that the processes in the compressor and turbine are isentropic in the case of an ideal cycle. Thus

$$T_{2s} = T_1\left(\frac{P_{2s}}{P_1}\right)^{(\gamma-1)/\gamma}, \quad T_{4s} = T_3\left(\frac{P_{4s}}{P_3}\right)^{(\gamma-1)/\gamma}$$

Since $P_{2s} = P_3$ and $P_{4s} = P_1$, the above expressions may be simplified further as

$$T_{2s} = T_1\,r_p^{(\gamma-1)/\gamma}, \quad T_{4s} = \frac{T_3}{r_p^{(\gamma-1)/\gamma}}$$

where, $r_p = P_{2s}/P_1$ is the pressure ratio of the cycle. Therefore, the net power generated is given as

$$\dot{W}_{x,net} = \dot{m}\,C_p\,T_1\left[\frac{T_3}{T_1}\left(1 - \frac{1}{r_p^{(\gamma-1)/\gamma}}\right) - r_p^{(\gamma-1)/\gamma} + 1\right] \tag{11.1}$$

The thermal efficiency of the cycle is thus

$$\eta = \frac{\dot{W}_{x,net}}{\dot{Q}_H} = 1 - \frac{1}{r_p^{(\gamma-1)/\gamma}} \tag{11.3}$$

The efficiency of the ideal cycle thus depends on the pressure ratio, r_p, alone. Consequently, today's gas turbine powerplants routinely operate with r_p values as high as 35. It can be seen from Fig. 11.8 that increasing r_p increases the maximum temperature in the cycle, T_3. Currently, metallurgical considerations limit this value to about 1500 K. It is important to realize that this value is itself about 300-350°C higher than the melting temperature of the turbine blade metal! With this in mind, if the maximum temperature in the cycle is fixed, and the pressure ratio is increased, the resulting cycle is shown in Fig. 11.8 as $1\text{-}2s'\text{-}3\text{-}4s'$. It may be inferred that, while this cycle has a higher efficiency, the net specific work output is smaller (since the heat that can be added is smaller). Consequently, the mass flow rate of the working substance and hence size of the components has to be increased in order to generate the required amount of power.

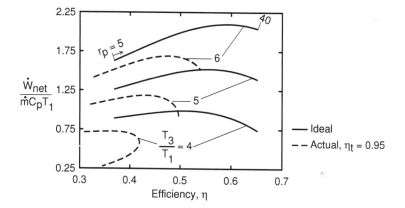

Figure 11.9: Variation of the specific net work output and efficiency of a basic Brayton cycle for $T_3 \div T_1 = 4$, 5 and 6. For each case, the pressure ratio r_p ranges from 5 to 40. Note that the efficiency of the ideal cycle is independent of $T_3 \div T_1$.

The variation of the specific net work output and efficiency for the ideal basic Brayton cycle is shown in Fig. 11.9 for several values of the pressure ratio and shown from the expression for specific work, Eqn. 11.1, that, the net specific work is a maximum when

$$r_p = \sqrt{\left(\frac{T_3}{T_1}\right)^{\gamma/(\gamma-1)}}$$

For this value of r_p, it is easy to establish that $T_{2s} = T_{4s}$. That is, the exit temperatures from the compressor and turbine are equal.

The limiting case for which the net work output becomes zero occurs when the pressure ratio is such that the compressor exit temperature becomes equal to the maximum allowed temperature. Consequently, no heat addition is possible and so the net work output is zero. The cycle in this case may be thought of as a "motoring" cycle that alternately compresses and expands the air between the same pressures.

■ EXAMPLE 11.7

An ideal Brayton cycle operates with a pressure ratio of 30 and a peak temperature of 1300 K. Air is the working substance and it enters the compressor at 100 kPa and 300 K. Determine the heat supplied, power output, the thermal and the second law efficiency of the cycle. Also determine the rate of exergy destruction in the individual components. The ambient temperature may be taken to be 300 K.

Solution : For air, $M = 28.8$ kg/kmol and $\gamma = 7/5$. Therefore, $R = 8314 \div 28.8 = 288.68$ J/kg.K and $C_p = \gamma R \div (\gamma - 1) = 1010.38$ J/kg.K.

Given $P_1 = 100$ kPa, $T_1 = 300$ K, $r_p = 30$ and $T_3 = 1300$ K. Using the expressions given above, we may evaluate $T_{2s} = 793$ K, $T_{4s} = 492$ K.

Compressor power required may be determined as

$$\frac{\dot{W}_{x,comp}}{\dot{m}} = C_p\left(T_{2s} - T_1\right) = 498.117\,\text{kJ/kg}$$

Power generated by the turbine is given as

$$\frac{\dot{W}_{x,turbine}}{\dot{m}} = C_p\left(T_3 - T_{4s}\right) = 816.387\,\text{kJ/kg}$$

The compressor consumes 61 percent of the power generated by the turbine. This should be contrasted with the pump work calculated in the Rankine cycle. Net power

$$\frac{\dot{Q}_H}{\dot{m}} = C_p \left(T_3 - T_{2s} \right) = 512.263 \, \text{kJ/kg}$$

and heat rejected is

$$\frac{\dot{Q}_C}{\dot{m}} = C_p \left(T_{4s} - T_1 \right) = 194 \, \text{kJ/kg}$$

The efficiency of the cycle is

$$\eta = \frac{\dot{W}_{x,net}}{\dot{Q}_H} = \frac{318.27}{512.263} = 62.13\%$$

The efficiency of the ideal basic Brayton cycle may be seen to be much higher than that of the ideal, basic Rankine cycle.

Rate at which exergy is supplied

$$\frac{\dot{X}_{supplied}}{\dot{m}} = \frac{\dot{W}_{x,comp}}{\dot{m}} + \frac{\dot{Q}_H}{\dot{m}} \left(1 - \frac{T_0}{T_H} \right) = 892.17 \, \text{kJ/kg}$$

where we have taken T_H = 1300 K. Rate at which exergy is recovered may be evaluated as

$$\frac{\dot{X}_{recovered}}{\dot{m}} = \frac{\dot{W}_{x,turbine}}{\dot{m}} = 816.387 \, \text{kJ/kg}$$

Note that, the exergy recovered during the heat rejection process is zero since heat is rejected to the ambient. The second law efficiency is thus

$$\eta_{II} = \frac{\text{Exergy recovered}}{\text{Exergy supplied}} = 91.5\%$$

The rate of exergy destruction in the compression and turbine are zero, as these are isentropic. The rate of exergy destruction during the heat addition process is given as

$$T_0 \frac{\dot{\sigma}}{\dot{m}} = \frac{T_0}{\dot{m}} \left[\dot{m} \left(s_3 - s_{2s} \right) - \frac{\dot{Q}_H}{T_H} \right]$$

$$= T_0 \left[\left(C_p \ln \frac{T_3}{T_{2s}} - R \ln \cancelto{0}{\frac{P_3}{P_{2s}}} \right) - \frac{\dot{Q}_H}{\dot{m}} \frac{1}{T_H} \right]$$

$$= 31.6136 \, \text{kJ/kg}$$

$$m \qquad \dot{m} \qquad T_C \Bigg]$$

$$= T_0 \left[\left(C_p \ln \frac{T_1}{T_{4s}} - R \ln \frac{P_1}{P_{4s}}^{\,0} \right) + \frac{\dot{Q}_C}{\dot{m}} \frac{1}{T_C} \right]$$

$$= 44.05 \, \text{kJ/kg}$$

The total exergy destruction rate in the cycle is 75.6636 kW per unit mass flow rate of air.

As mentioned before, Eqn. 11.3 shows that the efficiency of the ideal basic Brayton cycle depends only on the pressure ratio. In reality, two parameters, namely, the pressure ratio, r_p and the the temperature ratio, $T_3 \div T_1$, control the performance of the cycle. This is demonstrated next for an actual cycle.

Let the isentropic efficiency of the compressor and turbine be η_c and η_t respectively. Equation 11.1 may then be written as

$$\dot{W}_{x,net} = \dot{m} \, C_p \, T_1 \left[\frac{T_3}{T_1} \, \eta_t \left(1 - \frac{1}{r_p^{(\gamma-1)/\gamma}} \right) - \frac{1}{\eta_c} \left(r_p^{(\gamma-1)/\gamma} - 1 \right) \right] \qquad (11.4)$$

The heat supplied becomes, after modifying Eqn. 11.2,

$$\dot{Q}_H = \dot{m} \, C_p \, T_1 \left[\frac{T_3}{T_1} - 1 - \frac{1}{\eta_c} \left(r_p^{(\gamma-1)/\gamma} - 1 \right) \right] \qquad (11.5)$$

The thermal efficiency of the actual cycle is thus

$$\eta = \frac{\dot{W}_{x,net}}{\dot{Q}_H} = 1 - \frac{\frac{T_3}{T_1} \left(1 - \eta_t - \frac{\eta_t}{r_p^{(\gamma-1)/\gamma}} \right) - 1}{\frac{T_3}{T_1} - 1 - \frac{1}{\eta_c} \left(r_p^{(\gamma-1)/\gamma} - 1 \right)} \qquad (11.6)$$

Variation of the specific net work output and efficiency of the actual basic Brayton cycle is shown in Fig. 11.9 for the same values of r_p and $T_3 \div T_1$ as before and with $\eta_t = 0.95$ and $\eta_c = 0.9$. It is clear that the efficiency of the actual cycle depends upon both r_p and $T_3 \div T_1$ and in general, increases with both, except for small values of the latter.

Solution : The compressor power in this case may be evaluated as $498.117 \div 0.9 = 553.463$ kJ/kg and the power generated by the turbine is equal to $0.95 \times 816.387 = 775.57$ kJ/kg. Net power produced is 222.105 kW per unit mass flow rate.

From the definition of the isentropic efficiency of the compressor,

$$\eta_{comp} = \frac{T_{2s} - T_1}{T_2 - T_1}$$

we can evaluate $T_2 = 848$ K. Heat supplied may be evaluated as

$$\frac{\dot{Q}_H}{\dot{m}} = C_p (T_3 - T_2) = 456.692 \, \text{kJ/kg}$$

The efficiency of the cycle is

$$\eta = \frac{\dot{W}_{x,net}}{\dot{Q}_H} = \frac{222.106}{456.692} = 48.63\%$$

Rate at which exergy is supplied

$$\frac{\dot{X}_{supplied}}{\dot{m}} = \frac{\dot{W}_{x,comp}}{\dot{m}} + \frac{\dot{Q}_H}{\dot{m}} \left(1 - \frac{T_0}{T_H}\right) = 904.765 \, \text{kJ/kg}$$

where we have taken $T_H = 1300$ K. Rate at which exergy is recovered may be evaluated as

$$\frac{\dot{X}_{recovered}}{\dot{m}} = \frac{\dot{W}_{x,turbine}}{\dot{m}} = 775.57 \, \text{kJ/kg}$$

The second law efficiency is thus

$$\eta_{II} = \frac{\text{Exergy recovered}}{\text{Exergy supplied}} = 85.7\%$$

The decrease in the second law efficiency is due to the additional exergy destruction in the turbine and compressor.

Intercooling It was shown in section 9.6.1 that, for a given pressure ratio, a steady flow reversible isothermal compression process requires less power than an isentropic compression process. Hence, the efficiency of the basic Brayton cycle can be improved by replacing the isentropic process with a reversible isothermal process. However, the latter is impossible to achieve because of the excessive cooling requirement. A better (and more practicable) strategy is to accomplish the overall compression process in multiple stages with the air being cooled to its initial

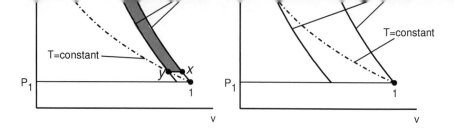

Figure 11.10: Work interaction for two stage compression with intercooling between the stages. The shaded region represents to the saving in the power required.

temperature at the end of each stage. The compression process in each stage is still isentropic. A two stage compression process that utilizes this strategy is illustrated in Fig. 11.10. The initial and final pressures are the same between the two illustrations in this figure, but the intermediate pressure, $P_x = P_y$, is different. It is closer to the initial pressure in Fig. 11.10 (left) and the final pressure in Fig. 11.10 (right). Since the power required in this case is the area under the curve in $v - P$ coordinates (Fig. 9.8), the saving in compressor power for each case is shown shaded in Fig. 11.10. It is evident from these illustrations that, as the intermediate pressure is varied, the saving in compressor power also varies, suggesting that there is likely to be an optimum value at which the required compressor power is a minimum. In order to determine this optimum value for the intermediate pressure, we start with SFEE applied to the compression process:

$$\frac{\dot{W}_{x,comp}}{\dot{m}} = C_p\left(T_x - T_1\right) + C_p\left(T_f - T_y\right),$$

where the subscript f denotes the final state. Note that $T_y = T_1$. Since $1\text{-}x$ and $y\text{-}f$ are isentropic processes, we may rewrite the above expression as

$$\frac{\dot{W}_{x,comp}}{\dot{m}} = C_p T_1 \left[\left(\frac{P_x}{P_1}\right)^{(\gamma-1)/\gamma} - 1\right] + C_p T_1 \left[\left(\frac{P_2}{P_y}\right)^{(\gamma-1)/\gamma} - 1\right].$$

Upon differentiating this expression with respect to the intermediate pressure, P_x, (keeping in mind $P_x = P_y$), it may be shown that

$$\frac{P_x}{P_1} = \frac{P_2}{P_y} \Rightarrow P_x = P_y = (P_1 P_2)^{1/2}.$$

This shows that, the pressure ratio is the same in both the stages and is equal to $r_p^{1/2}$ where $r_p = P_2 \div P_1$. In addition, since the temperature at the beginning of each stage

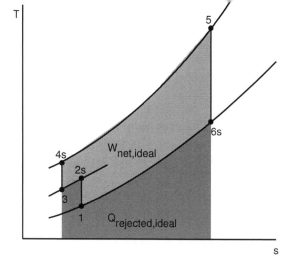

Figure 11.11: $T - s$ diagram of a Brayton cycle with intercooling

The $T - s$ diagram of a Brayton cycle with two stage compression and intercooling between the stages is shown in Fig. 11.11. It is clear from this figure that the net work produced increases. However, the heat added is also more now since the air enters the combustor at a lower temperature now when compared to single stage compression. Consequently, the overall efficiency of the cycle, in general, decreases with the use of multistage compression alone.

EXAMPLE 11.9

Redo Example 11.7, after incorporating two stage compression with intercooling.

Solution : Since it is given that $r_p = 30$, the pressure ratio in each stage is $\sqrt{30}$. The temperature at the end of each stage may be evaluated as

$$T_{2s} = T_{4s} = T_1 \left(r_p^{1/2} \right)^{(\gamma-1)/\gamma} = 488\,\text{K}$$

Total compressor power required is

$$\frac{\dot{W}_{x,comp}}{\dot{m}} = 2\,C_p\,(T_{2s} - T_1) = 379.903\,\text{kJ/kg}$$

This represents a 23.73 percent saving in the required compressor power, which is expected. Power generated by the turbine remains the same, 816.387 kJ/kg. Net

and heat rejected during the cycle is

$$\frac{\dot{Q}_C}{\dot{m}} = C_p\left(T_{2s} - T_3\right) + C_p\left(T_{6s} - T_1\right) = 383.94\,\text{kJ/kg}$$

The thermal efficiency works out to be 53.21 percent, which is less than that of the single stage compression case by 14.36 percent. Since the temperature at entry into the combustor is less now than before, the heat that has to be added to attain the same peak temperature as before is higher.

Rate at which exergy is supplied

$$\frac{\dot{X}_{supplied}}{\dot{m}} = \frac{\dot{W}_{x,comp}}{\dot{m}} + \frac{\dot{Q}_H}{\dot{m}}\left(1 - \frac{T_0}{T_H}\right) = 1011\,\text{kJ/kg}$$

where we have taken $T_H = 1300$ K. Rate at which exergy is recovered may be evaluated as

$$\frac{\dot{X}_{recovered}}{\dot{m}} = \dot{W}_{x,turbine} = 816.387\,\text{kJ/kg}$$

The second law efficiency is thus

$$\eta_{II} = \frac{\text{Exergy recovered}}{\text{Exergy supplied}} = 80.75\%$$

The rate of exergy destruction in the compression and turbine are zero, as these are isentropic. The rate of exergy destruction during the heat addition process is given as

$$
\begin{aligned}
T_0 \frac{\dot{\sigma}}{\dot{m}} &= \frac{T_0}{\dot{m}}\left[\dot{m}\left(s_5 - s_{4s}\right) - \frac{\dot{Q}_H}{T_H}\right] \\
&= T_0\left[C_p \ln \frac{T_5}{T_{4s}} - \frac{\dot{Q}_H}{\dot{m}}\frac{1}{T_H}\right] \\
&= 107.66\,\text{kJ/kg}
\end{aligned}
$$

The rate of exergy destruction during the heat rejection process is given as

$$
\begin{aligned}
T_0 \frac{\dot{\sigma}}{\dot{m}} &= \frac{T_0}{\dot{m}}\left[\dot{m}\left(s_3 - s_{2s}\right) + \dot{m}\left(s_1 - s_{6s}\right) + \frac{\dot{Q}_C}{T_C}\right] \\
&= T_0\left[C_p\left(\ln \frac{T_3}{T_{2s}} + \ln \frac{T_1}{T_{6s}}\right) + \frac{\dot{Q}_C}{\dot{m}}\frac{1}{T_C}\right] \\
&= 86.52\,\text{kJ/kg}
\end{aligned}
$$

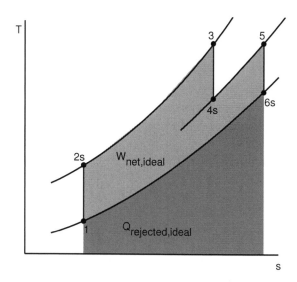

Figure 11.12: $T - s$ diagram of a Brayton cycle with reheat

Reheat   Just as multistage compression with intercooling requires less power when compared to single stage compression, it is straightforward to show that multistage expansion with reheating between the stages produces more work when compared to single stage expansion. It is left as an exercise to the reader to repeat the analysis of the previous section for a two stage expansion process and prove that the pressure ratio in each stage must be the same for maximizing the turbine power output. Consequently, the work produced in each stage is the same. The $T - s$ diagram of a Brayton cycle with reheat is shown in Fig. 11.12. Although the total work developed with reheat is higher, the heat supplied is also higher and so a gain in efficiency is unlikely, just as in the case of multistage compression with intercooling.

■ EXAMPLE 11.10

Redo Example 11.7 with two stage expansion and reheating in-between the stages to the same peak temperature.

Solution : Since the overall pressure ratio across the cycle is 30, the pressure ratio across each turbine stage is $\sqrt{30}$. It is also given that $T_3 = T_5 = 1300$ K. Since processes *3-4s* and *5-6s* are isentropic, we may evaluate the temperature at the end

The total work produced by the turbines is given as

$$\frac{\dot{W}_{x,turbine}}{\dot{m}} = 2\,C_p\,(T_3 - T_{4s}) = 1010.38\,\text{kJ/kg}$$

This represents an increase of 23.76 percent over the case with single stage expansion. The compressor power is the same as before and so the net power produced is 1010.38 - 498.117 = 512.263 kJ per unit mass flow rate. Total heat supplied is given as

$$\frac{\dot{Q}_H}{\dot{m}} = C_p\,(T_3 - T_{2s}) + C_p\,(T_5 - T_{4s}) = 1017.45\,\text{kJ/kg}$$

which is nearly twice as much as before. Heat rejected during the cycle is

$$\frac{\dot{Q}_C}{\dot{m}} = C_p\,(T_{6s} - T_1) = 505.19\,\text{kJ/kg}$$

The efficiency of the cycle works out to be 50.35 percent which is a 12 percent reduction.

Rate at which exergy is supplied

$$\frac{\dot{X}_{supplied}}{\dot{m}} = \frac{\dot{W}_{x,comp}}{\dot{m}} + \frac{\dot{Q}_H}{\dot{m}}\left(1 - \frac{T_0}{T_H}\right) = 1280.77\,\text{kJ/kg}$$

where we have taken $T_H = 1300$ K. Rate at which exergy is recovered may be evaluated as

$$\frac{\dot{X}_{recovered}}{\dot{m}} = \frac{\dot{W}_{x,turbine}}{\dot{m}} = 1010.38\,\text{kJ/kg}$$

The second law efficiency is thus

$$\eta_{II} = \frac{\text{Exergy recovered}}{\text{Exergy supplied}} = 78.89\%$$

The rate of exergy destruction during the heat addition process is given as

$$
\begin{aligned}
T_0\frac{\dot{\sigma}}{\dot{m}} &= \frac{T_0}{\dot{m}}\left[\dot{m}\,(s_3 - s_{2s}) + \dot{m}\,(s_5 - s_{4s}) - \frac{\dot{Q}_H}{T_H}\right] \\
&= T_0\left[C_p\left(\ln\frac{T_3}{T_{2s}} + \ln\frac{T_5}{T_{4s}}\right) - \frac{\dot{Q}_H}{\dot{m}}\frac{1}{T_H}\right] \\
&= 62.196\,\text{kJ/kg}
\end{aligned}
$$

$$= T_0 \left(C_p \ln \frac{T_1}{T_{6s}} + \frac{\dot{Q}_C}{\dot{m}} \frac{1}{T_C} \right)$$

$$= 207.887 \,\text{kJ/kg}$$

The total exergy destruction rate in the cycle is 270.083 kW per unit mass flow rate of air.

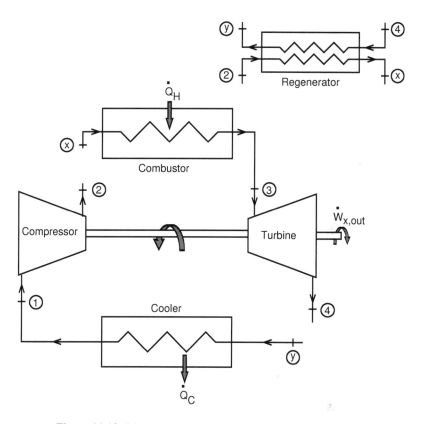

Figure 11.13: Block diagram of a Brayton cycle with regeneration

Regeneration It was seen that reheat increases the work produced but the efficiency of the cycle decreases since the additional heat supplied more than offsets the gain in the work developed. This cannot be viewed as a penalty since heat has to be supplied during the reheat stage. In the case of multistage compression with intercooling, the required compressor power decreases but the heat supplied

expansion with reheating can generate more power for a given heat input.

It may be inferred by comparing Figs. 11.8 and 11.12 that, for the same pressure ratio and peak temperature, the temperature at the exit of the second stage turbine is higher when reheat is employed. Furthermore, this temperature is also much higher than the temperature at the exit of the second stage compressor (Fig. 11.11). Hence, there is scope for heating the air before it enters the combustion chamber by utilizing the exhaust heat from the second stage turbine. This is quite similar to the regenerative feedwater heating utilized in the Rankine cycle and, in fact, bears the same name. The difference, however, is that, in the case of the Brayton cycle, there is no penalty in the power generated due to regeneration unlike the Rankine cycle.

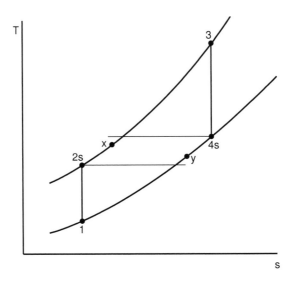

Figure 11.14: $T - s$ diagram of a Brayton cycle with regeneration

The block diagram of a basic Brayton cycle with regeneration is shown in Fig. 11.13. The regenerator can be seen to be just a heat exchanger. States x and y respectively denote the exit states of the compressor and turbine air stream. The $T - s$ diagram of a basic Brayton cycle with regeneration is shown in Fig. 11.14. The potential for regeneration is dependent directly on the difference between the temperatures at the turbine exit and the compressor exit. For instance, in the basic Brayton cycle considered in Example 11.7, there is no scope for regeneration, since the compressor exit temperature (793 K) is considerably higher than the turbine exit temperature (492 K). In Example 11.9, the compressor exit temperature decreases to 488 K, but

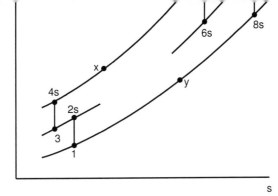

Figure 11.15: $T - s$ diagram of a Brayton cycle with intercooling, reheat and regeneration

the scope for regeneration is still negligibly small. However, in Example 11.10, with two stage expansion and reheating between the stages, the turbine exit temperature is 800 K. Consequently, the scope for regeneration is quite high, when multistage compression with intercooling and multistage expansion with reheating is employed. Accordingly, a regenerator effectiveness, ϵ, is defined as follows:

$$\epsilon = \frac{h_x - h_2}{h_4 - h_2}$$

Given a value for ϵ, the enthalpy of the air before entering the combustor, h_x may be determined.

EXAMPLE 11.11

Redo the previous example, after including two stage compression with intercooling and a regenerator with effectiveness 0.9.

Solution : With reference to Fig. 11.15, temperatures at the states shown in the table may be retrieved from Examples 11.7, 11.9 and 11.10.

Since the regenerator effectiveness is given to be 0.9, we have

$$0.9 = \frac{h_x - h_{4s}}{h_{8s} - h_{4s}} = \frac{T_x - T_{4s}}{T_{8s} - T_{4s}}$$

Upon substituting the known values, we get T_x = 769 K. SFEE applied to the heat

$$\Rightarrow \quad T_y \quad = \quad T_{8s} + T_{4s} - T_x = 519\,\text{K}$$

State	T (K)
1	300
2s	488
3	300
4s	488
5	1300
6s	800
7	1300
8s	800
x	769
y	519

The total compressor power required is 379.904 kW per unit mass flow rate. The total power generated by the turbines is 1010.38 kW per unit mass flow rate. Net power produced is thus 1010.38 - 379.903 = 630.477 kW per unit mass flow rate. Heat supplied is given as

$$\frac{\dot{Q}_H}{\dot{m}} = C_p\,(T_5 - T_x) + C_p\,(T_7 - T_{6s}) = 1041.702\,\text{kJ/kg}$$

and the heat rejected is

$$\frac{\dot{Q}_C}{\dot{m}} = C_p\,(T_{2s} - T_3) + C_p\,(T_y - T_1) = 411.226\,\text{kJ/kg}$$

The thermal efficiency of the cycle is, therefore 60.52 percent. Although this is about 1.5 percent less than that of the ideal cycle considered in Example 11.7, the specific power generated is almost twice that of the latter, which is a big advantage in terms of the size of the powerplant.

Rate at which exergy is supplied

$$\frac{\dot{X}_{supplied}}{\dot{m}} = \frac{\dot{W}_{x,comp}}{\dot{m}} + \frac{\dot{Q}_H}{\dot{m}}\left(1 - \frac{T_0}{T_H}\right) = 1181.213\,\text{kJ/kg}$$

where we have taken T_H = 1300 K. Rate at which exergy is recovered may be evaluated as

$$\frac{\dot{X}_{recovered}}{\dot{m}} = \frac{\dot{W}_{x,turbine}}{\dot{m}} = 1010.38\,\text{kJ/kg}$$

The second law efficiency has increased considerably with the addition of regeneration. This suggests that the rate of exergy destruction during the heat addition as well as the heat rejection process has decreased. The rate of exergy destruction during the heat addition process is given as

$$
\begin{aligned}
T_0 \frac{\dot{\sigma}}{\dot{m}} &= \frac{T_0}{\dot{m}} \left[\dot{m}(s_5 - s_x) + \dot{m}(s_7 - s_{6s}) - \frac{\dot{Q}_H}{T_H} \right] \\
&= T_0 \left[C_p \left(\ln \frac{T_5}{T_x} + \ln \frac{T_7}{T_{6s}} \right) - \frac{\dot{Q}_H}{\dot{m}} \frac{1}{T_H} \right] \\
&= 65.91 \, \text{kJ/kg}
\end{aligned}
$$

This is less than the corresponding value obtained in Example 11.9 with intercooling but without regeneration. The rate of exergy destruction during the heat rejection process is given as

$$
\begin{aligned}
T_0 \frac{\dot{\sigma}}{\dot{m}} &= \frac{T_0}{\dot{m}} \left[\dot{m}(s_3 - s_{2s}) + \dot{m}(s_1 - s_y) + \frac{\dot{Q}_C}{T_C} \right] \\
&= T_0 \left[C_p \left(\ln \frac{T_3}{T_{2s}} + \ln \frac{T_1}{T_y} \right) + \frac{\dot{Q}_C}{\dot{m}} \frac{1}{T_C} \right] \\
&= 97.608 \, \text{kJ/kg}
\end{aligned}
$$

This is less than the corresponding value in Example 11.10 with reheat but without regeneration.

Rate of exergy destruction in the regenerator may be evaluated as

$$
\begin{aligned}
T_0 \frac{\dot{\sigma}}{\dot{m}} &= \frac{T_0}{\dot{m}} \left[\dot{m}(s_x - s_{4s}) + \dot{m}(s_y - s_{8s}) \right] \\
&= T_0 C_p \left(\ln \frac{T_x}{T_{4s}} + \ln \frac{T_y}{T_{8s}} \right) \\
&= 6.689 \, \text{kJ/kg}
\end{aligned}
$$

The total exergy destruction rate in the cycle is 170.207 kW per unit mass flow rate. Although this is higher than the value calculated in Example 11.7, it is important to note that both the first law and the second law efficiencies are almost the same and the specific work output now is twice the earlier value.

A comparison of the performance metrics of the basic air standard Brayton cycle

Cycle	Thermal efficiency η_{th}	Second law efficiency η_{II}	Specific work (kJ/kg) $\frac{\dot{W}_{x,net}}{\dot{m}}$
Basic	62.13	91.5	318.27
Intercooling	53.21	80.75	436.485
Reheat	50.35	78.89	512.263
Reheat with regeneration and intercooling	60.52	85.54	630.476

Calculation with variable properties The calculations in all the examples above have assumed that C_p is a constant. In reality, C_p increases with temperature and the effect due to this departure becomes more significant at higher temperatures. Since the peak temperature in the Brayton cycle examples is 1300 K, it is likely that the effect of variable C_p is significant. With this in mind, the previous example is solved again using the data given in Table L[†].

The table below lists the property values at each state in the cycle shown in Fig. 11.15. In each row, the value used to retrieve from Table L is shown in italics; values that are retrieved are shown in bold; other values are given or calculated using known quantities. Temperature values retrieved from Table L are not required for any calculations, but are given here for the sake of comparison with the corresponding values from the previous example. Using the values given in the table, we can complete the analysis of the cycle.

$$\frac{\dot{W}_{x,comp}}{\dot{m}} = (h_{2s} - h_1) + (h_{4s} - h_3) = 376.9\,\text{kJ/kg}$$

$$\frac{\dot{W}_{x,turbines}}{\dot{m}} = (h_5 - h_{6s}) + (h_7 - h_{8s}) = 1037.04\,\text{kJ/kg}$$

$$\frac{\dot{Q}_H}{\dot{m}} = (h_5 - h_x) + (h_7 - h_{6s}) = 1075.92\,\text{kJ/kg}$$

$$\frac{\dot{Q}_C}{\dot{m}} = (h_{2s} - h_3) + (h_y - h_1) = 415.78\,\text{kJ/kg}$$

$$\eta = \frac{\dot{W}_{x,net}}{\dot{Q}_H} = 61.36\%$$

[†] see Appendix A for details

2s	$100\sqrt{30}$	486	488.64	-	$1.386\times30^{-?}$	1.70203
3	$100\sqrt{30}$	300	300.19	1.70203	1.386	1.21386
4s	3000	486	488.64	-	$1.386\times30^{1/2}$	1.21386
5	3000	1300	1395.97	3.27345	330.9	2.29715
6s	$100\sqrt{30}$	850	877.45	-	$330.9\div30^{1/2}$	2.29715
7	$100\sqrt{30}$	1300	1395.97	3.27345	330.9	2.7853
8s	100	850	877.45	-	$330.9\div30^{1/2}$	2.7853
x	3000	815	838.569	2.73836	-	1.76205
y	100	524	527.521	2.26738	-	2.26738

$$\frac{\dot{X}_{supplied}}{\dot{m}} = \frac{\dot{W}_{x,comp}}{\dot{m}} + \frac{\dot{Q}_H}{\dot{m}}\left(1 - \frac{T_0}{T_H}\right) = 1204.53\,\text{kJ/kg}$$

$$\frac{\dot{X}_{recovered}}{\dot{m}} = \frac{\dot{W}_{x,turbines}}{\dot{m}} = 1037.04\,\text{kJ/kg}$$

$$\eta_{II} = \frac{\text{Exergy recovered}}{\text{Exergy supplied}} = 86.1\%$$

The rate of exergy destruction during the heat addition process is given as

$$T_0\frac{\dot{\sigma}}{\dot{m}} = T_0\left[(s_5 - s_x) + (s_7 - s_{6s}) - \frac{\dot{Q}_H}{\dot{m}}\frac{1}{T_H}\right]$$
$$= 58.686\,\text{kJ/kg}$$

The rate of exergy destruction during the heat rejection process is given as

$$T_0\frac{\dot{\sigma}}{\dot{m}} = T_0\left[(s_3 - s_{2s}) + (s_1 - s_y) + \frac{\dot{Q}_C}{\dot{m}}\frac{1}{T_C}\right]$$
$$= 99.724\,\text{kJ/kg}$$

Rate of exergy destruction in the regenerator may be evaluated as

$$T_0\frac{\dot{\sigma}}{\dot{m}} = T_0\left[(s_x - s_{4s}) + (s_y - s_{8s})\right]$$
$$= 9.081\,\text{kJ/kg}$$

It is left as an exercise to redo the examples taking into account internal irreversibilities of the turbine(s) and the compressor(s).

Air and fuel (gasoline) vapor are taken into the engine during the intake stroke. The mixture is then compressed during the compression stroke. It is then ignited using a spark plug. The combustion of the air- fuel mixture essentially occurs at constant volume. The combustion products then expand during the power stroke, generating work. The products of combustion are then expelled from the cylinder during the exhaust stroke and the engine is ready to receive a fresh mixture of air and fuel vapor. This entire sequence is executed in four strokes (one stroke corresponding to a travel of the piston from one dead center to the other) and hence such an engine is called a four stroke engine. Although the sequence of processes appears to be cyclic on the $P - v$ diagram, the working substance does not execute a cyclic process.

The equivalent air standard cycle, referred to as the Otto cycle, is also shown in Fig. 11.16. Here, it is assumed that a fixed mass of air, m, executes a cyclic process. Consequently, the intake and the exhaust strokes are absent. Heat addition and heat rejection are assumed to occur at constant volume. Note that there is no explicit heat rejection process in the sequence of process executed in an actual engine, but it is required in the air standard cycle, as demanded by the K-P statement.

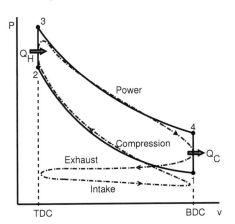

Figure 11.16: $P - v$ diagram of an air standard Otto cycle (solid line) and the processes in a spark ignition (SI) engine (chain line). TDC and BDC denote Top Dead Center and Bottom Dead Center respectively.

First law applied to process *1-2* gives

$$\Delta U = \cancel{Q_{1-2}}^{0} - W_{1-2} \quad \Rightarrow \quad W_{1-2} = m\left(u_2 - u_1\right)$$

where we have assumed the process to be isentropic and set $Q = 0$. Similarly,

$$Q_{2-3} = Q_H \;\; = \;\; m\left(u_3 - u_2\right)$$

however, W_{1-2} is compression work and Q_{4-1} is heat rejected. Net work output during each cycle is given as

$$W_{net} = m\left(u_3 - u_4\right) - m\left(u_2 - u_1\right)$$

Thermal efficiency of the cycle is given as

$$\eta = \frac{W_{net}}{Q_{in}} = \frac{\left(u_3 - u_4\right) - \left(u_2 - u_1\right)}{u_3 - u_2} = 1 - \frac{u_4 - u_1}{u_3 - u_2}$$

If we assume the working substance (air) to be calorically perfect (cold air standard analysis), the above two expressions may be written as

$$\frac{W_{net}}{m\, C_v\, T_1} = \frac{T_3}{T_1}\left(1 - \frac{1}{r^{\gamma-1}}\right)$$

$$\eta = 1 - \frac{1}{r^{\gamma-1}}$$

where $r = V_1 \div V_2$ is the compression ratio†.

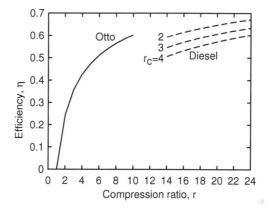

Figure 11.17: Variation of the thermal efficiency of the ideal air standard Otto and Diesel cycle with compression ratio

It is clear that the performance of the cycle is controlled by two parameters – the compression ratio, r and the ratio $T_3 \div T_1$. However, in the case of a calorically

---

†V_2 is referred to as the clearance volume and $V_1 - V_2$ is referred to as the displacement or the swept volume.

air-fuel mixture – and not air alone, is taken in, high compression ratios can result in premature auto-ignition of the mixture. Since the combustion is intended to start at the spark plug and proceed outward, the pressure rise due to auto-ignition can interfere with this and cause an uneven pressure rise, or knocking. It is evident from this figure that the maximum value of the efficiency is about 60 percent for the cold air standard analysis.

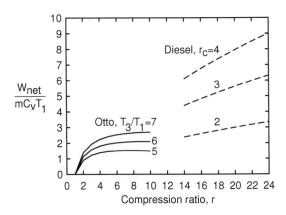

Figure 11.18: Variation of the specific work output of the ideal air standard Otto and Diesel cycle with compression ratio

The variation of the dimensionless specific work, $W_{net} \div (mC_v T_1)$, is shown in Fig. 11.18 for several representative values of $T_3 \div T_1$ (assuming T_1 to be 300 K) and for values of r up to 10. The specific work increases initially and levels off afterwards. It can also be seen that, for a given value of r, specific work is higher for higher values of $T_3 \div T_1$.

◼ EXAMPLE 11.12

At the beginning of an ideal air standard Otto cycle the air is at 100 kPa and 298 K. The compression ratio is 9 and the peak temperature is 2200 K. Determine (a) the thermal efficiency, (b) the mean effective pressure and (c) the second law efficiency. The ambient temperature may be taken to be 298 K.

Solution : Property values at each state calculated using values from Table L are given in the table below. In each row, the value used to retrieve from Table L is shown in italics; values that are retrieved are shown in bold; other values are given or calculated using known quantities. Since the cycle is ideal, processes *1-2* and *3-4*

and $P_3 = P_2 \times T_3 \div T_2$. Also

$$P_4 = P_3 \frac{T_4}{T_3} \frac{v_3}{v_4} = P_3 \frac{T_4}{T_3} \frac{1}{r}$$

State	P	T	u	s°	v_r	s (Eqn. A.1)
	(kPa)	(K)	(kJ/kg)	(kJ/kg.K)		(kJ/kg.K)
1	100	298	212.64	1.69528	631.9	1.69528
2	2108.05	698	511.06	-	631.9÷9	1.69528
3	6644.28	2200	1872.4	3.9191	2.012	2.7159
4	374.5	1116	858.95	-	2.012×9	2.7159

(a) Work supplied during the compression process is given as

$$\frac{W_{1-2}}{m} = u_2 - u_1 = 298.42\,\text{kJ/kg}$$

and work done during the expansion process is

$$\frac{W_{3-4}}{m} = u_3 - u_4 = 1013.45\,\text{kJ/kg}$$

The net work produced during every cycle is thus 715.03 kJ/kg. Heat added during each cycle may be evaluated as

$$\frac{Q_H}{m} = u_3 - u_2 = 1361.34\,\text{kJ/kg}$$

and heat rejected during each cycle is

$$\frac{Q_C}{m} = u_4 - u_1 = 646.31\,\text{kJ/kg}$$

Thermal efficiency of the cycle is thus

$$\eta = \frac{W_{net}}{Q_H} = \frac{715.03}{1361.34} = 52.52\%$$

(b) The mean effective pressure (MEP) is defined as

$$\text{MEP} = \frac{W_{net}}{V_1 - V_2} = \frac{\frac{W_{net}}{m}}{\frac{RT_1}{P_1}\left(1 - \frac{1}{r}\right)}$$

Upon substituting the numerical values we get this to be 940.39 kPa. The MEP is a quantity similar to the average temperature of heat addition that was calculated for

$$\frac{X_{supplied}}{m} = (u_2 - u_1) + P_0(v_2 - v_1) + \frac{Q_H}{m}\left(1 - \frac{T_0}{T_H}\right)$$

$$= (u_2 - u_1) - \frac{P_0}{P_1}RT_1\left(1 - \frac{1}{r}\right) + \frac{Q_H}{m}\left(1 - \frac{T_0}{T_H}\right)$$

$$= 1399.32\,\text{kJ/kg}$$

where we have taken $T_H = 2200\,\text{K}$ and also used Eqn. 10.4. Exergy recovered during the cycle is given as

$$\frac{X_{recovered}}{m} = (u_3 - u_4) + P_0(v_3 - v_4)$$

$$= (u_3 - u_4) - \frac{P_0}{P_3}RT_3(r - 1)$$

$$= 937.5\,\text{kJ/kg}$$

where we have taken $T_C = 300\,\text{K}$ (the lowest temperature during process 3-4). Equation 10.4 has been used for evaluating the exergy recovered from the engine in such a manner as to get a positive value. The second law efficiency is thus

$$\eta_{II} = \frac{\text{Exergy recovered}}{\text{Exergy supplied}} = 67\%$$

11.2.3 Diesel cycle

Internal combustion engines that utilize diesel as the fuel are called compression ignition (CI) engines. In these engines, air is inducted into the engine during the intake stroke and it is then compressed during the compression stroke. In contrast to SI engines, compression ratios in CI engines are more than 14 and typically around 20. Consequently, the temperature of the air at the end of the compression process is much higher than in the case of the SI engine and in fact, higher than the auto-ignition temperature of the diesel fuel. The fuel is sprayed into the cylinder at the end of the compression process and it ignites on its own without requiring a spark plug – hence the name compression ignition engine. However, the heat release is more gradual than in the case of a SI engine and is such that the pressure remains constant while the temperature increases. The fuel supply is cut-off once the temperature reaches a suitable value and the combustion gases now expand and produce work. At the end of the power stroke, the combustion gases are pushed out of the cylinder and the entire sequence is repeated with a fresh charge of air.

The $P - v$ diagram of the air standard Diesel cycle, which is an idealized representation of the aforementioned sequence of processes that take place in a CI engine is shown in Fig. 11.19.

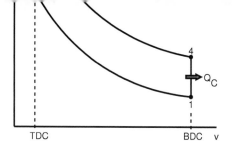

Figure 11.19: $P - v$ diagram of an air standard Diesel cycle

First law applied to process *1-2* gives

$$\Delta U = \cancelto{0}{Q_{1-2}} - W_{1-2} \quad \Rightarrow \quad W_{1-2} = m\,(u_2 - u_1)$$

where we have assumed the process to be isentropic and set $Q = 0$. Similarly,

$$
\begin{aligned}
Q_{2-3} = Q_H &= m\,(u_3 - u_2) + m\,P_2\,(v_3 - v_2) \\
W_{3-4} &= m\,(u_3 - u_4) \\
Q_{4-1} &= m\,(u_4 - u_1)
\end{aligned}
$$

where we have used $P_3 = P_2$. Net work output during each cycle is given as

$$W_{net} = m\,P_2\,(v_3 - v_2) + m\,(u_3 - u_4) - m\,(u_2 - u_1)$$

Thermal efficiency of the cycle is given as

$$
\begin{aligned}
\eta = \frac{W_{net}}{Q_{in}} &= \frac{P_2\,(v_3 - v_2) + (u_3 - u_4) - (u_2 - u_1)}{u_3 - u_2} \\
&= \frac{(h_3 - h_2) - (u_4 - u_1)}{u_3 - u_2}
\end{aligned}
$$

If we assume the working substance (air) to be calorically perfect (cold air standard analysis), the above two expressions may be written as

$$
\begin{aligned}
\frac{W_{net}}{m\,C_v\,T_1} &= \gamma\,(r_c - 1)\,r^{\gamma-1} + 1 - r_c^{\gamma} \\
\eta &= 1 - \frac{1}{\gamma\,r^{\gamma-1}}\left(\frac{r_c^{\gamma} - 1}{r_c - 1}\right)
\end{aligned}
$$

where $r = V_1 \div V_2$ is the compression ratio, as before and $r_c = V_3 \div V_2$ is the so-called cut-off ratio. These are the parameters that control the performance of the

that the expression for the efficiency of the cold air standard Diesel cycle becomes identical to its counterpart for the Otto cycle as $r_c \to 1$. However, it is evident from Fig. 11.19 that as $r_c \to 1$, $Q_H \to 0$ and so the cycle becomes a motoring cycle with zero net work output. This is corroborated by the expression for W_{net} given above, which goes to zero for $r_c = 1$. Hence, the Diesel cycle does not approach the Otto cycle as $r_c \to 1$, as the expression for the thermal efficiency seems to suggest.

The specific work for the cold air standard Diesel cycle is plotted in Fig. 11.18 for several values of r_c. It is clear that the specific work of the Diesel cycle for $r = 20$ and $r_c > 2$, which are typical values for production Diesel engines, is comparable to the highest value attained in an Otto cycle. The fact that the thermal efficiency and specific work of the Diesel cycle for typical values of the operational parameters is comparable to the best values that could be attained with the Otto cycle, explains the popularity of Diesel engines.

▪ EXAMPLE 11.13

An ideal air standard Diesel cycle operates with a compression ratio of 18.75 and a cut-off ratio of 2.4283. At the beginning of the cycle, the air is at 100 kPa and 298 K. Determine (a) the thermal efficiency, (b) the mean effective pressure and (c) the second law efficiency. The ambient temperature may be taken to be 298 K.

Solution : Property values at each state calculated using values from Table L are given in the table below. In each row, the value used to retrieve from Table L is shown in italics; values that are retrieved are shown in bold; other values are given or calculated using known quantities.

State	P	T	u	$s°$	v_r	s (Eqn. A.1)
	(kPa)	(K)	(kJ/kg)	(kJ/kg.K)		(kJ/kg.K)
1	100	_298_	**212.64**	**1.69528**	**631.9**	1.69528
2	5700.5	**906**	**679.35**	-	_631.9÷18.75_	1.69528
3	5700.5	_906×2.4283_	**1872.4**	**3.9191**	**2.012**	2.76
4	393.63	**1173**	**909.39**	-	_2.012×18.75÷2.4283_	2.76

Since it is given that the cycle is ideal, processes _1-2_ and _3-4_ are isentropic. In the table,

$$P_2 = P_1 \frac{T_2}{T_1} \frac{v_1}{v_2} = r\, P_1 \frac{T_2}{T_1}$$

(a) Work supplied during the compression process is given as

$$\frac{W_{1-2}}{m} = u_2 - u_1 = 466.71 \text{ kJ/kg}$$

and work done during the cycle is

$$\begin{aligned}
\frac{W_{2-3}}{m} + \frac{W_{3-4}}{m} &= P_2 \left(v_3 - v_2\right) + \left(u_3 - u_4\right) \\
&= RT_2 \left(r_c - 1\right) + \left(u_3 - u_4\right) \\
&= 1334.46 \text{ kJ/kg}
\end{aligned}$$

The net work produced during every cycle is thus 867.75 kJ/kg. Heat added during each cycle may be evaluated as

$$\begin{aligned}
\frac{Q_H}{m} &= \left(u_3 - u_2\right) + P_2 \left(v_3 - v_2\right) \\
&= \left(u_3 - u_2\right) + RT_2 \left(r_c - 1\right) \\
&= 1564.5 \text{ kJ/kg}
\end{aligned}$$

and heat rejected during each cycle is

$$\frac{Q_C}{m} = u_4 - u_1 = 696.75 \text{ kJ/kg}$$

Thermal efficiency of the cycle is thus

$$\eta = \frac{W_{net}}{Q_H} = \frac{867.75}{1564.5} = 55.47\%$$

(b) The mean effective pressure (MEP) is defined as

$$\text{MEP} = \frac{W_{net}}{V_1 - V_2} = \frac{\frac{W_{net}}{m}}{\frac{RT_1}{P_1}\left(1 - \frac{1}{r}\right)}$$

Upon substituting the numerical values we get this to be 1071.59 kPa.

(c) Exergy supplied during the cycle is given as

$$\begin{aligned}
\frac{X_{supplied}}{m} &= \left(u_2 - u_1\right) + P_0 \left(v_2 - v_1\right) + \frac{Q_H}{m}\left(1 - \frac{T_0}{T_H}\right) \\
&= \left(u_2 - u_1\right) - \frac{P_0}{P_1} RT_1 \left(1 - \frac{1}{r}\right) + \frac{Q_H}{m}\left(1 - \frac{T_0}{T_H}\right) \\
&= 1738.313 \text{ kJ/kg}
\end{aligned}$$

$$
\begin{aligned}
m \quad &= \quad RT_2 \left(1 - \frac{P_0}{P_2}\right)(r_c - 1) + (u_3 - u_4) - \frac{P_0}{P_3} RT_3 (r - 1) \\
&= \quad 1131.31 \text{ kJ/kg}
\end{aligned}
$$

where we have taken $T_C = 300$ K (the lowest temperature during process 3-4). Equation 10.4 has been used for evaluating the exergy recovered from the turbine in such a manner as to get a positive value. The second law efficiency is thus

$$
\eta_{II} = \frac{\text{Exergy recovered}}{\text{Exergy supplied}} = 65.08\%
$$

11.2.4 Dual cycle

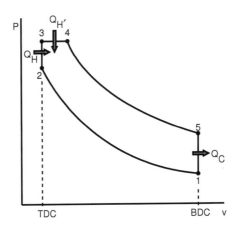

Figure 11.20: $P - v$ diagram of an air standard dual cycle

As the name suggests, the dual cycle is a combination of the air standard Otto and Diesel cycle, wherein a part of the heat addition occurs at constant volume and the rest occurs at constant pressure (Fig. 11.20). The thermodynamic analysis of the cycle proceeds along the same lines as the other two cycles and is hence is not discussed here.

11.3 Vapor compression refrigeration cycle

The vapor compression refrigeration cycle was described in section 8.2. The block diagram of the cycle is illustrated in Fig. 8.3. In this section, thermodynamic aspects

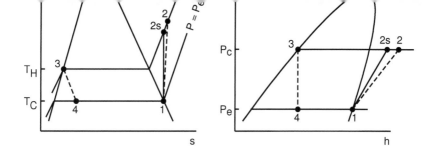

Figure 11.21: $T - s$ diagram of an ideal vapor compression refrigeration cycle

of the cycle are discussed in detail.

The $T - s$ diagram of an ideal vapor compression refrigeration cycle, labelled *1-2-3-4*, is shown in Fig. 11.21 (left). This cycle, just like the Rankine and the Brayton cycles, operates between two isobars, corresponding to the evaporator pressure, P_e and the condenser pressure, P_c. The temperature of the refrigerated space is T_C and the temperature of the ambient to which heat is rejected is T_H. In an ideal cycle, these are respectively the saturation temperatures corresponding to the evaporator and condenser pressures. A reversible Carnot cycle will operate with reservoir temperatures T_H and T_C. The term ideal is used, even though processes *1-2* and *3-4* are internally irreversible, because the the refrigerant leaves the evaporator as a saturated vapor and the condenser as a saturated liquid. It may be seen that processes *2-3* and *4-1* are isobaric processes while process *3-4* is an isenthalpic process. Accordingly, the cycle is customarily depicted using $P - h$ coordinates in the refrigeration community, since then, three out of the four processes in the cycle are parallel to the axes. This is illustrated in Fig. 11.21 on the right.

SFEE applied to the individual process in the cycle leads to (after neglecting heat losses, PE and KE changes),

$$\frac{\dot{W}_{x,comp}}{\dot{m}} = h_2 - h_1$$

$$\frac{\dot{Q}_H}{\dot{m}} = h_2 - h_3$$

$$h_4 = h_3$$

$$\frac{\dot{Q}_C}{\dot{m}} = h_1 - h_4$$

It is customary to express the heat removal rate from the refrigerated space in *tons*. One *ton* of refrigeration is defined as the heat removal rate required to convert 1 ton (910 kg) of liquid water at 0°C into ice at 0°C in 24 hours and is equal to 211 kJ/min.

■ **EXAMPLE 11.14**

An ideal vapor compression cycle using refrigerant R134a operates with the refrigerated space at -5°C. The refrigerant leaves the condenser as a saturated liquid at 30°C. Refrigerant enters the compressor as saturated vapor at the rate 0.0041 m³/s. Determine the power input to the compressor, the rate of heat removal from the refrigerated space in tons and the COP. Also determine the second law efficiency of the cycle and the rate of exergy destruction in the individual components. Take the ambient temperature to be 25°C and the isentropic efficiency of the compressor to be 100 percent.

Solution : With reference to Fig. 11.21, $v_1 = v_g = 0.08282$ m³/kg, $h_1 = h_g = 247.55$ kJ/kg and $s_1 = s_g = 0.9345$ kJ/kg.K, from Table D. The mass flow rate of the refrigerant is thus, $\dot{m} = 0.0041 \div 0.08282 = 0.0495$ kg/s. Also, $h_3 = h_f(30°C) = 93.58$ kJ/kg, $s_3 = 0.3479$ kJ/kg.K from Table D and $h_4 = h_3 = 93.58$ kJ/kg.

State 2s is superheated at $P = P_{sat}(30°C) = 770.6$ kPa and since *1-2s* is an isentropic process, $s_{2s} = s_1 = 0.9345$ kJ/kg.K. From Table F, we can get $h_{2s} = 271.46$ kJ/kg, by interpolation[†]. The compressor power required is therefore

$$\dot{W}_{x,comp} = \dot{m}(h_{2s} - h_1) = 1.184\,\text{kW}$$

The rate of heat removal from the refrigerated space may be evaluated as

$$\dot{Q}_C = \dot{m}(h_1 - h_4) = 7.622\,\text{kW} = 2.167\,\text{tons}$$

The COP may be evaluated as

$$\text{COP} = \frac{\dot{Q}_C}{\dot{W}_{x,comp}} = \frac{7.622}{1.184} = 6.44$$

The rate at which exergy is supplied to the cycle is given as

$$\dot{X}_{supplied} = \dot{W}_{x,comp} = 1.184\,\text{kW}$$

[†]The miniREFPROP application available for free at
https://trc.nist.gov/refprop/MINIREF/MINIREF.HTM is a convenient alternative.

$$= \dot{m}\,(h_{2s} - h_3)\left(1 - \frac{T_0}{T_H}\right) + \dot{Q}_C\left(\frac{T_0}{T_C} - 1\right)$$

$$= 0.9985\,\text{kW}$$

where $T_H = 30°\text{C}$ and $T_C = -5°\text{C}$. The second law efficiency is thus

$$\eta_{II} = \frac{\text{Exergy recovered}}{\text{Exergy supplied}} = 84.33\%$$

There is no exergy destruction in the compressor since the process is isentropic. For the evaporator,

$$\dot{\sigma}_{evaporator} = \dot{m}\,(s_1 - s_4) - \frac{\dot{Q}_C}{T_C}$$

$$= \dot{m}\,(s_1 - s_4) - \frac{\dot{m}\,(h_1 - h_4)}{T_C}$$

$$= \dot{m}\,\frac{h_1 - h_4}{T_C} - \frac{\dot{m}\,(h_1 - h_4)}{T_C}$$

$$= 0$$

where we have used the fact that, for process 4-1, in which the temperature and pressure are constant, $T\,ds = dh - v\,\cancel{dP}^{\,0}$ and so $\Delta s = \Delta h \div T$.

Rate of exergy destruction in the condenser is given as

$$T_0\,\dot{\sigma}_{condenser} = T_0\left[\dot{m}\,(s_3 - s_{2s}) + \frac{\dot{Q}_H}{T_H}\right] = 6.825\,\text{W}$$

Rate of exergy destruction in the throttling valve is given as

$$T_0\,\sigma_{thr} = T_0\,\dot{m}\,(s_4 - s_3) = 0.1844\,\text{kW}$$

Here, we have calculated $s_4 = 0.3604$ kJ/kg.K after evaluating the dryness fraction $x_4 = 0.2393$ from $h_4 = h_f + x_4\,(h_g - h_f)$.

The total exergy destruction in the cycle is almost entirely due to the exergy destruction that occurs during the throttling process.

EXAMPLE 11.15

Repeat the previous example with the isentropic efficiency for the compressor being 73 percent.

1	247.55	0.9345
2s	271.46	0.9345
3	93.58	0.3479
4	93.58	0.3604

Since it is given that $\eta_c = 0.73$, we can get $h_2 = 280.3$ kJ/kg, from

$$\eta_c = \frac{h_{2s} - h_1}{h_2 - h_1}$$

From Table F, we can get $s_2 = 0.96285$ kJ/kg.K by interpolation. The compressor power required is therefore

$$\dot{W}_{x,comp} = \dot{m}\,(h_2 - h_1) = 1.6211\,\text{kW}$$

The rate of heat removal in the condenser may be evaluated as

$$\dot{Q}_H = \dot{m}\,(h_2 - h_3) = 9.4264\,\text{kW}$$

The rate of heat removal from the refrigerated space may be evaluated as

$$\dot{Q}_C = \dot{m}\,(h_1 - h_4) = 7.622\,\text{kW} = 2.167\,\text{tons}$$

The COP may be evaluated as

$$\text{COP} = \frac{\dot{Q}_C}{\dot{W}_{x,comp}} = \frac{7.622}{1.184} = 6.44$$

The rate at which exergy is supplied to the cycle is given as

$$\dot{X}_{supplied} = \dot{W}_{x,comp} = 1.6211\,\text{kW}$$

and the rate at which exergy is recovered is

$$\dot{X}_{recovered} = \dot{Q}_H \left(1 - \frac{T_0}{T_H}\right) + \dot{Q}_C \left(\frac{T_0}{T_C} - 1\right)$$
$$= 1.0088\,\text{kW}$$

where $T_H = 30°$C and $T_C = -5°$C. The second law efficiency is thus

$$\eta_{II} = \frac{\text{Exergy recovered}}{\text{Exergy supplied}} = 62.23\%$$

$$T_0 \, \dot{\sigma}_{condenser} = T_0 \left[\dot{m} \, (s_3 - s_2) + \frac{\dot{Q}_H}{T_H} \right] = 0.019 \, \text{W}$$

Rate of exergy destruction in the throttling valve is given as

$$T_0 \, \sigma_{thr} = T_0 \, \dot{m} \, (s_4 - s_3) = 0.1844 \, \text{kW}$$

It can be seen that the exergy destruction in the compressor is now substantial.

11.3.1 Actual cycle

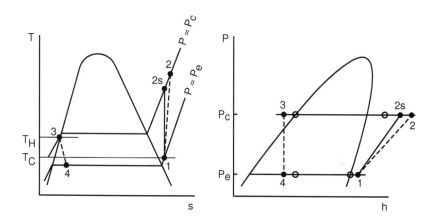

Figure 11.22: $T - s$ diagram of an actual vapor compression refrigeration cycle. The refrigerant is superheated at the exit of the evaporator and sub-cooled at the exit of the condenser. Open circles denote states corresponding to the ideal cycle.

It is quite difficult in practice to ensure that the refrigerant leaves the evaporator as a saturated vapor and the condenser as a saturated liquid. In reality, the refrigerant is always slightly superheated as it leaves the evaporator and slightly sub-cooled when it leaves the condenser. This is illustrated in $T - s$ and $P - h$ coordinates in Fig. 11.22. If the temperature of the refrigerated space, T_C and that of the ambient, T_H, remain the same, then these departures from the ideal cycle represent external irreversibilities. Furthermore, internal irreversibilities are always present during the compression process, which cause the exit state from the compressor to move to 2 from 2s. The combined effect of these irreversibilities is to reduce the performance of the cycle. This is demonstrated next.

flow rate) remaining the same.

Solution : With reference to Fig. 11.22, state *1* is now superheated. From Table F, corresponding to 200 kPa and -5°C, we can get $h_1 = 248.85$ kJ/kg and $s_1 = 0.954$ kJ/kg.K. We can evaluate h_{2s} to be 280.948 kJ/kg in the same manner as in the previous example. Therefore,

$$h_2 = h_1 + \frac{h_{2s} - h_1}{\eta_{comp}} = 292.82 \text{ kJ/kg}$$

By interpolation in Table F, we can get $s_2 = 0.9907$ kJ/kg.K. The compressor power required is therefore

$$\dot{W}_{x,comp} = \dot{m}(h_2 - h_1) = 2.177 \text{ kW}$$

State *3* is a sub-cooled liquid at 30°C. Hence, $h_3(900 \text{ kPa}, 30°C) = u_f(30°C) + v_f(30°C) \times 900 = 93.688$ kJ/kg and $s_3 \approx s_f(30°C) = 0.3479$ kJ/kg.K. Also $h_4 = h_3 = 93.688$ kJ/kg and state *4* is a saturated mixture at 200 kPa. Hence, the dryness fraction $x_4 = 0.2682$ from $h_4 = h_f + x_4(h_g - h_f)$ and s_4 may be evaluated as 0.3646 kJ/kg.K.

Therefore, the rate at which heat is removed from the refrigerated space is given as

$$\dot{Q}_C = \dot{m}(h_1 - h_4) = 7.681 \text{ kW} = 2.184 \text{ tons}$$

The COP may be evaluated as

$$\text{COP} = \frac{\dot{Q}_C}{\dot{W}_{x,comp}} = \frac{7.681}{2.177} = 3.53$$

This is 45 percent less than the COP for the ideal cycle in the previous example.

The rate at which heat is removed in the condenser is given as

$$\dot{Q}_H = \dot{m}(h_2 - h_3) = 9.857 \text{ kW}$$

The rate at which exergy is supplied to the cycle is given as

$$\dot{X}_{supplied} = \dot{W}_{x,comp} = 2.177 \text{ kW}$$

and the rate at which exergy is recovered is

$$\dot{X}_{recovered} = \dot{Q}_H \left(1 - \frac{T_0}{T_H}\right) + \dot{Q}_C \left(\frac{T_0}{T_C} - 1\right)$$
$$= 1.0225 \text{ kW}$$

This is also 45 percent less than the second law efficiency for the ideal cycle.

Rate of exergy destruction in the compressor may be calculated as

$$T_0 \dot{\sigma}_{compressor} = T_0 \dot{m} (s_2 - s_1) = 0.5414 \, \text{kW}$$

Rate of exergy destruction in the evaporator,

$$T_0 \dot{\sigma}_{evaporator} = T_0 \left[\dot{m} (s_1 - s_4) - \frac{\dot{Q}_C}{T_C} \right] = 0.1534 \, \text{kW}$$

Rate of exergy destruction in the condenser is given as

$$T_0 \dot{\sigma}_{condenser} = T_0 \left[\dot{m} (s_3 - s_2) + \frac{\dot{Q}_H}{T_H} \right] = 0.2124 \, \text{kW}$$

Rate of exergy destruction in the throttling valve is given as

$$T_0 \sigma_{thr} = T_0 \dot{m} (s_4 - s_3) = 0.2463 \, \text{kW}$$

The total rate of exergy destruction is now seen to be distributed among all the components.

The performance of the vapor compression cycle may be improved by using the cascade cycle or the multistage vapor compression cycle. These cycles and topics such as selection of refrigerants for a particular application are not discussed here, as they are beyond the scope of this book. Interested readers may consult the books suggested at the end for a discussion of these topics.

PSYCHROMETRY

Psychrometry is the study of the changes in the state and content of moist air or humid air in applications such as heating, ventilation and air-conditioning (HVAC), drying and cooling towers used in thermal power plants. It is well known that atmospheric air usually contains some amount of water vapor, which is colloquially referred to as humidity. Consider, for instance air at a room temperature of 25°C and 1 atm (101325 Pa). The air contains a certain amount of water vapor which is at the same temperature but at its partial pressure, P_v. The state of the water vapor is shown in Fig. 12.1.

The water vapor in the room air is superheated on account of the fact that its partial pressure is quite small. Furthermore, in psychrometric applications, the air is usually at a pressure of 1 atm or slightly higher and the range of temperature is typically between 0°C and about 60°C. It is clear from Fig. 12.1, that, for these conditions, the water vapor in the air may be treated as an ideal gas. Thus, moist air itself may be treated conveniently as a mixture of two ideal gases – dry air (denoted using the subscript a) and water vapor (denoted using the subscript v). This enormously simplifies the thermodynamic analysis of psychrometric applications.

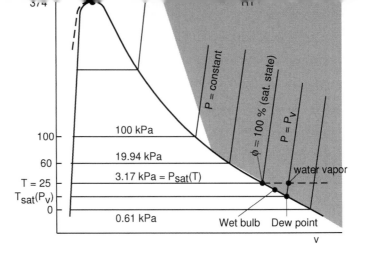

Figure 12.1: $T - v$ diagram of water with relevance to psychrometric applications

12.1 Moist air

Since the moist air is treated as a mixture of two ideal gases, we may write

$$P = \frac{n\mathcal{R}T}{V} \, ; \quad P_a = \frac{n_a\mathcal{R}T}{V} \, ; \quad P_v = \frac{n_v\mathcal{R}T}{V}$$

for the moist air, dry air and water vapor respectively. Here, n represents the number of moles and $\mathcal{R}$ is the universal gas constant, as before. Note that $P = P_a + P_v$ and $n = n_a + n_v$. If we divide the second and third expression above by the first one, we get $P_v = y_v P$ and $P_a = y_a P$, where y represents the mole fraction, as usual.

The humidity ratio, ω is defined as the ratio of the mass of the water vapor to the mass of dry air in a given sample. Thus,

$$\omega = \frac{m_v}{m_a} \frac{\text{kg vapor}}{\text{kg dry air}} \tag{12.1}$$

The above expression may be expanded and rewritten as follows:

$$\omega \quad = \quad \frac{m_v}{m_a} = \frac{n_v M_v}{n_a M_a} = \frac{M_v}{M_a} \frac{y_v}{y_a} = \frac{M_v}{M_a} \frac{y_v}{1 - y_v}$$

vapor (M_v), we are finally led to

$$\omega = 0.622 \frac{P_v}{P - P_v} \frac{\text{kg vapor}}{\text{kg dry air}} \qquad (12.2)$$

Consider again the moist air sample for which the state of the water vapor is as shown in Fig. 12.1. If, now water is added to this sample while keeping its temperature and pressure constant, the water evaporates. Consequently, P_v increases and the state point moves to the left along the isotherm. This continues until the amount of water vapor present is such that $P_v = P_{sat}(25^\circ C)$, at which point the sample of moist air is said to be saturated. The state point of the water vapor lies on the saturated vapor line at $T = 25^\circ C$. No more addition of water into the sample is possible. The ratio of the mole fraction of the water vapor in a given sample of moist air to the maximum mole fraction of water vapor that can be present at the same temperature and pressure of moist air is defined as the relative humidity, ϕ. Thus

$$\phi = \left.\frac{y_v}{y_{v,sat}}\right|_{T,\,P} = \left.\frac{y_v\,P}{y_{v,sat}\,P}\right|_{T,\,P} = \frac{P_v}{P_{sat}(T)} \qquad (12.3)$$

On the other hand, if the same sample of moist air is cooled at constant pressure – keeping both P and P_v constant, *i.e.*, without adding or removing water, then the state point of the water vapor moves downwards along the isobar $P = P_v$. When the sample temperature becomes equal to $T_{sat}(P_v)$, it becomes saturated and the state point of the water lies on the saturated vapor line. This temperature is called the dew point temperature, T_{dp}. Hence,

$$T_{dp} = T_{sat}(P_v) \qquad (12.4)$$

It can be inferred from Fig. 12.1, that, as moist air is heated at constant pressure it becomes drier while it becomes more humid when it is cooled at constant pressure, although the mass of the water vapor itself remains the same during these two processes. Relative humidity, thus brings out the fact that the perception of humidity depends on both the amount of water vapor in the air as well as the temperature. Practical HVAC systems heat/cool and humidify/de-humidify air to achieve an ideal combination of temperature and humidity.

■ EXAMPLE 12.1

Ambient air at $25^\circ C$ and 1 atm is observed to have a dew point temperature of $15^\circ C$. Determine the relative humidity and the humidity ratio.

Furthermore,

$$\omega = 0.622 \frac{P_v}{P - P_v} = 0.622 \times \frac{1.706}{101.325 - 1.706} = 0.01065 \frac{\text{kg vapor}}{\text{kg dry air}}$$

▉ EXAMPLE 12.2

Ambient air at a geographic location is at 25°C, 1 atm and 90 percent relative humidity. Determine the maximum amount of liquid water that can be extracted per unit volume of ambient air.

Solution : From Table A, we can get $P_{sat}(25°C) = 3.169$ kPa. Using Eqn. 12.3, we can get $P_v = \phi \times P_{sat}(25°C) = 2.8521$ kPa. Therefore,

$$m_v = \frac{P_v V}{R_v T} = \frac{2.8521 \times 10^3 \times 1}{\frac{8314}{18} \times 298} = 20.72\,\text{g}$$

We can extract 20.72 cc of liquid water from 1 m³ of air.

▉ EXAMPLE 12.3

One kg of air at 25°C, 1 atm and 70 percent relative humidity is compressed to 5 atm in a polytropic process $Pv^{1.25} = constant$. It is then cooled at constant volume to 25°C. Determine the humidity ratio and relative humidity of the air after the compression process. Also, determine the amount of water (if any) that condenses in the tank and the final pressure.

Solution : Initially, $P_{v,1} = \phi_1 \times P_{sat}(25°C) = 0.7 \times 3.169 = 2.2183$ kPa. Therefore,

$$\omega_1 = 0.622 \frac{P_{v,1}}{P_1 - P_{v,1}} = 0.622 \times \frac{2.2183}{101.325 - 2.2183}$$

$$= 0.0139 \frac{\text{kg vapor}}{\text{kg dry air}}$$

The temperature at the end of the compression may be calculated as

$$T_2 = T_1 \left(\frac{P_2}{P_1}\right)^{(n-1)\div n} = 411\,\text{K}$$

After the compression, since no water is added or removed, $\omega_2 = \omega_1 = 0.0139$ kg vapor per kg dry air. From Eqn. 12.2 we can get

$$P_{v,2} = \frac{\omega_2 P_2}{0.622 + \omega_2} = 11.07\,\text{kPa}$$

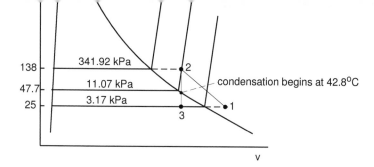

Therefore,

$$\phi_2 = \frac{P_{v,2}}{P_{sat}(138°C)} = \frac{11.07}{341.92} = 3.24\%$$

Although the amount of water vapor has remained the same, the relative humidity decreases as a result of the compression process. This suggests that lines of constant ϕ are closely spaced at higher temperatures and broadly spaced at lower temperatures.

Since the cooling takes place at constant volume, condensation begins at that temperature when the specific volume of the vapor becomes equal to the specific volume of the saturated vapor at that temperature (as shown in the figure). The specific volume of the vapor after the compression process may be evaluated using the ideal gas equation of state as

$$v_{v,2} = \frac{R_v T_2}{P_{v,2}} = \frac{8314}{18} \times \frac{411}{11.07 \times 10^3} = 17.15 \, \text{m}^3/\text{kg}$$

From Table A, it can be seen that $v_{v,2} = v_g(42.8°C)$. Condensation thus begins at 42.8°C. When the air is finally cooled to 25°C, the air is still saturated and since the final state of the water is a saturated mixture state, we may write its specific volume as

$$v_f + x \, (v_g - v_f)|_{25°C} = 17.15$$

The dryness fraction x may be evaluated as 0.3955. The mass of liquid water is

$$m_w = (1 - x) \, m_{v,1} = (1 - x) \, \omega_1 \, m_a = (1 - x) \, \omega_1 \frac{m}{1 + \omega_1} = 8.287 \, \text{g}$$

This is the amount water that condenses. Here, we have used the fact the mass of the moist air, $m = m_a + m_v = m_a(1 + \omega)$.

$$P_{a,3} = P_{a,2} \frac{T_3}{T_2} = (P_2 - P_{v,2}) \frac{T_3}{T_2} = 359.308 \, \text{kPa}$$

The mixture pressure is $P_3 = P_{a,3} + P_{v,3} = 362.477$ kPa, where $P_{v,3} = P_{sat}(25°\text{C})$ = 3.169 kPa.

■ EXAMPLE 12.4

Repeat the previous example, assuming that the cooling takes place at constant pressure.

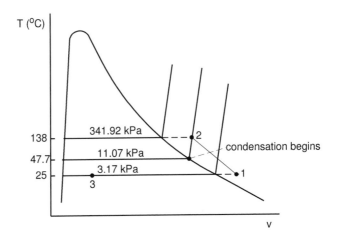

Solution : In this case, condensation begins when the dew point temperature, $T_{dp} = T_{sat}(11.07 \, \text{kPa}) = 47.7°\text{C}$ is reached. When the mixture reaches 25°C, it is still saturated and so $P_{v,3} = P_{sat}(25°\text{C}) = 3.169$ kPa. Hence

$$\omega_3 = 0.622 \, \frac{P_{v,3}}{P_3 - P_{v,3}} = 0.622 \times \frac{3.169}{5 \times 101.325 - 3.169}$$

$$= 3.915 \times 10^{-3} \, \frac{\text{kg vapor}}{\text{kg dry air}}$$

Note that, $P_3 = P_2 = 5$ atm in this case. In contrast to the previous example, we are able to use Eqn. 12.2 to evaluate ω_3 in this case, since the the final pressure is known. The amount of water that condenses is given as

$$m_w = m_a \, (\omega_1 - \omega_3) = \frac{m}{1 + \omega_1} \, (\omega_1 - \omega_3) = 9.847 \, \text{g}$$

12.2 Application of first law to psychrometric processes

Application of first law to psychrometric process is quite straightforward, since moist air is being treated as just a mixture of two ideal gases, namely, air and water vapor. The change in internal energy of air may be calculated in the usual manner for a calorically perfect gas. The specific internal energy of the water vapor is customarily approximated as being equal to that of the saturated vapor at the same temperature, i.e., $u_v(T, P) \approx u_g(T)$.

■ **EXAMPLE 12.5**

Determine the heat removed during the cooling process in the last two examples.

Solution : Application of First law to the constant volume cooling process gives

$$\Delta E = \Delta U = Q - \overbrace{W}^{\text{0, constant volume}}$$

$$Q = \Delta U = m_a C_{v,a} (T_3 - T_2) + m_{v,1} (u_3 - u_2)|_{H_2O}$$

$$= \frac{m}{1 + \omega_1} \left[C_{v,a} (T_3 - T_2) + \omega_1 (u_3 - u_2)|_{H_2O} \right]$$

We can get $u_2 = u_g(138°C) = 2548$ kJ/kg and $u_3 = u_f + x (u_g - u_f) = 1016.04$ kJ/kg using Table A. With $C_{v,a} = 717.86$ J/kg.K, and after substituting the known values into the above expression, we get $Q = -101$ kJ.

Application of First law to the constant pressure cooling process gives

$$\Delta E = \Delta U = Q - W = Q - P_3 (V_3 - V_2)$$

$$Q = \Delta H = m_a h_{a,3} + m_{v,3} h_{v,3} + m_w h_{w,3} - m_a h_{a,2} - m_{v,2} h_{v,2}$$

$$= \frac{m}{1 + \omega_1} \left[C_{p,a} (T_3 - T_2) + \omega_3 h_{v,3} - \omega_2 h_{v,2} \right] + m_w h_{w,3}$$

We have, from Table A, $h_{v,3} = h_g(25°C) = 2546.3$ kJ/kg, $h_{v,2} = h_g(138°C) = 2731.16$ kJ/kg and $h_{w,3} = h_f(25°C) = 104.75$ kJ/kg. With $C_{p,a} = 1.005$ kJ/kg.K and after substituting the known values into the above expression, we get $Q = -138.59$ kJ.

amount of water that condenses (if any) and the heat removed.

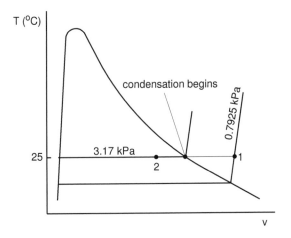

Solution : Initially, $P_{v,1} = \phi_1 \times P_{sat}(25°C) = 0.25 \times 3.169 = 0.79225$ kPa. Therefore,

$$\omega_1 = 0.622 \frac{P_{v,1}}{P_1 - P_{v,1}} = 0.622 \times \frac{0.79225}{101.325 - 0.79225}$$

$$= 4.902 \times 10^{-3} \frac{\text{kg vapor}}{\text{kg dry air}}$$

Since the compression process is isothermal, the final pressure is five times the initial pressure and so $P_2 = 5 \times 101.325$ kPa. The partial pressure of the water vapor can increase only up to 3.169 kPa, which is four times the initial value. Hence, condensation will take place.

The initial specific volume of the water vapor may be evaluated as

$$v_{v,1} = \frac{R_v T_1}{P_{v,1}} = \frac{8314}{18} \times \frac{298}{0.79225 \times 10^3} = 173.74 \, \text{m}^3/\text{kg}$$

The final specific volume is thus $v_{v,2} = v_{v,1} \div 5 = 34.748$ m³/kg. The dryness fraction at the final state may be evaluated from

$$v_f + x \left(v_g - v_f \right)|_{25°C} = 34.748$$

as 0.8014. The mass of liquid water is

$$m_w = (1 - x) \, m_{v,1} = (1 - x) \, \omega_1 \, m_a = (1 - x) \, \omega_1 \frac{m}{1 + \omega_1} = 1.94 \, \text{g}$$

$$\Delta E = \Delta U = Q - W = Q - P_1 V_1 \ln \frac{V_2}{V_1} = Q - P_1 V_1 \ln \frac{1}{5}$$

$$\Rightarrow Q = \Delta U_a^{\;\;0} + m_{v,1}(u_2 - u_1) + P_1 \frac{m_{v,1}}{v_{v,1}} \ln 0.2$$

$$= \frac{m}{1 + w_1} w_1 \left[(u_2 - u_1) + P_1 \frac{1}{v_{v,1}} \ln 0.2\right]$$

We can get $u_1 = u_g(25°C) = 2408.9$ kJ/kg and $u_2 = u_f + x(u_g - u_f) = 1951.3$ kJ/kg using Table A. Upon substituting the known values into the above expression, we get $Q = -4.4736$ kJ.

The following examples illustrate unit processes in HVAC, namely, heating, cooling, humidification and de-humidification. Since these are steady flow applications, the enthalpy of the moist air is used in the calculations. The enthalpy of a sample of moist air may be written as $H = m_a h_a + m_v h_v$. Here, $h_a = C_{p,a} T$ where $C_{p,a} = 1.005$ kJ/kg.°C and T is in °C and h_v is usually approximated to be $h_g(T)$.

■ EXAMPLE 12.7

Ambient air at 5°C, 1 atm pressure and 85 percent relative humidity is to be heated as it flows steadily in a duct so that the relative humidity becomes 25

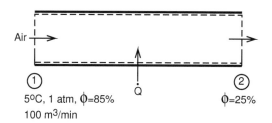

percent. If the volume rate is 100 m³/min at the inlet, determine the rate of heat addition. Assume that the pressure remains constant.

Solution : Since it is given that $\phi_1 = 0.85$, we can get

$$P_{v,1} = \phi_1 P_{sat}(5°C) = 0.85 \times 0.873 = 0.74205 \text{ kPa}$$

and

$$w_1 = 0.622 \frac{P_{v,1}}{P_1 - P_{v,1}}$$

water is added or removed). Also, $\phi_2 = 0.25$. Hence

$$P_{sat}(T_2) = \frac{P_{v,2}}{\phi_2} = 2.9682\,\text{kPa}$$

From Table A, we can get $T_2 \approx 24°C$. SFEE applied to the duct gives

$$0 = \dot{Q} - \overset{0}{\cancel{\dot{W}_x}} + \dot{m}\,(h_1 - h_2)$$

$$\dot{Q} = \dot{m}_a\,(h_{a,2} - h_{a,1}) + \dot{m}_v\,(h_{v,2} - h_{v,1})$$

$$= \dot{m}_a\,[C_{p,a}\,(T_2 - T_1) + \omega_1\,(h_g(T_2) - h_g(T_1))]$$

Mass flow rate of dry air may be evaluated from the given volumetric flow rate as follows:

$$\dot{m}_a = \frac{\dot{V}_1}{v_{a,1}} = \frac{\dot{V}_1}{\dfrac{R_a T_1}{P_{a,1}}} = \frac{\dot{V}_1}{\dfrac{R_a T_1}{P_1 - P_{v,1}}}$$

$$= \frac{100}{\dfrac{8314}{28.97} \times \dfrac{278}{(101.325 - 0.74205) \times 10^3}} = 126.072\,\frac{\text{kg}}{\text{min}}$$

From Table A, $h_g(5°C) = 2509.7$ kJ/kg and $h_g(24°C) = 2544.5$ kJ/kg. Upon substituting the known values into the expression above, we get $\dot{Q} = 40.458$ kW.

◼ **EXAMPLE 12.8**

Moist air at 35°C, 1 atm and 80 relative humidity enters an air-conditioning duct where it is cooled and de-humidified as shown in the figure. Both the condensate and air that is saturated leave at 22°C. The volume flow rate of air at the inlet is 120 m³/min. Assuming steady state operation and neglecting heat loss, determine (a) the mass flow rate of the condensate and (b) the required cooling capacity in tons. Assume that the pressure remains constant.

Solution : Since it is given that $\phi_1 = 0.8$, we can get

$$P_{v,1} = \phi_1\,P_{sat}(35°C) = 0.8 \times 5.627 = 4.5016\,\text{kPa}$$

and

$$\omega_1 = 0.622\,\frac{P_{v,1}}{P_1 - P_{v,1}}$$

$$= 0.622 \times \frac{4.5016}{101.325 - 4.5016} = 0.0289\,\frac{\text{kg vapor}}{\text{kg dry air}}$$

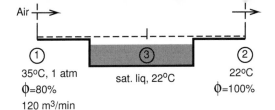

Air

① 35°C, 1 atm
ϕ=80%
120 m³/min

③ sat. liq, 22°C

② 22°C
ϕ=100%

Mass flow rate of dry air may be evaluated from the given volumetric flow rate as follows:

$$\dot{m}_a = \frac{\dot{V}_1}{v_{a,1}} = \frac{\dot{V}_1}{\frac{R_a T_1}{P_{a,1}}} = \frac{\dot{V}_1}{\frac{R_a T_1}{P_1 - P_{v,1}}}$$

$$= \frac{120}{\frac{8314}{28.97} \times \frac{308}{(101.325 - 4.5016) \times 10^3}} = 131.447 \, \frac{\text{kg}}{\text{min}}$$

At the exit, the air is saturated and so $P_{v,2} = P_{sat}(22°C) = 2.645$ kPa. It follows that

$$\omega_2 = 0.622 \frac{P_{v,2}}{P_2 - P_{v,2}}$$

$$= 0.622 \times \frac{2.645}{101.325 - 2.645} = 0.0167 \, \frac{\text{kg vapor}}{\text{kg dry air}}$$

where we have taken $P_2 = P_1$.

(a) Mass balance of water across the control volume shown gives

$$\dot{m}_w = \dot{m}_3 = \dot{m}_{v_1} - \dot{m}_{v,2} = \dot{m}_a \left(\omega_1 - \omega_2 \right) = 1.6037 \, \frac{\text{kg}}{\text{min}}$$

This is the mass flow rate of the condensate.

(b) SFEE applied to the control volume shown gives

$$0 = \dot{Q} - \cancel{\dot{W}_x}^{0} + \dot{m}_1 h_1 - \dot{m}_2 h_2 - \dot{m}_3 h_3$$

$$\dot{Q} = \dot{m}_a \left(h_{a,2} - h_{a,1} \right) + \dot{m}_{v,2} h_{v,2} - \dot{m}_{v,1} h_{v,1} + \dot{m}_3 h_3$$

$$= \dot{m}_a \left[C_{p,a} \left(T_2 - T_1 \right) + \omega_2 \, h_g(T_2) - \omega_1 \, h_g(T_1) \right] + \dot{m}_w h_f(T_3)$$

EXAMPLE 12.9

Moist air at 20°C, 1 atm, 40 percent relative humidity enters an insulated duct with a volumetric flow rate of 50 m³/min. Steam (saturated vapor) at 100°C

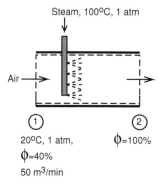

Steam, 100°C, 1 atm

Air →

20°C, 1 atm,
ϕ=40%
50 m³/min

ϕ=100%

is injected into the duct in order to humidify the air as shown in the figure. If it is desired to have a relative humidity of 100 percent at the exit, determine the exit temperature and required mass flow rate of steam. Assume steady state operation and also that the pressure remains constant.

Solution : At the inlet, $\phi_1 = 0.4$, and so

$$P_{v,1} = \phi_1 P_{sat}(20°C) = 0.4 \times 2.339 = 0.9356 \,\text{kPa}$$

and

$$
\begin{aligned}
\omega_1 &= 0.622 \frac{P_{v,1}}{P_1 - P_{v,1}} \\
&= 0.622 \times \frac{0.9356}{101.325 - 0.9356} = 5.8 \times 10^{-3} \frac{\text{kg vapor}}{\text{kg dry air}}
\end{aligned}
$$

Mass flow rate of dry air may be evaluated from the given volumetric flow rate as follows:

$$
\begin{aligned}
\dot{m}_a &= \frac{\dot{V}_1}{v_{a,1}} = \frac{\dot{V}_1}{\frac{R_a T_1}{P_{a,1}}} = \frac{\dot{V}_1}{\frac{R_a T_1}{P_1 - P_{v,1}}} \\
&= \frac{50}{\frac{8314}{28.97} \times \frac{293}{(101.325 - 0.9356) \times 10^3}} = 59.694 \frac{\text{kg}}{\text{min}}
\end{aligned}
$$

SFEE applied to the control volume shown gives

$$0 = \cancelto{0}{\dot{Q}} - \cancelto{0}{\dot{W}_x} + \dot{m}_1 h_1 + \dot{m}_s h_s - \dot{m}_2 h_2$$

$$0 = \dot{m}_a (h_{a,2} - h_{a,1}) + \dot{m}_{v,2} h_{v,2} - \dot{m}_{v,1} h_{v,1} - \dot{m}_s h_s$$

$$0 = \dot{m}_a [C_{p,a} (T_2 - T_1) + w_2 h_g(T_2) - w_1 h_g(T_1) - (w_2 - w_1) h_g(100°C)]$$

$$0 = C_{p,a} (T_2 - T_1) + w_2 h_g(T_2) - w_1 h_g(T_1) - (w_2 - w_1) h_g(100°C)$$

Upon substituting the known values into the above expression, it becomes

$$1.005 \times (T_2 - 20) + w_2 h_g(T_2) - 14.71576 - (w_2 - 5.8 \times 10^{-3}) \times 2675.7 = 0$$

where T is in °C. Since $\phi_2 = 1$, we also have

$$w_2 = 0.622 \frac{P_{v,2}}{P_2 - P_{v,2}} = 0.622 \frac{P_{sat}(T_2)}{101.325 - P_{sat}(T_2)}$$

where P is in kPa. The only unknown is the exit temperature and this has to be obtained iteratively as shown below:

T_2 (°C)	w_2 kg vapor / kg dry air	$1.005 \times (T_2 - 20)$ $+w_2 h_g(T_2) - 14.71576$ $-(w_2 - 5.8 \times 10^{-3}) \times 2675.7$
30	0.0272	7.5784
25	0.02	3.2403
22	0.0167	0.56

The exit temperature may be taken as 22°C. The required mass flow rate of steam comes out to be

$$\dot{m}_s = \dot{m}_a (w_2 - w_1) = 0.6506 \frac{kg}{min}$$

40°C, 1 atm and 15 percent relative humidity enters the cooler at a rate of 1.344

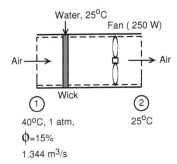

Water, 25°C
Fan (250 W)
Air
Air
Wick
(1)
(2)
40°C, 1 atm,
ϕ=15%
1.344 m³/s
25°C

m³/s. Saturated liquid water at 25°C enters the wick where it evaporates. A fan driven by an electric motor rated at 250 W pulls the air through the cooler and sends it out at 25°C. Assuming steady state operation, calculate (a) the mass flow rate of liquid water that must be supplied to the cooler and (b) the relative humidity of air at the exit. The pressure may be assumed to remain constant and heat loss to the surroundings may be neglected.

Solution : At the inlet, $\phi_1 = 0.15$, and so

$$P_{v,1} = \phi_1 P_{sat}(40°C) = 0.15 \times 7.381 = 1.10715 \, \text{kPa}$$

and

$$\omega_1 = 0.622 \, \frac{P_{v,1}}{P_1 - P_{v,1}}$$

$$= 0.622 \times \frac{1.10715}{101.325 - 1.10715} = 6.872 \times 10^{-3} \, \frac{\text{kg vapor}}{\text{kg dry air}}$$

Mass flow rate of dry air may be evaluated from the given volumetric flow rate as follows:

$$\dot{m}_a = \frac{\dot{V}_1}{v_{a,1}} = \frac{\dot{V}_1}{\frac{R_a T_1}{P_{a,1}}} = \frac{\dot{V}_1}{\frac{R_a T_1}{P_1 - P_{v,1}}}$$

$$= \frac{1.344}{\frac{8314}{28.97} \times \frac{313}{(101.325 - 1.10715) \times 10^3}} = 1.5 \, \frac{\text{kg}}{\text{s}}$$

Let the mass flow rate of liquid water be $\dot{m}_w$ kg/s. A mass balance of water gives

$$\dot{m}_{v,2} = \dot{m}_{v,1} + \dot{m}_w \quad \Rightarrow \quad \dot{m}_w = \dot{m}_a \, (\omega_2 - \omega_1)$$

$$U = \cancel{Q} - W_x + m_1 h_1 + m_w h_w - m_2 h_2$$

$$-\dot{W}_x = \dot{m}_a (h_{a,2} - h_{a,1}) + \dot{m}_{v,2} h_{v,2} - \dot{m}_{v,1} h_{v,1} - \dot{m}_w h_w$$

$$= \dot{m}_a \left[C_{p,a} (T_2 - T_1) + w_2 h_g(T_2) - w_1 h_g(T_1) - (w_2 - w_1) h_f(25°C) \right]$$

$$\Rightarrow \quad w_2 = \frac{-\dot{W}_x \div \dot{m}_a + C_{p,a} (T_1 - T_2) + w_1 \left[h_g(T_1) - h_f(25°C) \right]}{h_g(T_2) - h_f(25°C)}$$

If we substitute the known values into the above expression, we get $w_2 = 0.0132$ kg vapor / kg dry air.

(a) Required mass flow rate of water may be evaluated as

$$\dot{m}_w = \dot{m}_a (w_2 - w_1) = 9.492 \, \frac{g}{s}$$

(b) The relative humidity of the air at the exit may be evaluated using

$$\phi_2 = \frac{101.325 \times w_2}{P_{sat}(25°C) \times (0.622 - w_2)} = 69.33\%$$

12.3 Wet bulb and the adiabatic saturation temperature

In the previous section, definitions of two fundamental quantities specific to psychrometry, namely, relative humidity and humidity ratio were given. Their use in psychrometric applications was also illustrated through several examples. In practical applications, it is often essential to measure one of these quantities, since the other one can be calculated using either Eqn. 12.2 or 12.3.

The sling psychrometer shown in Fig. 12.2 is a simple device that can be used for this purpose. It consists of two thermometers – one called the dry bulb thermometer, which is the commonly used thermometer and another one called the wet bulb thermometer, the bulb of which is covered with a moist wick (hence the name). The wick is thoroughly moistened before use, and the assembly is then whirled around the handle. The liquid water in the wick evaporates as a result of convective heat transfer and the temperature of the surrounding air in the immediate vicinity of the wet bulb decreases. Evaporation of the water continues until the air around the bulb (which initially contained a certain of quantity of moisture that we wish to determine), becomes saturated. At this point, the reading of the wet bulb thermometer reaches

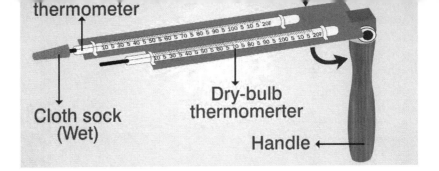

Figure 12.2: Sling psychrometer. Source: `https://sciencestruck.com`

a minimum and does not change any more. This is noted to be the wet bulb temperature, T_{wb} and the reading of the dry bulb thermometer, T_{db} is also noted down. The subscript db is superfluous and will be dropped. A model describing the thermodynamic process in the vicinity of the wet bulb will be developed next with the objective of relating the two measured temperatures to the humidity ratio of the air.

Consider the adiabatic saturator shown in Fig. 12.3. Moist air at pressure P_1 whose humidity ratio we wish to determine enters an insulated duct at a temperature T_1 which is equal to the measured dry bulb temperature. The air flows over a reservoir of liquid water at a temperature $T_3 < T_1$, causing the water to evaporate. Since the enthalpy of the incoming air is utilized for the evaporation, the temperature of the air decreases and its moisture content increases as a result of the evaporation. If the duct

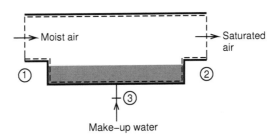

Figure 12.3: Schematic of an adiabatic saturator

is made sufficiently long, then the air will become saturated and more evaporation will not be possible. At this point, the temperature of the air is the same as that of

as the adiabatic saturation temperature, T_{as}. With $P_1 = 1$ atm, T_{as} is an excellent *approximation* of the wet bulb temperature.

Let the mass flow rate of liquid water be $\dot{m}_w$ kg/s. A mass balance of water gives

$$\dot{m}_{v,2} = \dot{m}_{v,1} + \dot{m}_w \quad \Rightarrow \quad \dot{m}_w = \dot{m}_a\,(\omega_2 - \omega_1)$$

SFEE applied to the control volume shown gives

$$0 = \overset{0}{\cancel{\dot{Q}}} - \overset{0}{\cancel{\dot{W}_x}} + \dot{m}_1 h_1 + \dot{m}_w h_w - \dot{m}_2 h_2$$

$$0 = \dot{m}_a\,(h_{a,2} - h_{a,1}) + \dot{m}_{v,2} h_{v,2} - \dot{m}_{v,1} h_{v,1} - \dot{m}_w h_w$$

$$0 = \dot{m}_a\,[C_{p,a}\,(T_2 - T_1) + \omega_2\,h_g(T_2) - \omega_1\,h_g(T_1) - (\omega_2 - \omega_1)\,h_f(T_3)]$$

$$0 = C_{p,a}\,(T_2 - T_1) + \omega_2\,h_g(T_2) - \omega_1\,h_g(T_1) - (\omega_2 - \omega_1)\,h_f(T_2)$$

$$\Rightarrow \quad \omega_1 = \frac{C_{p,a}\,(T_2 - T_1) + \omega_2\,[h_g(T_2) - h_f(T_2)]}{h_g(T_1) - h_f(T_2)} \tag{12.5}$$

With $T_1 = T_{db}$ and $T_2 = T_{wb}$, the above expression may be used to determine ω_1, since $\omega_2 = 0.622\,P_{sat}(T_2) \div [P_1 - P_{sat}(T_2)]$ is also known. Note that, for a given state at temperature T and a certain humidity, $T > T_{wb} > T_{dp}$, as shown in Fig. 12.1.

EXAMPLE 12.11

The dry bulb and wet bulb temperatures in a room are measured to be 25°C and 18°C respectively. Assuming the pressure to be 1 atm, determine the (a) humidity ratio, (b) relative humidity and (c) dew point temperature.

Solution : From Table A, we can retrieve the following values: $P_{sat}(25°C) = 3.169$ kPa, $h_g(25°C) = 2546.3$ kJ/kg, $P_{sat}(18°C) = 2.064$ kPa, $h_f(18°C) = 75.47$ kJ/kg and $h_g(18°C) = 2533.5$ kJ/kg.

(a) Thus, $\omega_{wb} = 0.01293$ kg vapor / kg dry air. From Eqn. 12.5, we can get $\omega = 0.01$ kg vapor / kg dry air.

It follows that $\phi = P_v \div P_{sat}(25°C) = 50.6$ percent.

(c) From Table B, $T_{sat}(1.603\,\text{kPa}) = T_{dp} \approx 14°C$.

■ EXAMPLE 12.12

Determine the wet bulb temperature of ambient air at 25°C, 1 atm and 75 percent relative humidity.

Solution : Since it is given that $\phi = 0.75$,

$$P_{v,1} = \phi\, P_{sat}(25°C) = 0.75 \times 3.169 = 2.37675\,\text{kPa}$$

and

$$\omega = 0.622\,\frac{P_v}{P - P_v}$$

$$= 0.622 \times \frac{2.37675}{101.325 - 2.37675} = 0.015\,\frac{\text{kg vapor}}{\text{kg dry air}}$$

Also $h_g(25°C) = 2546.3$ kJ/kg. Equation 12.5 may be written as

$$C_{p,a}\,(T_{wb} - T) + \omega_{wb}\,[h_g(T_{wb}) - h_f(T_{wb})] - \omega\,[h_g(T) - h_f(T_{wb})] = 0$$

where $\omega_{wb} = 0.622\,P_{sat}(T_{wb}) \div [P - P_{sat}(T_{wb})]$. T_{wb} has to be determined iteratively as shown below. Since $20 < T_{dp} < 21$, the initial guess can be 23°C.

T_{wb} (°C)	ω_{wb} kg vapor kg dry air	$1.005\,(T_{wb} - 25)$ $+\omega_{wb}\,[h_g(T_{wb}) - h_f(T_{wb})]$ $-0.015\,[2546.3 - h_f(T_{wb})]$
23	0.01774	4.637
22	0.01667	0.99
21	0.01566	-2.511

Hence, $T_{wb} \approx 22°C$.

12.4 Psychrometric chart

In psychrometry, the specific enthalpy of moist air is customarily defined as

$$h^* = \frac{H}{m_a} = \frac{m_a h_a + m_v h_v}{m_a} = h_a + \omega h_v \quad \frac{\text{kJ}}{\text{kg dry air}} \qquad (12.6)$$

$$P_v = \frac{\omega}{0.622 + \omega} P$$

$$\phi = \frac{P_v}{P_{sat}(T)}$$

$$h^* = h_a + \omega\, h_v = C_{p,a}\, T(^\circ C) + \omega\, h_g(T) \quad \frac{kJ}{kg\ dry\ air}$$

$$v_a = \frac{V}{m_a} = \frac{\omega\, R_v\, T}{P_v} = \frac{R_v\, T}{P}(0.622 + \omega) \quad \frac{m^3}{kg\ dry\ air}$$

$$h^*(T_{wb}) + (\omega - \omega_{wb})\, h_f(T_{wb}) = h^*(T)$$

$$T_{dp} = T_{sat}(P_v)$$

The above representation brings out the fact that all the quantities of interest depend only on two independent properties T and ω. Hence, these relations may be graphically represented in a $\omega - T$ diagram with T as the abscissa and ω as the ordinate. Such a diagram is popularly called the psychrometric chart. The pressure of the moist air, P, is assumed to remain constant, which is quite reasonable in psychrometric applications.

Psychrometric chart corresponding to $P = 1$ atm is given at the end of the book. All the quantities listed above, namely, P_v, ϕ, h^*, T_{wb} and T_{dp} may be retrieved from the chart for a given T and ω. The following observations regarding the dependence of these quantities on T and ω can be made:

- In the expression for the specific enthalpy, h^*, the quantity $h_g(T)$ increases by 100 kJ/kg from 2500 kJ/kg in the temperature range $0.01^\circ C \leq T \leq 55^\circ C$, which is less than a 5 percent change. Hence, it is reasonable to assume that $h_g(T)$ is practically a constant in the range of temperatures of interest. Consequently, lines of $h^* = C_{p,a}\, T + \omega\, h_g(T) = $ constant, where T is in $^\circ C$ are straight lines with a negative slope in the $T - \omega$ coordinates space and may be conveniently represented in the second quadrant.

- In the expression for the vapor pressure, P_v, the denominator, $0.622 + \omega$ depends very weakly on ω, since ω is only of the order of 0.01 in psychrometric applications. Consequently, P_v depends almost linearly on ω alone. For the same reason, the specific volume of dry air, v_a, depends weakly on ω and depends linearly on T. In other words, $v_a = $ constant lines are nearly vertical lines in the psychrometric chart.

The following examples demonstrate the use of the psychrometric chart. A close-up view of the chart depicting the different states accompanies each example. The pair of lines with arrows pointing *towards* a state denote the two variables whose values are being used to fix that state. Lines with arrows pointing away from a state indicate values that are being retrieved from the chart, using interpolation, if necessary.

▌ EXAMPLE 12.13

Redo Example 12.11 using the psychrometric chart.

Solution : The initial state may be located on the chart at $T = 25°C$ and $T_{wb} = 18°C$. We may retrieve, $\phi = 50$ percent, $\omega = 0.01$ kg vapor/kg dry air, $T_{dp} = 14°C$ and $P_v = 12$ mm of mercury $= 1.6$ kPa.

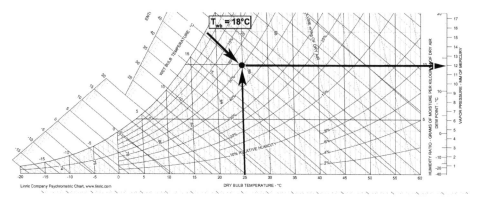

▌ EXAMPLE 12.14

Redo Example 12.7 using the psychrometric chart.

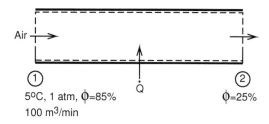

Solution : The inlet state may be located on the chart using the given $\phi_1 = 85$ percent and $T_1 = 5°C$, as shown. From the chart, we can retrieve $h_1^* = 17$ kJ/kg dry air and

Mass flow rate of dry air may be evaluated as $\dot{m}_a = \dot{V}_1 \div v_{a,1} = 100 \div 0.795$ = 125.786 kg dry air/min. SFEE applied to the control volume shown gives

$$0 = \dot{Q} - \cancel{\dot{W}_x}^{0} + \dot{m}_a (h_1^* - h_2^*)$$
$$\dot{Q} = 125.786 \times (36 - 17) = 39.83 \, \text{kW}$$

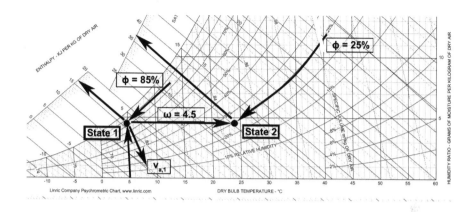

EXAMPLE 12.15

Redo Example 12.8 using the psychrometric chart.

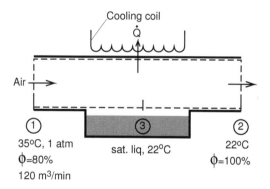

Solution : State 1 may be located in the chart using the given relative humidity, ϕ_1 = 80 percent and temperature, 35°C. From the chart, we can retrieve, $\omega_1 = 0.029$ kg

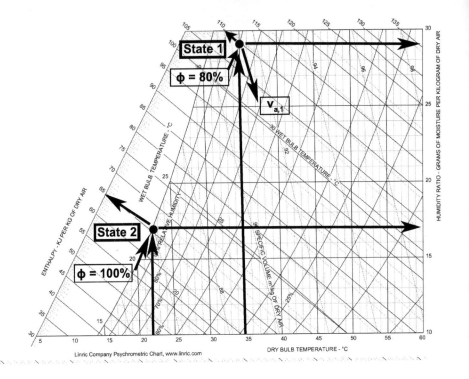

Mass flow rate of dry air may be evaluated from the given volumetric flow rate as follows:

$$\dot{m}_a = \frac{\dot{V}_1}{v_{a,1}} = \frac{120}{0.915} = 131.15 \; \frac{\text{kg}}{\text{min}}$$

(a) Mass balance of water across the control volume shown gives

$$\dot{m}_w = \dot{m}_a \left(\omega_1 - \omega_2\right) = 1.5869 \; \frac{\text{kg}}{\text{min}}$$

This is the mass flow rate of the condensate.

(b) SFEE applied to the control volume shown gives

$$\begin{aligned}
0 &= \dot{Q} - \overset{0}{\overbrace{\dot{W}_x}} + \dot{m}_a h_1^* - \dot{m}_a h_2^* - \dot{m}_3 h_3 \\
\dot{Q} &= \dot{m}_a (h_2^* - h_1^*) + \dot{m}_w h_f(T_3)
\end{aligned}$$

Redo Example 12.9 using the psychrometric chart.

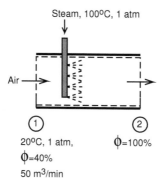

Steam, 100°C, 1 atm

Air →

①   ②

20°C, 1 atm, $\phi = 100\%$
$\phi = 40\%$
50 m³/min

Solution : The initial state may be located on the chart at $\phi_1 = 40$ percent and $T_1 =$ 20°C. Thus, $\omega_1 = 0.006$ kg vapor/kg dry air, $v_{a,1} = 0.839$ m³/kg dry air and $h_1^* = 35$ kJ/kg dry air. Mass flow rate of dry air may be evaluated from the given volumetric

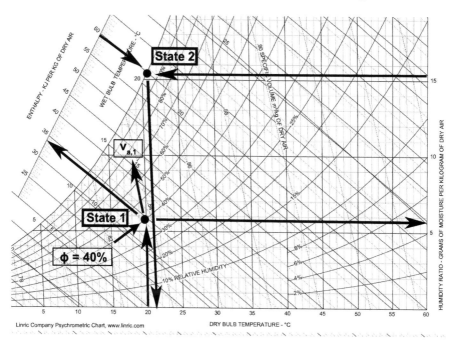

A mass balance of water gives

$$\dot{m}_s = \dot{m}_a \left(w_2 - w_1\right)$$

SFEE applied to the control volume shown gives

$$
\begin{aligned}
0 &= \cancel{\dot{Q}}^{0} - \cancel{\dot{W}_x}^{0} + \dot{m}_a h_1^* + \dot{m}_s h_s - \dot{m}_a h_2^* \\
0 &= \dot{m}_a \left[(h_2^* - h_1^*) - (w_2 - w_1)\, h_g(100^\circ \mathrm{C})\right] \\
0 &= (h_2^* - 35) - (w_2 - 0.006)\, 2675.7
\end{aligned}
$$

The last expression may be rewritten as $h_2^* - 2675.7\,w_2 = 18.95$. State 2 lies on the $\phi = 100$ percent line, to the right of and above state 1. We can locate the state as shown by trial and error, at $h_2^* = 60$ kJ/kg dry air and $w_2 = 0.0155$ kg vapor/kg dry air. The temperature T_2 may be retrieved as 20.5°C. The mass flow rate of steam comes out to be 0.566 kg/min.

■ EXAMPLE 12.17

Redo Example 12.10 using the psychrometric chart.

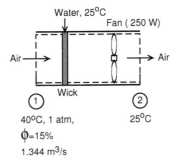

Water, 25°C
Fan (250 W)
Air →
→ Air
Wick
① ②
40°C, 1 atm,   25°C
φ=15%
1.344 m³/s

Solution : The initial state may be located on the chart at $\phi_1 = 15$ percent and $T_1 = 40°\mathrm{C}$. Thus, $w_1 = 0.007$ kg vapor/kg dry air, $v_{a,1} = 0.895$ m³/kg dry air and $h_1^* = 58$ kJ/kg dry air. Mass flow rate of dry air may be evaluated from the given volumetric flow rate as follows:

$$\dot{m}_a = \frac{\dot{V}_1}{v_{a,1}} = \frac{1.344}{0.895} = 1.5017 \ \frac{\mathrm{kg}}{\mathrm{s}}$$

Let the mass flow rate of liquid water be m_w kg/s. A mass balance of water gives $\dot{m}_w = \dot{m}_a \left(w_2 - w_1\right)$.

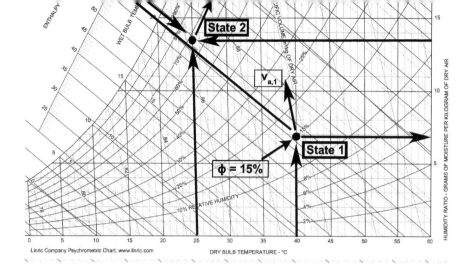

SFEE applied to the control volume shown gives

$$0 = \overset{0}{\cancel{\dot{Q}}} - \dot{W}_x + \dot{m}_a h_1^* + \dot{m}_w h_w - \dot{m}_a h_2^*$$
$$-\dot{W}_x = \dot{m}_a \left(h_2^* - h_1^* \right) - \dot{m}_w h_w$$
$$= \dot{m}_a \left[h_2^* - h_1^* - (\omega_2 - \omega_1) h_f(25°\text{C}) \right]$$

If we substitute the known values into the above expression, we get $h_2^* - 104.75\,\omega_2 = 57.433$. State 2 lies on the $T = 25°$C line, above state 1. We can locate the state by trial and error as shown, at $h_2^* = 59$ kJ/kg dry air, $\omega_2 = 0.0133$ kg vapor/kg dry air and $\phi_2 = 69$ percent relative humidity. The mass flow rate of liquid water comes out to be 9.46 g/min.

▆ EXAMPLE 12.18

Air at 30°C, 1 atm and 80 percent relative humidity enters an insulated air-conditioning duct where it is first cooled and de-humidified and then heated to 22°C and 40 percent relative humidity. Determine the mass flow rate of the condensate, cooling requirement in tons and the rate of heat addition in kW per unit volume flow rate of air that enters.

Solution : Let the states at the inlet and exit be denoted *1* and *3* respectively. The intermediate state, in between the cooling and heating sections be denoted 2. The

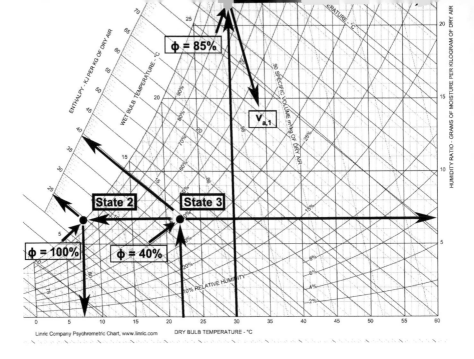

Linric Company Psychrometric Chart, www.linric.com   DRY BULB TEMPERATURE - °C

initial state may be located on the chart at ϕ_1 = 80 percent and T_1 = 30°C. Thus, ω_1 = 0.0215 kg vapor/kg dry air, $v_{a,1}$ = 0.89 m³/kg dry air and h_1^* = 85 kJ/kg dry air.

Mass flow rate of dry air may be evaluated from the given volumetric flow rate as follows:

$$\dot{m}_a = \frac{\dot{V}_1}{v_{a,1}} = \frac{1.0}{0.89} = 1.1236 \ \frac{\text{kg}}{\text{min}}$$

The final state may be located on the chart at ϕ_3 = 40 percent and T_3 = 22°C. Thus, ω_3 = 0.00675 kg vapor/kg dry air, and h_3^* = 38.5 kJ/kg dry air.

In the heating section, the humidity ratio remains the same as no water is added or removed, and so $\omega_2 = \omega_3$. Furthermore, since water condenses in the cooling section, the moist air that leaves this section is saturated and so ϕ_2 = 100 percent. State 2 may thus be located in the chart as shown. We can retrieve h_2^* = 24 kJ/kg dry air. Note that the temperature T_2 = 7.5°C.

SFEE applied to the cooling section gives

$$0 = \dot{Q}_C + \dot{m}_a(h_1^* - h_2^*) - \dot{m}_w\, h_f(7.5°\text{C})$$

We can calculate $\dot{Q}_C = 68.02$ kJ/min $= 0.322$ tons.

SFEE applied to the heating section leads to

$$\dot{Q}_H = \dot{m}_a(h_3^* - h_2^*) = 0.2715\,\text{kW}$$

12.5 Other applications

12.5.1 Adiabatic mixing of two streams

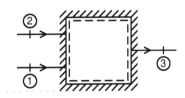

Figure 12.4: Adiabatic mixing of two moist air streams

Consider the adiabatic mixing chamber shown in Fig. 12.4, where two incoming moist air streams are mixed and the mixed stream exits the chamber. Assuming steady state operation and constant pressure, application of mass balance for the dry air and water, give, respectively,

$$\dot{m}_{a,1} + \dot{m}_{a,2} = \dot{m}_{a,3} \tag{12.7}$$

$$\dot{m}_{v,1} + \dot{m}_{v,2} = \dot{m}_{v,3}$$

$$\Rightarrow \quad \omega_1\, \dot{m}_{a,1} + \omega_2\, \dot{m}_{a,2} = \omega_3\, \dot{m}_{a,3}$$

$$\omega_3 = \frac{\omega_1\, \dot{m}_{a,1} + \omega_2\, \dot{m}_{a,2}}{\dot{m}_{a,1} + \dot{m}_{a,2}} \tag{12.8}$$

Application of SFEE to the control volume shown gives,

$$0 = \overset{0}{\cancel{\dot{Q}}} - \overset{0}{\cancel{\dot{W}_x}} + \dot{m}_{a,1}\, h_1^* + \dot{m}_{a,2}\, h_2^* - \dot{m}_{a,3}\, h_3^*$$

$$h_3^* = \frac{\dot{m}_{a,1}\, h_1^* + \dot{m}_{a,2}\, h_2^*}{\dot{m}_{a,1} + \dot{m}_{a,2}} \tag{12.9}$$

These two equations may be rearranged to give

$$\frac{\omega_3 - \omega_1}{\omega_2 - \omega_3} = \frac{h_3^* - h_1^*}{h_2^* - h_3^*} = \frac{\dot{m}_{a,2}}{\dot{m}_{a,1}}$$

$$\Rightarrow \quad \frac{h_1^* - h_3^*}{\omega_1 - \omega_3} = \frac{h_2^* - h_3^*}{\omega_2 - \omega_3}$$

The last expression shows that state points *1*, *2* and *3* lie in a straight line in the psychrometric chart.

■ EXAMPLE 12.19

In the previous example, consider replacing the heating section with a mixing section to accomplish the same objective. Determine the volume flow rate and relative humidity of air at 30°C that would be required. Neglect any heat loss in the mixing section and assume that the pressure remains constant at 1 atm.

Solution : With reference to Fig. 12.4. a stream of saturated air, $\dot{m}_a = 1.1236$ kg/min, 7.5°C (state *1*), $\omega = 0.00675$ kg vapor/kg dry air and air at 30°C (state *2*) are mixed adiabatically and the resulting mixture leaves at 22°C and 40 percent relative humidity (state *3*).

From Eqn. 12.8, since $\omega_1 = \omega_3$, we get $\omega_2 = \omega_1 = 0.00675$ kg vapor/kg dry air. From the chart, for $T_2 = 30°C$ and $\omega_2 = 0.00675$ kg vapor/kg dry air, we have $\phi_2 = 25$ percent, $v_{a,2} = 0.87$ m³/kg dry air and $h_2^* = 47$ kJ/kg dry air. SFEE applied to the mixing process gives

$$0 = \cancel{\dot{Q}}^{0} - \cancel{\dot{W}}_x^{0} + \dot{m}_{a,1} h_1^* + \dot{m}_{a,2} h_2^* - (\dot{m}_{a,1} + \dot{m}_{a,2}) h_3^*$$

$$\dot{m}_{a,2} = \frac{\dot{m}_{a,1} (h_3^* - h_1^*)}{h_2^* - h_3^*}$$

$$= \frac{1.1236 \times (38.5 - 24)}{(47 - 38.5)} = 1.9167 \ \frac{\text{kg}}{\text{min}}$$

The volume flow rate of the incoming air may be evaluated as $\dot{V}_{a,2} = \dot{m}_{a,2} \times v_{a,2} = 1.6676$ m³/min.

relative humidity and a volume flow rate of 175 m³/min. The mixing takes place adiabatically. Determine the exit temperature, relative humidity and volume flow rate. Assume that the pressure remains constant.

Solution : The following property values may be retrieved from the chart for the two incoming streams:

State	ω	v_a	h^*
	$\dfrac{\text{kg vapor}}{\text{kg dry air}}$	$\dfrac{\text{m}^3}{\text{kg dry air}}$	$\dfrac{\text{kJ}}{\text{kg dry air}}$
1	0.0255	0.91	100
$(35°C, T_{wb} = 30°C)$			
2	0.002	0.805	15
$(10°C, \phi = 25\%)$			

The incoming mass flow rates of dry air may be evaluated to be

$$\dot{m}_{a,1} = \frac{\dot{V}_1}{v_{a,1}} = \frac{50}{0.91} = 54.945 \ \frac{\text{kg}}{\text{min}}$$

$$\dot{m}_{a,2} = \frac{\dot{V}_2}{v_{a,2}} = \frac{175}{0.805} = 217.391 \ \frac{\text{kg}}{\text{min}}$$

From Eqn. 12.8, we can get $\omega_3 = 6.74 \times 10^{-3}$ kg vapor/kg dry air. From Eqn. 12.9, we can get $h_3^* = 32.15$ kJ/kg dry air. From the chart, we can retrieve $T_3 = 16°C$, $v_{a,3} = 0.825$ m³/kg dry air and $\phi_3 = 57$ percent. The volume flow rate of the exiting stream is thus $\dot{V}_{a,3} = \dot{m}_{a,3} \times v_{a,3} = 224.677$ m³/min.

It is left as an exercise to the reader to locate the three states in the chart and verify that they indeed lie along a straight line.

12.5.2 Drying

EXAMPLE 12.21

Wet paddy at 25°C that contains one-third moisture by weight is to be dried completely before being stored in order prevent rotting while in storage. Hot air is to be used for this purpose. Ambient air at 25°C, 1 atm and 50 percent relative humidity with a volume flow rate of 250 m³/min is heated to 60°C. The hot air

Solution : Let the state of the ambient air be denoted *1*. From the chart, we can retrieve, $\omega_1 = 0.01$ kg vapor/kg dry air, $v_{a,1} = 0.8575$ m^3/kg dry air and $h_1^* = 50.5$ kJ/kg dry air. Mass flow rate of dry air may be evaluated from the given volumetric flow rate as follows:

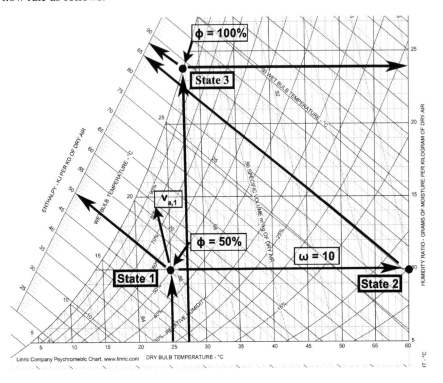

$$\dot{m}_a = \frac{\dot{V}_1}{v_{a,1}} = \frac{250}{0.8575} = 291.545 \; \frac{\text{kg}}{\text{min}}$$

For the hot air at the exit of the heating section (state 2), $\omega_2 = \omega_1 = 0.01$ kg vapor/kg dry air. With $T_2 = 60°$C, we can retrieve, $h_2^* = 85$ kJ/kg dry air. Application of SFEE to the heating section gives

$$\dot{Q}_H = \dot{m}_a(h_2^* - h_1^*) = 167.64 \, \text{kW}$$

If the air absorbs the maximum possible amount of water vapor, then its state at the exit of the drying section will be fully saturated (state *3*).

$$0 = \dot{m}_a \left(h_2^* - h_3^* \right) + \dot{m}_w \, h_f(25°C)$$
$$0 = 291.545 \times \left(h_2^* - h_3^* \right) + 291.545 \times (\omega_3 - 0.01) \times 104.75$$
$$0 = (85 - h_3^*) + (\omega_3 - 0.01) \times 104.75$$

This may be written as $h_3^* - 104.75\,\omega_3 = 83.9525$. The solution has to be obtained iteratively: If a value is guessed for T_3, then all the remaining quantities in the above expression become known and hence the right hand side of the expression may be evaluated. The guess is corrected until the expression is satisfied. The final state is as shown on the chart: $T_3 = 27.5°C$, $h_3^* = 86.5$ kJ/kg dry air and $\omega_3 = 0.0233$ kg vapor/kg dry air.

The mass flow rate of wet paddy is:

$$\dot{m}_{paddy} = 3 \times \dot{m}_w = 3 \times 291.545 \times (0.0233 - 0.01) = 11.633 \ \frac{\text{kg}}{\text{min}}$$

12.5.3 Cooling tower

As discussed earlier in Chapter 11, a condenser is used in a steam power plant for the working fluid to reject heat to the ambient. It is customary to use water from a nearby river or lake for absorbing the heat from the working fluid. This water is usually at a temperature of 30°C at entry to the condenser and its temperature increases by 10-15°C after passing through the same. This warm water is then taken to a cooling tower where it is cooled to 30°C by losing heat to the atmosphere and then returned to the condenser. In this section, we will look at the psychrometric aspects of the operation of the cooling tower (Fig. 12.5).

This type of cooling tower is known as an induced draft cooling tower, since the fan at the top draws the atmospheric air through the tower. In some installations, where the fan is absent, the draft inside the tower occurs due to natural convection and these are called natural draft cooling towers. The warm water from the condenser is sent downwards in a shower and it meets the upward draft of the atmospheric air which is drawn in through the openings provided along the circumference of the cooling tower. Since the atmospheric air is at a lower temperature, some of the liquid water evaporates and the moisture laden air leaves through the top. The rest of the water cools down as a result of evaporative cooling and collects at the bottom of the tower. Make-up water is added to compensate for the water lost due to evaporation and the cooled water is returned to the condenser.

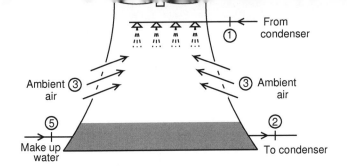

Figure 12.5: Illustration of a cooling tower used in steam power plants

EXAMPLE 12.22

Cooling water enters the condenser of a steam power plant at 30°C and leaves at 45°C with a mass flow rate of 3.75×10^7 kg/h. It is then taken to a cooling tower where it is cooled before being returned. Ambient air at 25°C, 1 atm and 45 percent relative humidity enters the cooling tower and leaves at 31°C, fully saturated. Make-up water at 25°C is also provided. Determine the mass flow rates of the dry air and make-up water. Neglect heat losses and fan power and assume that the pressure remains constant at 1 atm.

Solution : With reference to Fig. 12.5, the following property values may be retrieved from the psychrometric chart and Table A:

State		
1	$h_1 = h_f(45°C) = 188.42$ kJ/kg	
2	$h_2 = h_f(30°C) = 125.67$ kJ/kg	
3	$h_3^* = 47$ kJ/kg dry air	$\omega_3 = 0.009$ kg vapor/kg dry air
4	$h_4^* = 105$ kJ/kg dry air	$\omega_4 = 0.029$ kg vapor/kg dry air
5	$h_5 = h_f(25°C) = 104.75$ kJ/kg	

Mass balance of water across the cooling tower gives

$$\dot{m}_{w,1} + \dot{m}_{w,5} - \dot{m}_{w,2} + \dot{m}_{v,3} - \dot{m}_{v,4} = 0$$
$$\dot{m}_{w,5} = \dot{m}_{v,4} - \dot{m}_{v,3}$$
$$\dot{m}_{w,5} = \dot{m}_a(\omega_4 - \omega_3)$$

$$0 \;=\; (h_3^* - h_4^*) + \frac{\dot{m}_{w,1}}{\dot{m}_a}(h_1 - h_2) + (\omega_4 - \omega_3)h_5$$

Upon substituting the known values into the above expression, we get $\dot{m}_a = 4.21 \times 10^7$ kg/h. The mass flow rate of make-up water may be evaluated to be 8.42×10^5 kg/h.

COMBUSTION THERMODYNAMICS

In our discussion of power producing cycles such as Rankine, Brayton and so on, it was assumed that a certain amount of heat was supplied to the cycle. The details of how this heat was generated was immaterial. It is well known that this heat is generated by the combustion of a fuel (solid, liquid or gas) usually with atmospheric air. In this chapter, the thermodynamic aspects of this combustion process are studied in detail.

13.1 Combustion stoichiometry

Combustion is an oxidation reaction that is accompanied by heat release (exothermic). For instance,

$$H_2 + \frac{1}{2}O_2 \rightarrow H_2O$$

is a combustion reaction in which 1 kmol of H_2 combines with 1/2 kmol of O_2 to form 1 kmol of water vapor. The required oxygen may be provided as a pure oxygen stream or, as is commonly done, from atmospheric air. In the thermodynamic context, it will be assumed that combustion takes place. The question of whether

pure O_2 is supplied in the above reaction, we have

$$H_2 + \frac{1}{2}(O_2 + 3.76N_2) \rightarrow H_2O + 1.88N_2$$

It is assumed throughout in this chapter that N_2 is inert and does not participate in the chemical reaction. In reality, this is not true; as the temperature of the products increases, N_2 oxidizes to form oxides, which are of course, pollutants owing to their harmful effects on the environment and humans.

Consider the complete combustion of liquid n-octane (C_8H_{18}) with air. Here, complete combustion of a hydrocarbon fuel with air implies that the combustion products are fully oxidized, $i.e.$, CO_2 and H_2O are the only products and there is no excess oxygen in the products. The balanced combustion equation can be written as

$$C_8H_{18\,(l)} + 12.5(O_2 + 3.76N_2) \rightarrow 8CO_2 + 9H_2O + (12.5)(3.76)N_2$$

Thus, each kmole of fuel requires 12.5 kmoles of O_2 or (12.5)(4.76)=59.5 kmoles of air for complete combustion. This amount of air is termed theoretical or stoichiometric air. Therefore, the stoichiometric air-fuel ratio on a molar basis is 59.5 kmol of air per kmol of fuel. Air-fuel ratio on a mass basis is

$$= 59.5\frac{\text{kmol of air}}{\text{kmol of fuel}} \times 28.97\frac{\text{kg}}{\text{kmol of air}} \times \frac{1}{114}\frac{\text{kmol of fuel}}{\text{kg}}$$

$$= 15\frac{\text{kg of air}}{\text{kg of fuel}}$$

For any hydrocarbon fuel with composition of the form C_nH_{2n+2}, the complete combustion equation can be written as

$$C_nH_{2n+2} + \left(\frac{3n+1}{2}\right)(O_2+3.76N_2) \rightarrow nCO_2+(n+1)H_2O+\left(\frac{3n+1}{2}\right)(3.76)N_2$$

Air-fuel ratio on a molar basis for the above reaction is

$$\overline{AF} = \left(\frac{3n+1}{2}\right)(4.76)\frac{\text{kmol of air}}{\text{kmol of fuel}}$$

On a mass basis, air-fuel ratio is given as

$$AF = \left(\frac{3n+1}{2}\right)(4.76)\frac{\text{kmol of air}}{\text{kmol of fuel}} \times 28.97\frac{\text{kg}}{\text{kmol of air}}$$

$$\times \frac{1}{12n + 2n + 2}\frac{\text{kmol of fuel}}{\text{kg}}$$

This is a remarkable result since it is independent of n. Although this result has been obtained for large n, it can be used as a good engineering approximation even for small values of n. It is left as an exercise to the reader to derive this for other families of hydrocarbon fuels such as $C_n H_{2n}$ and $C_n H_{2n-2}$.

13.1.1 Excess air and equivalence ratio

In many practical applications, the amount of air used is in excess of the stoichiometric amount of air for several reasons. The most important reason is that the presence of the additional amount of O_2 and N_2 serve to dilute the combustion products thereby reducing the temperature to manageable levels. Also, the presence of excess O_2 tends to stabilize the combustion and improves the combustion efficiency, up to a certain extent. There is also an attendant effect on the composition of the product gas mixture, namely, that the availability of O_2 leads to the formation of pollutants such as oxides of nitrogen. This idea of excess air leads naturally to the definition of the equivalence ratio,

$$\Phi = \frac{\text{Actual Fuel} - \text{air ratio}}{\text{Stoichiometric Fuel} - \text{air ratio}}$$

Fuel-air ratio is the reciprocal of the air-fuel ratio. An equivalence ratio of unity represents a stoichiometric mixture. Equivalence ratios less than one and greater than one represent a fuel lean mixture and fuel rich mixture respectively.

EXAMPLE 13.1

A dry analysis of the products from the combustion of methane with air is as follows: CO_2 = 9%, CO = 2%, O_2 = 2.11% and N_2 = 86.89% by volume. Calculate the excess air, equivalence ratio and the dew point temperature of the products assuming the pressure to be 100 kPa.

Solution: Since the product analysis is on a volumetric basis, it is the same as molar basis. Also, note that the combustion of methane would have produced water vapor, but the analysis is on a dry basis. If we assume 100 kmole of dry products, a kmol of fuel and b kmol of air, then the complete combustion equation can be written as

$$aCH_4 + b(O_2 + 3.76N_2) \rightarrow 9CO_2 + 2a\,H_2O + 86.89\,N_2 + 2CO + 2.11\,O_2$$

Carbon atom balance gives $a = 11$ and N atom balance gives $b = 23.11$. Oxygen atom balance is automatically satisfied. Therefore, the balanced combustion reaction is

$$11CH_4 + 23.11(O_2 + 3.76N_2) \rightarrow 9CO_2 + 22H_2O + 86.89\,N_2 + 2CO + 2.11\,O_2$$

The stoichiometric air-fuel ratio is thus equal to $2 \times 4.76 = 9.52$ kmol of air/kmol of fuel. The excess air is $(10.06 - 9.52) \div 9.52 = 5.67$ percent. The equivalence ratio is $9.52 \div 10.06 = 0.946$.

The partial pressure of the water vapor in the products may be calculated as

$$P_{H_2O} = y_{H_2O}P = \frac{22}{9 + 22 + +86.89 + 2 + 2.11} \times 100 = 18.03 \, \text{kPa}$$

The dew point temperature, $T_{dp} = T_{sat}(P_{H_2O}) = 57.8°C$, from Table B.

◼ EXAMPLE 13.2

An ultimate analysis of Indian coal on a dry basis gives the following composition by mass percent: C 43.7, H 3.8, N 0.9, O 14.5, S 0.7, with the rest being ash. Assuming the ash and sulphur to be inert, determine the mass flow rate of air required per kg of coal for complete combustion.

Solution : The composition of 100 kg of fuel on a molar basis may be determined as:

	C	H	N	O	S
Mass (kg)	43.7	3.8	0.9	14.5	0.7
Moles (kmol)	3.6417	3.8	0.06428	0.90625	0.021875

The balanced chemical reaction for the complete combustion of this fuel may be written as:

$$C_{3.6417}H_{3.8}N_{0.06428}O_{0.90625}S_{0.021875} + 4.138575 \, (O_2 + 3.76N_2)$$
$$\rightarrow 3.6417 \, CO_2 + 1.9 \, H_2O + 15.529612 \, N_2 + 0.021875 \, S$$

The required mass flow rate of air per kg of coal may be evaluated as

$$4.138575 \times 4.76 \, \frac{\text{kmoles of air}}{100 \, \text{kg of fuel}} \times 28.8 \, \frac{\text{kg of air}}{\text{kmol air}} = 5.6735 \, \frac{\text{kg of air}}{\text{kg of fuel}}$$

◼ EXAMPLE 13.3

Indian rice husk, a bio-mass fuel, has the following percentage composition on a mass basis: C 35, H 5.5, N 1.53, O 36, S 0.08. It is burnt with 185 percent theoretical air, with air at 298 K, 100 kPa and 75 percent relative humidity. Determine the required mass flow rate of air per kg of fuel and the change in the

	C	H	N	O	S
Mass (kg)	35	5.5	1.53	36	0.08
Moles (kmol)	2.91667	5.5	0.1093	2.25	0.0025

The balanced chemical reaction for the complete combustion of this fuel with dry air may be written as:

$$C_{2.91667}H_{5.5}N_{0.1093}O_{2.25}S_{0.0025} + 3.16667\,(O_2 + 3.76N_2)$$
$$\rightarrow 2.9166\,CO_2 + 2.75\,H_2O + 11.9613\,N_2 + 0.0025\,S$$

With 185 percent theoretical air (dry), the above reaction becomes

$$C_{2.91667}H_{5.5}N_{0.1093}O_{2.25}S_{0.0025} + 1.85 \times 3.16667\,(O_2 + 3.76N_2)$$
$$\rightarrow 2.9166\,CO_2 + 2.75\,H_2O + 0.0025\,S + 0.85 \times 3.16667\,O_2$$
$$+(11.9613 + 0.85 \times 3.16667 \times 3.76)\,N_2$$

The partial pressure of the water vapor in the products may be evaluated as

$$P_{H_2O} = \frac{2.75}{30.44275} \times 100 = 9.033\,\text{kPa}$$

The dew point temperature of the products is $T_{dp} = T_{sat}(9.033\,\text{kPa}) = 44°\text{C}$, using Table B.

With the given data, the partial pressure of the water vapor in the supplied air may be evaluated as $P_{H_2O} = 0.75 \times P_{sat}(298\,\text{K}) = 2.37675$ kPa, using Table A. The mole fraction is thus $y_{H_2O} = 2.37675 \div 100 = 0.0237675$. Assuming that the ratio of the mole fractions of N_2 and O_2 in the air remains at 3.76, we can get

$$y_{O_2} + y_{N_2} + y_{H_2O} = 1$$
$$4.76\,y_{O_2} + 0.0237675 = 1$$
$$\Rightarrow \quad y_{O_2} = 0.2051$$

Hence, each kmol of O_2 in the humid air is accompanied by 3.76 kmoles of N_2 and $0.0237675 \div 0.2051 = 0.11588$ kmoles of water vapor. The reaction for complete combustion with 185 percent theoretical air may now be written as

$$C_{2.91667}H_{5.5}N_{0.1093}O_{2.25}S_{0.0025} + 1.85 \times 3.16667\,(O_2 + 3.76N_2 + 0.11588\,H_2O)$$

The partial pressure of water vapor in the products may be evaluated as 11.0176 kPa and the dew point temperature of the products is $T_{dp} = 47.7°C$, using Table B. The humidity in the incoming air thus elevates the dew point temperature of the products by 3.7°C.

▌ EXAMPLE 13.4

Wood, which contains C, O and H is burnt with air. The composition of the products (usually called producer gas) is, $N_2 - 39$, $CO - 17$, $H_2 - 11.5$, $CO_2 - 13.5$, $CH_4 - 9$, and $H_2O - 10$, by volume percent. Determine the composition of the fuel in weight percentage and the percent theoretical air supplied.

Solution : The combustion reaction for this case may be written as

$$C_a O_b H_d + x\,(O_2 + 3.76 N_2) \quad \rightarrow \quad 13.5\,CO_2 + 17\,CO + 9\,CH_4$$
$$+11.5\,H_2 + 10\,H_2O + 39\,N_2$$

Performing atom balance gives $a = 39.5$, $d = 79$, $b + 2x = 54$ and $3.76x = 39$. Therefore, $x = 10.3723$ and $b = 33.2553$. The composition of the fuel by weight percentage may be evaluated as follows:

	C	H	O
Moles (kmol)	39.5	79	33.2533
Mass (kg)	474	79	532.0528
Mass (%)	43.685	7.281	49.034

Stoichiometric combustion of the fuel is represented by

$$C_{39.5}O_{33.2553}H_{79} + y\,(O_2 + 3.76 N_2) \rightarrow 39.5\,CO_2 + 39.5\,H_2O + 3.76\,y\,N_2$$

Atom balance of O gives, $y = 42.62235$. The percent theoretical air is thus $10.3723 \div 42.62235 = 24.336$ percent.

▌ EXAMPLE 13.5

A hydrocarbon fuel with chemical formula $C_n H_{1.87n}$ undergoes combustion with a certain amount of air. It is noted that combustion of 1 kmol of fuel produces 49.047 kmoles of N_2 and 1.3044 kmoles of O_2. Determine the composition of the fuel and the equivalence ratio.

Solution : Combustion of 1 kmol of the given fuel with stoichiometric air can be written as

$$C_n H_{1.87n} + b\,(O_2 + 3.76\,N_2) \rightarrow n\,CO_2 + \frac{1.87n}{2}\,H_2O + b\,N_2$$

This equation may be modified as follows for an equivalence ratio of Φ :

$$C_n H_{1.87n} + \frac{1.4675\,n}{\Phi}\,(O_2 + 3.76\,N_2) \rightarrow n\,CO_2 + \frac{1.87n}{2}\,H_2O$$
$$+ \frac{5.5178\,n}{\Phi}\,N_2 + \frac{(1.4675\,n)(1-\Phi)}{\Phi}\,O_2$$

Using the given product composition, we can write

$$\frac{5.5178\,n}{\Phi} = 49.047; \qquad \frac{(1.4675\,n)(1-\Phi)}{\Phi} = 1.3044$$

from which we can get, in succession, $n = 8$ and $\Phi = 0.9$.

13.2 First law analysis of combustion systems

We have already defined in section 4.2, the enthalpy per unit mass as $h = u + Pv$, where u is the internal energy. The connection between u and the modes of energy storage in molecules such as translation, rotation and vibration for ideal gases, was discussed in section 5.2. It was also shown that the energy stored in these modes is dependent only on the temperature of the gas. However, the energy stored in the chemical bonds that make up the molecules was not accounted for in this development. This energy is somewhat like 'dead storage' in a reservoir or dam and does not play a role so long as there are no chemical reactions. Once chemical reactions begin to take place, new product species are formed from the reactant species and it is important to account for the energy used to break the bonds in the reactant species and form new bonds in the product species.

13.2.1 Enthalpy of formation and sensible enthalpy

With this in mind, for any chemical species, the enthalpy at a given temperature T is defined to be the sum of the enthalpy of formation, $\bar{h}_f^\circ(T_{ref})$ and a sensible enthalpy, $\Delta\bar{h}(T)$. The enthalpy of formation is evaluated at the standard reference state which is at temperature $T_{ref} = 298$ K and pressure 1 atm. Subscript f refers to formation and superscript $^\circ$ refers to standard state. Thus

$$\bar{h}(T) = \bar{h}_f^\circ + \Delta\bar{h}(T) \tag{13.1}$$

The sensible enthalpy depends only on the temperature and it is easy to see that the enthalpy that we have used so far is actually the sensible enthalpy. However, in combustion calculations, the sensible enthalpy has to be calculated from tabulated

standard state from the basic elements which make up the species at the standard state. It is usually determined from experiments involving the combination of elements to form the particular species. When the species is not formed from the elements but from some other substances, then the heat required to form these substances from their constituent elements is considered to determine the enthalpy of formation of the particular species. The enthalpy of formation accounts for the energy contained in the chemical bonds between atoms that make up the molecules of the species over and above the energy of the bonds of the elements. By definition, enthalpies of formation of elements in their natural states at 298 K, like, O_2, N_2 in gaseous state and C in solid state are zero.

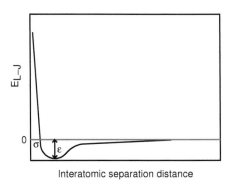

Figure 13.1: Lennard-Jones 6-12 potential curve

The potential energy due to forces between two atoms can be modelled with the well-known Lennard-Jones 6-12 potential,

$$E_{L-J} = 4\epsilon \left[\left(\frac{\sigma}{r} \right)^{12} - \left(\frac{\sigma}{r} \right)^{6} \right].$$

Here, σ is the atomic radius, ϵ is the bond energy and r is the interatomic separation distance. Values of the Lennard-Jones parameters (σ and ϵ) for various species are available in standard textbooks on Combustion. Note that the force between the atoms is attractive for $r > \sigma$ and repulsive for $r < \sigma$. This is illustrated in Fig. 13.1. Hence, an amount of energy ϵ has to be spent to break a bond between two atoms or, an amount of energy ϵ is released when the bond between two atoms is formed. The bond energies for some typical bonds are given in Table 13.1.

It should be noted that bond energy is defined as the the energy required to break

$H - H$	0.436×10^6
$C - C$	0.348×10^6
$C - H$	0.413×10^6
$O - H$	0.456×10^6
$C = O$	0.804×10^6
$C - O$	0.36×10^6
$C \equiv O$	1.072×10^6
$C \equiv C$	0.814×10^6
$O - O$	0.139×10^6
$(O = O)_{O_2}$	0.498×10^6

Source: *Combustion Science and Engineering*,
K. Annamalai and I. K. Puri, CRC Press, 2007

the bond. Also, bond energies for the double bond are higher than those of the single bond and those for the triple bond are even higher. The bond energy constitutes the major part of the energy contained in the molecule. The enthalpy of formation can therefore be approximately estimated using bond energies. This is demonstrated in the following worked example.

■ EXAMPLE 13.6

Determine the enthalpy of formation of gaseous methane, acetylene (C_2H_2), ethylene (C_2H_4), CO_2 and CO. Assume the enthalpy of formation of $C_{(g)}$ from the element $C(s)$ to be 0.7184×10^6 kJ/kmol.

Solution: Gaseous methane contains four $C - H$ bonds. Four H atoms are formed from the dissociation of two H_2 molecules. Carbon in the gaseous phase is formed from carbon in the solid state. Therefore, the enthalpy of formation of $CH_4 = -4 \times 0.413 \times 10^6 + 2 \times 0.436 \times 10^6 + 0.7184 \times 10^6$ kJ/kmol = -61600 kJ/kmol. The actual value is -74850 kJ/kmol.

Acetylene contains one $C \equiv C$ and two $C - H$ bonds. Two H atoms are formed from the dissociation of a H_2 molecule. Proceeding as before, enthalpy of formation of $C_2H_2 = -2 \times 0.413 \times 10^6 - 0.814 \times 10^6 + 0.436 \times 10^6 + 2 \times 0.7184 \times 10^6$ kJ/kmol = 232800 kJ/kmol, which is reasonably close to the actual value of 226736 kJ/kmol.

Ethylene contains one $C = C$ and four $C - H$ bonds. Four H atoms are formed from the dissociation of two H_2 molecules. Therefore, enthalpy of formation of $C_2H_4 = -4 \times 0.413 \times 10^6 - 0.615 \times 10^6 + 2 \times 0.436 \times 10^6 + 2 \times 0.7184 \times 10^6$

state. Hence, the enthalpy of formation of $CO_2 = -2 \times 0.804 \times 10^6 + 0.498 \times 10^6 + 0.7184 \times 10^6$ kJ/kmol = -391600 kJ/kmol. The actual value is -393510 kJ/kmol.

In CO, the carbon atom is joined to the oxygen atom by a triple bond. Hence, the enthalpy of formation of $CO = (-2 \times 1.072 \times 10^6 + 0.498 \times 10^6 + 2 \times 0.7184 \times 10^6)/2$ kJ/kmol = -104600 kJ/kmol. The actual value is -110530 kJ/kmol.

Note that the enthalpies of formation of the species considered above are large and negative. The values are negative since heat is released during the formation of the species from its constituent elements.

In the case of a molecule such as benzene, energy due to resonance also has to be accounted for. This arises since benzene can resonate between five distinct molecular structures without changing the number or nature of the bonds between the atoms. The interested reader can consult any advanced combustion book for details.

13.2.2 Enthalpy of combustion and calorific value

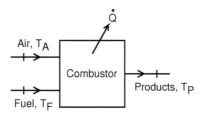

Figure 13.2: Schematic of a combustor operating at steady state

Consider a combustor operating at steady state in which the air enters at temperature T_A and fuel enters at temperature T_F as shown in Fig. 13.2. Combustion takes place and after removal of heat the products leave at temperature T_P. If we apply the steady flow energy equation to the combustor, we get

$$0 = \dot{Q} - \cancel{\dot{W}_x}^{0} + \dot{m}_F h_F + \dot{m}_A h_A - \dot{m}_P h_P$$

Since combustion reactions are always written on a molar basis, it would be advantageous to write the above expression on a molar basis as well. Thus,

$$0 = \dot{Q} + \dot{n}_F \bar{h}_F + \dot{n}_A \bar{h}_A - \dot{n}_P \bar{h}_P$$

$$0 = \frac{\dot{Q}}{\dot{n}_F} + \bar{h}_F + \frac{\dot{n}_A}{\dot{n}_F} \bar{h}_A - \frac{\dot{n}_P}{\dot{n}_F} \bar{h}_P$$

$$0 = \frac{\dot{Q}}{\dot{n}_F} + \bar{h}_{reactants} - \bar{h}_{products}$$

If the combustion reaction is written for 1 kmol of fuel, then the ratios inside the sum are simply the stoichiometric coefficients of the respective species in the chemical reaction.

For an exothermic reaction, the enthalpy of the reactants is greater than that of the products and hence heat is released in the combustion reaction. This is illustrated graphically in Fig. 13.3.

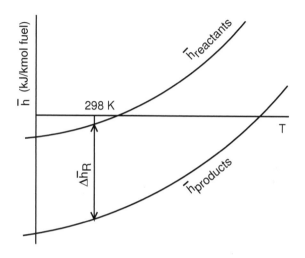

Figure 13.3: Illustration of enthalpy of combustion

The enthalpy of reaction (or combustion) $\Delta \bar{h}_R$ is defined as $\Delta \bar{h}_R = \bar{h}_{products} - \bar{h}_{reactants}$ with $T_P = T_A = T_F$, and has units of kJ/kmol of fuel. For combustion reactions which are exothermic, $\Delta \bar{h}_R$ is negative. This is because the heat released in the chemical reactions has to be *removed* to maintain the products at the same temperature as the reactants. $\Delta \bar{h}_R$ is usually evaluated at the reference temperature (298 K). In combustion applications, since the reactants are fuel plus air, the calorific value of the fuel is numerically equal to $\Delta \bar{h}_R$. However, here a distinction needs to be made depending upon whether the water in the product stream is in vapor or liquid form. The resulting value of $\Delta \bar{h}_R$ is called the lower calorific value (LCV) in the

Determine the higher and lower calorific value of methane at 298 K. Assume the enthalpy of formation of methane to be -74850 kJ/kmol.

Solution: Combustion of methane with a stoichiometric amount of air is represented by the equation

$$CH_4 + 2(O_2 + 3.76N_2) \rightarrow CO_2 + 2H_2O_{(v)} + 2 \times 3.76\,N_2$$

Equation 13.2 applied for this case, simplifies as follows (with $T_P = T_A = T_F = 298$ K):

$$0 = \frac{\dot{Q}}{\dot{n}_F} + \bar{h}^{\circ}_{f,CH_4} + \Delta\bar{h}^{\,0}_{CH_4} + \left[2\,(\bar{h}^{\circ}_{f,O_2}{}^0 + \Delta\bar{h}^{\,0}_{O_2}) + 2 \times 3.76\,(\bar{h}^{\circ}_{f,N_2}{}^0 + \Delta\bar{h}^{\,0}_{N_2}) \right]$$

$$- \left[\bar{h}^{\circ}_{f,CO_2} + \Delta\bar{h}^{\,0}_{CO_2} + 2\,(\bar{h}^{\circ}_{f,H_2O} + \Delta\bar{h}^{\,0}_{H_2O}) + 2 \times 3.76\,(\bar{h}^{\circ}_{f,N_2}{}^0 + \Delta\bar{h}^{\,0}_{N_2}) \right]$$

Therefore

$$LCV = -\frac{\dot{Q}}{\dot{n}_F} = -74850 - (-393520) - 2 \times (-241830) = 802330 \; \frac{kJ}{kmol \text{ of fuel}}$$

using the enthalpy of formation data given in Table G. On a mass basis, LCV = 802330 (kJ/kmol of CH_4) / 16 (kg/kmol of CH_4) = 50146 kJ/kg of CH_4.

The higher calorific value

$$HCV = -74850 - (-393520) - 2 \times \left(-241830 - 18\frac{kg}{kmol} \times 2441.55\frac{kJ}{kg} \right)$$

$$= 890226 \; \frac{kJ}{kmol \text{ of fuel}}$$

where we have retrieved Δh for the condensation at 298 K from Table A. On a mass basis, HCV = 55639 kJ/kg of CH_4.

■ EXAMPLE 13.8

Determine the LCV of the combustion products (producer gas) in Example 13.4, assuming that water vapor is not present.

Solution : The equation for the stoichiometric combustion of producer gas is

$$13.5\,CO_2 + 17\,CO + 9\,CH_4 + 11.5\,H_2 + 39\,N_2$$
$$+ 32.25\,(O_2 + 3.76N_2) \rightarrow 39.5\,CO_2 + 29.5\,H_2O + 160.26\,N_2$$

$$\frac{10.0}{90} CO_2 + \frac{17}{90} CO + \frac{9}{90} CH_4 + \frac{11.5}{90} H_2 + \frac{59}{90} N_2$$

$$+ \frac{32.25}{90} (O_2 + 3.76 N_2) \rightarrow \frac{39.5}{90} CO_2 + \frac{29.5}{90} H_2O + \frac{160.26}{90} N_2$$

Equation 13.2 applied for this case, simplifies as follows:

$$0 = -\frac{\dot{Q}}{\dot{n}_F} = -\frac{26}{90} \bar{h}_{f,CO_2}^\circ + \frac{17}{90} \bar{h}_{f,CO}^\circ + \frac{9}{90} \bar{h}_{f,CH_4}^\circ - \frac{29.5}{90} \bar{h}_{f,H_2O}^\circ$$

Therefore

$$LCV = -\frac{\dot{Q}}{\dot{n}_F} = 164.587 \ \frac{MJ}{kmol \ of \ fuel}$$

using the values for enthalpy of formation from Table G and taking $\bar{h}_{f,CH_4}^\circ$ from the previous example. The LCV of the producer gas is thus 6360.17 kJ/kg of fuel.

■ **EXAMPLE 13.9**

Liquid n-dodecane ($C_{12}H_{26}$) at 298 K and air at 700 K steadily enter the combustor of a gas turbine engine. Determine the required mass flow rate of air and the corresponding equivalence ratio, if the temperature of the products is (a) 1700 K and (b) 2200 K. Heat loss from the combustor may be neglected. The enthalpy of formation of liquid n-dodecane may be taken as -352100 kJ/kmol.

Solution : The chemical reaction for the complete combustion of n-dodecane with air may be written as

$$C_{12}H_{26,(l)} + 18.5 \, (O_2 + 3.76 \, N_2) \rightarrow 12CO_2 + 13H_2O + 18.5 \times 3.76 \, N_2$$

If the excess air supplied is x, then

$$C_{12}H_{26,(l)} + (1+x)(18.5) \, (O_2 + 3.76 \, N_2) \rightarrow \begin{array}{l} 12CO_2 + 13H_2O \\ +(x)(18.5) \, O_2 \\ +(1+x)(18.5 \times 3.76) \, N_2 \end{array}$$

Applying Eqn. 13.2 to the combustor, we get,

$$0 = \frac{\overset{0}{\cancel{\dot{Q}}}}{\dot{n}_F} + \bar{h}_{f,C_{12}H_{26,(l)}}^\circ + \Delta \bar{h}_{C_{12}H_{26}}^{\overset{0}{\cancel{\ }}}$$

$$+ \quad (1+x)(18.5 \times 3.76)\,(\overset{0}{\cancel{\bar{h}^\circ_{f,N_2}}} + \Delta \bar{h}_{N_2}) + (x)(18.5)(\overset{0}{\cancel{\bar{h}^\circ_{f,O_2}}} + \Delta \bar{h}_{O_2}) \Bigg]_{T_P}$$

The expression becomes, after retrieving the required values from Table G and rearranging,

$$x = \frac{12\Delta \bar{h}_{CO_2} + 13\Delta \bar{h}_{H_2O} + 69.56\Delta \bar{h}_{N_2} - 8577.391}{1063.46 - 69.56\Delta \bar{h}_{N_2} - 18.5\Delta \bar{h}_{O_2}} \Bigg|_{T_P}$$

(a) Upon setting $T_P = 1700$ K and solving this equation for x, we get $x = 1.2695$. The required mass flow rate of air may be evaluated as

$$\dot{m}_{air} = (1+x)(18.5)(4.76)\,\frac{\text{kmoles of air}}{\text{kmol of fuel}} \times 28.84\,\frac{\text{kg of air}}{\text{kmol of air}}$$
$$\times \frac{1}{170}\,\frac{\text{kmol of fuel}}{\text{kg of fuel}}$$
$$= 33.905\,\frac{\text{kg}}{\text{kg of fuel}}$$

The equivalence ratio $\Phi = 1 \div (1+x) = 0.44$.

(b) For $T_P = 2200$ K, we get $x = 0.4032$. Therefore, $\dot{m}_{air} = 20.9626$ kg / kg fuel and $\Phi = 0.7127$.

▮ EXAMPLE 13.10

Gasoline E-10 is a blend of 90 percent n-octane and 10 percent ethanol (C_2H_5OH) by volume. Determine its LCV. Take the enthalpy of formation liquid n-octane and liquid ethanol to be -249950 and -277650 kJ/kmol respectively and the density to be 0.703 and 0.790 kg/L respectively.

Solution : One litre of the blended fuel contains 0.9 litre of n-octane and 0.1 litre of ethanol, or, $0.9 \times 0.703 = 0.6327$ kg and $0.1 \times 0.79 = 0.079$ kg, or, on a molar basis, 5.55 moles and 1.7174 moles respectively. The respective mole fraction in the blended fuel is thus 0.7637 and 0.2363.

The chemical reaction for the complete combustion of the blended fuel with stoichiometric air may be written as

$$0.7637\,C_8H_{18\,(l)} + 0.2363\,C_2H_5OH_{(l)} + 10.25515\,(O_2 + 3.76\,N_2) \rightarrow 6.5822\,CO_2$$

$$LCV = 0.7637\, \tilde{n}_{f,C_8H_{18}} + 0.2363\, \tilde{n}_{f,C_2H_5OH} - [0.5822\, \tilde{n}_{f,CO_2} + 7.5822\, \tilde{n}_{f,H_2O}]$$

This may be evaluated as 4167335 kJ/kmol of fuel, using the values in Table G. On a mass basis, the LCV is 42553 kJ/kg of fuel.

13.2.3 Adiabatic flame temperature

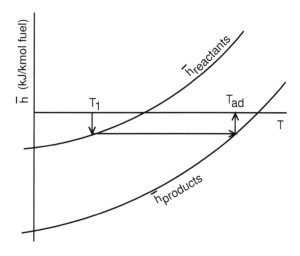

Figure 13.4: Illustration of the adiabatic flame temperature concept

If the combustor considered in the previous subsection is insulated, then the energy equation reduces to

$$0 = \frac{\overset{0}{\cancel{\dot{Q}}}}{\dot{n}_F} + \bar{h}_{reactants} - \bar{h}_{products}$$

$$\Rightarrow \qquad \bar{h}_{reactants} = \bar{h}_{products}$$

This implies that the products leave at a temperature at which the enthalpy of the product stream is equal to the enthalpy of the incoming reactants at temperature T_1. The products in this case will leave at a temperature which is different from that of the reactants. This is illustrated graphically in Fig. 13.4. The temperature of the products in this case is called the adiabatic flame temperature. This is an important concept since it represents the theoretical maximum product temperature that can be realized at the combustor exit.

of the products leaving the combustor.

Solution: SFEE (Eqn. 13.2) applied to this combustor gives

$$0 = \cancelto{0}{\frac{\dot{\varnothing}}{\dot{n}_F}} + \bar{h}^\circ_{f,CH_4} + \cancelto{0}{\Delta\bar{h}_{CH_4}} + \left[2\left(\cancelto{0}{\bar{h}^\circ_{f,O_2}} + \cancelto{0}{\Delta\bar{h}_{O_2}} \right) + 2 \times 3.76 \left(\cancelto{0}{\bar{h}^\circ_{f,N_2}} + \cancelto{0}{\Delta\bar{h}_{N_2}} \right) \right]$$

$$- \left[\bar{h}^\circ_{f,CO_2} + \Delta\bar{h}_{CO_2} + 2\left(\bar{h}^\circ_{f,H_2O} + \Delta\bar{h}_{H_2O} \right) + 2 \times 3.76 \left(\cancelto{0}{\bar{h}^\circ_{f,N_2}} + \Delta\bar{h}_{N_2} \right) \right]$$

Upon substituting the values for the enthalpies of formation given in Table G, we get,

$$\Delta\bar{h}_{CO_2} + 2\Delta\bar{h}_{H_2O} + 7.52\Delta\bar{h}_{N_2} = 802330$$

The temperature of the products has to be determined iteratively. It is reasonable to expect the value to lie between 298 K and 3000 K. Hence, we start with an initial guess equal to $(298+3000)\div2 = 1649$ K. Proceeding with the nearest entry in Table G which is 1700 K, the left hand side of the above equation comes out to be 529731. This suggests that our initial guess is on the lower side. The next guess can be taken to be $(1700 + 3000)\div2 = 2350$ K or 2400 K. The left hand side now comes out to be 833440, which is less than the right hand side. The next guess is 2300 K, for which the left hand side is 789177. With linear interpolation, we can finally estimate 2330 K as the temperature of the products[†].

T (K)	T (K)	$\Delta\bar{h}_{CO_2} + 2\Delta\bar{h}_{H_2O} + 7.52\Delta\bar{h}_{N_2}$
(guess)	(nearest entry in Table G)	(kJ/kmol of fuel)
1649	1700	529731
2350	2400	833440
2300	2300	789177

◼ EXAMPLE 13.12

Hydrogen is burnt with 150 percent theoretical air in an adiabatic combustor operating at 1 atm pressure. Assuming steady state operation and the reactant temperature to be 298 K, determine the product temperature.

[†] This is still somewhat approximate but accurate enough for our purpose.

Application of SFEE (Eqn. 13.2) gives

$$
0 = \cancelto{0}{\frac{\dot{Q}}{\dot{n}_F}} + \cancelto{0}{\bar{h}^\circ_{f,H_2}} + \cancelto{0}{\Delta\bar{h}_{H_2}} + 1.25 \times \left[(\cancelto{0}{\bar{h}^\circ_{f,O_2}} + \cancelto{0}{\Delta\bar{h}_{O_2}}) + 3.76\,(\cancelto{0}{\bar{h}^\circ_{f,N_2}} + \cancelto{0}{\Delta\bar{h}_{N_2}}) \right]
$$
$$
- \left[\bar{h}^\circ_{f,H_2O} + \Delta\bar{h}_{H_2O} + 0.25\,(\cancelto{0}{\bar{h}^\circ_{f,O_2}} + \Delta\bar{h}_{O_2}) + 0.75 \times 3.76\,(\cancelto{0}{\bar{h}^\circ_{f,N_2}} + \Delta\bar{h}_{N_2}) \right]
$$

Upon substituting for $\bar{h}^\circ_{f,H_2O}$ from Table G and rearranging, we get

$$
\Delta\bar{h}_{H_2O} + 0.25\,\Delta\bar{h}_{O_2} + 2.82\,\Delta\bar{h}_{N_2} = 241830
$$

The temperature of the products has to be determined iteratively, as shown below:

T (K) (guess)	T (K) (nearest entry in Table G)	$\Delta\bar{h}_{H_2O} + 0.25\,\Delta\bar{h}_{O_2} + 2.82\,\Delta\bar{h}_{N_2}$ (kJ/kmol of fuel)
2000	2000	245870
1900	1900	229711

The final value may be obtained as 1975 K by interpolation.

EXAMPLE 13.13

A stoichiometric mixture of n-octane vapor and air at 700 K and 1000 kPa undergoes combustion in the cylinder of a spark ignition IC engine. Assuming the combustion to occur at constant volume and neglecting heat losses, determine the final temperature and final pressure. The enthalpy of n-octane vapor at 700 K may be taken as -94880.75 kJ/kmol.

Solution : Since combustion occurs at constant volume, application of First law to the contents of the cylinder gives,

$$
\Delta E = \Delta U = \cancelto{0}{\dot{Q}} - \cancelto{0,\,V = const}{W}
$$
$$
\Rightarrow \quad U_{products} = U_{reactants}
$$

$$
\sum_{products} n_i\,\bar{u}_i = \sum_{reactants} n_i\,\bar{u}_i
$$

$$[8\,(\bar{h}_{CO_2} - \mathcal{R}T) + 9\,(\bar{h}_{H_2O} - \mathcal{R}T) + 47\,(\bar{h}_{N_2} - \mathcal{R}T)]_{T=T_P}$$
$$= [(\bar{h}_{C_8H_{18}(v)} - \mathcal{R}T) + 12.5\,(\bar{h}_{O_2} - \mathcal{R}T) + 47\,(\bar{h}_{N_2} - \mathcal{R}T)]_{T=T_R}$$

where T_P denotes the temperature of the products and T_R denotes the temperature of the reactants (= 700 K) and we have used the the fact that $\bar{h} = \bar{u} + \mathcal{R}T$. This may be rearranged to give

$$[8\,\Delta\bar{h}_{CO_2} + 9\,\Delta\bar{h}_{H_2O} + 47\,\Delta\bar{h}_{N_2} - 64\mathcal{R}T]_{T=T_P}$$
$$= [\bar{h}_{C_8H_{18}(v)} + 12.5\,\Delta\bar{h}_{O_2} + 47\,\Delta\bar{h}_{N_2} - 60.5\mathcal{R}T]_{T=T_R}$$
$$- [8\,\bar{h}^{\circ}_{f,CO_2} + 9\,\bar{h}^{\circ}_{f,H_2O}]$$

If we substitute the known values from Table G into the above expression, we get

$$8\,\Delta\bar{h}_{CO_2} + 9\,\Delta\bar{h}_{H_2O} + 47\,\Delta\bar{h}_{N_2} - 64\mathcal{R}T\big|_{T=T_P} = 5596206$$

T (K) (guess)	T (K) (nearest entry in Table G)	$8\,\Delta\bar{h}_{CO_2} + 9\,\Delta\bar{h}_{H_2O} + 47\,\Delta\bar{h}_{N_2} - 64\mathcal{R}T$ (kJ/kmol of fuel)
1649	1700	2332687
2350	2400	3806530
2650	2700	4459791
3000	3000	5121512
3500	3500	6234764
3200	3200	5564593

Once again, the final temperature has to be obtained iteratively. The results are shown in the table above. The final temperature may be taken to be 3213 K after interpolation. The final pressure is

$$P = 1000 \times \frac{3213}{700} = 4590\,\text{kPa}$$

In the above examples, we have assumed CO_2, H_2O and N_2 to be the only products arising from the combustion of the fuels, which is not realistic. At such high temperatures, CO_2 will decompose to form CO, O_2 and H_2O will decompose to form O_2, H_2 and N_2 will combine with the O_2 to form oxides of nitrogen such as NO. Consequently, the final temperature will be less than the values obtained here.

participating species owing to the different composition of the products and the reactants. In other words, the total enthalpy of each species is required – not just Δh. In the same manner and for the same reason, entropy balance for a combustion process requires absolute entropy of each species and not just Δs.

13.3.1 Absolute entropy of ideal gases and ideal gas mixtures

The third law of thermodynamics states that the absolute entropy of a pure crystalline substance is zero at 0 K, which provides a datum against which absolute entropy of any species may be evaluated. The absolute entropy of substances at the reference temperature and pressure of 298 K and 1 atm have been measured and/or theoretically calculated. In the same manner as the enthalpy of formation, the entropy of a substance at any other temperature and pressure is evaluated with respect to this reference state. Thus

$$\bar{s}(T,P) = \underbrace{\bar{s}(T_{ref}, P_{ref}) + \int_{T_{ref}}^{T} \frac{dh}{T} - \mathcal{R} \ln \frac{P}{P_{ref}}}_{=\bar{s}(T,P_{ref})=\bar{s}^{\circ}(T)}$$

where we have used Eqn. 9.14. The first two terms of this expression have been combined and denoted $\bar{s}^{\circ}$. This is given in Table H for several gases of interest. Hence,

$$\bar{s}(T,P) = \bar{s}^{\circ}(T) - \mathcal{R} \ln \frac{P}{P_{ref}} \tag{13.3}$$

In a mixture of gases, the absolute entropy of component i must be evaluated at its partial pressure, P_i and so the absolute entropy of the mixture is given as

$$\begin{aligned} \bar{s}(T,P) &= \sum_i y_i \left(\bar{s}_i^{\circ}(T) - \mathcal{R} \ln \frac{P_i}{P_{ref}} \right) \\ &= \sum_i y_i \left(\bar{s}_i^{\circ}(T) - \mathcal{R} \ln \frac{y_i P}{P_{ref}} \right) \end{aligned} \tag{13.4}$$

13.3.2 Entropy generation

Consider the combustor shown in Fig. 13.2 again. Application of Eqn. 9.29 to the combustor gives, after assuming steady state operation,

$$\dot{\sigma} = \dot{m}_P s_P - \dot{m}_F s_F - \dot{m}_A s_A + \frac{\dot{Q}_{surr}}{T_0}$$

$$\frac{}{\dot{n}_F} = \sum_{products} \frac{}{\dot{n}_F} \bar{s}_i(T_P, P_i) - \bar{s}_F - \sum_{air} \frac{}{\dot{n}_F} \bar{s}_i(T_A, P_i) + \frac{}{\dot{n}_F} T_0$$

$$\Rightarrow \bar{\sigma} = \bar{s}_{products} - \bar{s}_{reactants} + \frac{\bar{Q}_{surr}}{T_0} \qquad (13.5)$$

◼ EXAMPLE 13.14

Determine the entropy generated per kmol of fuel in Example 13.9 if the combustor operates at 3 MPa and the products leave at 1700 K. The specific entropy of liquid n-dodecane at entry may be taken as 490.660 kJ/kmol.K.

Solution : The mole fractions of the individual species in the reactant and product stream are given in the table below. Note that since the fuel enters as a liquid, it is not included in this calculation.

	Reactants			Products	
Species	n_i	y_i	Species	n_i	y_i
O_2	41.98575	0.21	O_2	23.48575	0.11381
N_2	157.86642	0.79	N_2	157.86642	0.76504
			CO_2	12	0.05815
			H_2O	13	0.063
				206.35217	1

The terms on the right hand side of Eqn. 13.5 may be evaluated as follows:

$$\bar{s}_{reactants} = \bar{s}_{C_{12}H_{26},(l)} + \left[41.98575 \left(\bar{s}_{O_2}^\circ - \mathcal{R}\ln \frac{y_{O_2} P}{P_{ref}}\right)\right.$$

$$+ \left. 157.86642 \left(\bar{s}_{N_2}^\circ - \mathcal{R}\ln \frac{y_{N_2} P}{P_{ref}}\right)\right]_{3\,MPa,\,700\,K}$$

$$= 39654.42 \frac{kJ}{kmol\ fuel.K}$$

$$\bar{s}_{products} = \sum_i n_i \left[\left(\bar{s}_i^\circ - \mathcal{R}\ln \frac{y_i P}{P_{ref}}\right)\right]_{3\,MPa,\,1700\,K}$$

$$= 47506.56 \frac{kJ}{kmol\ fuel.K}$$

■ EXAMPLE 13.15

Determine the entropy generated per kmol of fuel in Example 13.13. The specific entropy of n-octane vapor at 700 K may be taken as 657.5 kJ/kmol.K.

Solution : Considering the contents of the cylinder as the system, Eqn. 9.28 may be written (on a per kmol of fuel basis) as

$$\bar{\sigma} = \Delta \bar{s}_{sys} + \frac{\cancel{\phi}_{surr}^{\;0}}{T_0} = \bar{s}_{products} - \bar{s}_{reactants}$$

The combustion reaction for this case is given as

$$C_8H_{18\,(v)} + 12.5(O_2 + 3.76N_2) \rightarrow 8CO_2 + 9H_2O + 47\,N_2$$

The mole fractions of the individual species in the reactant and product may be calculated as follows:

	Reactants			Products		
Species	n_i	y_i		Species	n_i	y_i
O_2	12.5	0.2066		CO_2	8	0.125
N_2	47	0.7769		N_2	47	0.7344
C_8H_{18}	1	0.0165		H_2O	9	0.1406
	60.5	1			64	1

Therefore

$$\bar{s}_{reactants} = \sum_i n_i \left[\bar{s}_i^{\circ} - \mathcal{R} \ln \frac{y_i P}{P_{ref}} \right]_{1\,MPa,\,700\,K}$$

$$= 12884.01 \frac{kJ}{kmol\ fuel.K}$$

using Table H. In the same manner,

$$\bar{s}_{products} = \sum_i n_i \left[\bar{s}_i^{\circ} - \mathcal{R} \ln \frac{y_i P}{P_{ref}} \right]_{4.59\,MPa,\,3213\,K}$$

$$= 16347.884 \frac{kJ}{kmol\ fuel.K}$$

EXAMPLE 13.16

Propane enters a combustor along with 200 percent theoretical air steadily at 298 K and 100 kPa. If the products of combustion leave at the same temperature and pressure, determine the entropy generated. Take the enthalpy of formation and entropy of propane at 298 K to be -104700 kJ/kmol 270 kJ/kmol.K respectively.

Solution : Combustion of propane with stoichiometric air may be written as

$$C_3H_8 + 5(O_2 + 3.76N_2) \rightarrow 3CO_2 + 4H_2O + 18.8\,N_2$$

Therefore, the combustion reaction with 200 percent theoretical air is

$$C_3H_8 + 10\,(O_2 + 3.76N_2) \rightarrow 3\,CO_2 + 4\,H_2O + 5\,O_2 + 37.6\,N_2$$

Equation 13.2 applied to the combustor gives, after using the fact that the reactants and products are at 298 K,

$$0 \;=\; \frac{\dot{Q}}{\dot{n}_F} + \bar{h}_{reactants} - \bar{h}_{products}$$

$$\frac{\dot{Q}}{\dot{n}_F} \;=\; 3\,\bar{h}^\circ_{f,CO_2} + 4\,\bar{h}^\circ_{f,H_2O} - \bar{h}^\circ_{f,C_3H_8}$$

$$=\; -2043180 \; \frac{\text{kJ}}{\text{kmol fuel}}$$

The mole fractions of the species are calculated in the same manner as before.

	Reactants			Products	
Species	n_i	y_i	Species	n_i	y_i
O_2	10	0.2058	O_2	5	0.1008
N_2	37.6	0.7736	N_2	37.6	0.7581
C_3H_8	1	0.0206	CO_2	3	0.0605
			H_2O	4	0.0806
	48.6	1		49.6	1

Therefore

$$\bar{s}_{reactants} \;=\; \bar{s}_{C_3H_8} + 10\,\bar{s}_{O_2} + 37.6\,\bar{s}_{N_2}$$

$$=\; \left(\bar{s}^\circ_{C_3H_8} - \mathcal{R}\ln y_{C_3H_8}\right) + 10\left(\bar{s}^\circ_{O_2} - \mathcal{R}\ln y_{O_2}\right) + 37.6\left(\bar{s}^\circ_{N_2} - \mathcal{R}\ln\right.$$

$$=\; 9770 \; \frac{\text{kJ}}{\text{kmol fuel.K}}$$

$$+5 \left(\bar{s}^{\circ}_{O_2} - \mathcal{R}\ln y_{O_2}\right) + 37.6 \left(\bar{s}^{\circ}_{N_2} - \mathcal{R}\ln y_{N_2}\right)$$

$$= \; 9962.689 \; \frac{kJ}{kmol \; fuel.K}$$

Equation 13.5 becomes,

$$
\begin{aligned}
\bar{\sigma} \;\; &= \;\; \bar{s}_{products} - \bar{s}_{reactants} + \frac{\dot{Q}_{surr}}{\dot{n}_F} \frac{1}{T_0} \\
&= \;\; 9962.689 - 9770 + \frac{2043180}{298} \\
&= \;\; 7049 \; \frac{kJ}{kmol \; fuel.K}
\end{aligned}
$$

GAS PHASE CHEMICAL EQUILIBRIUM

The notion of mechanical equilibrium was introduced in section 2.3 and it was also mentioned that the system must be in equilibrium for its state to be fixed using values for measurable properties. The notion of thermal equilibrium was introduced in section 8.4 in connection with the Carnot engine, and it was stated that the best possible engine is one which operates with reversible processes *i.e.,* processes connecting equilibrium – mechanical and thermal, states. In this chapter, the concept of chemical equilibrium, which determines the chemical composition, is introduced and discussed. The development is confined to chemical equilibrium in the gas phase alone.

14.1 Introduction

A thermodynamic system that is in a state of non-equilibrium as a result of an externally imposed constraint, will attain a state of equilibrium once the constraint is removed. This is demonstrated next using the system shown in Fig. 14.1.

An insulated vessel which is divided into two parts by an insulated piston that is

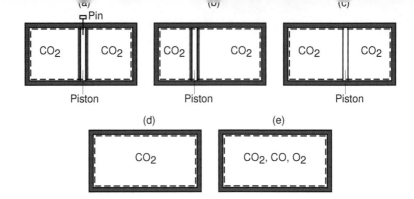

Figure 14.1: System in (a) non-equilibrium, (b) mechanical equilibrium, (c) mechanical and thermal equilibrium (d) chemical non-equilibrium and (e) mechanical, thermal and chemical equilibrium

at a different pressure and temperature. The system (shown using the dashed line) is thus initially in a state of mechanical non-equilibrium owing to the presence of the pin and thermal non-equilibrium owing to the insulation that covers the piston.

Let the mechanical constraint, namely, the pin, be removed first. As a result, the piston will oscillate back and forth and finally settle down when the pressure on both sides of it are equal. The pressure in the system no longer changes, *i.e.,* $dP = 0$ and mechanical equilibrium is established (Fig. 14.1b). Let the thermal constraint, namely, the insulation covering the piston be removed next. Once again, after a transient period during which the temperatures on either side of the piston change, the system attains an equilibrium state and $dT = 0$ (Fig. 14.1c). The piston itself may be removed now without causing any change in the pressure or temperature of the system (assuming that the volume occupied by the piston as well as its thermal capacity are negligibly small). We now have an insulated vessel containing CO_2 that is in mechanical and thermal equilibrium, as shown in Fig. 14.1(d). The question that naturally arises now is, whether this is the final equilibrium state or not. Note that, so far in this book, we have assumed it to be so. We now depart from this and develop the notion of chemical equilibrium. It is intuitive to expect that just as mechanical equilibrium determines the final pressure and thermal equilibrium the final temperature, chemical equilibrium determines the final composition.

In the development above, leading up to the state in Fig. 14.1(d), a chemical constraint (similar to the mechanical and thermal constraint), that chemical reactions do not take place, has been present. If the final temperature of the CO_2 is not

reach a new equilibrium state in which the final composition (amounts of CO_2, CO and O_2) is different from the initial composition, which had CO_2 alone. It may be recalled that such a situation prevailed in several examples in the previous chapter, but such additional reactions were not allowed to take place. Accordingly, the product composition evaluated in those examples are non-equilibrium compositions arising as a consequence of this artificial constraint. If the constraint is removed, then the CO_2 decomposes to form CO and O_2 according to the forward reaction. Simultaneously, the CO and O_2 recombine to form CO_2 through the backward reaction. When the system attains chemical equilibrium, the production and consumption of the individual species are balanced and the composition of the mixture will not change any more (Fig. 14.1e). The pressure and temperature will also change in the case of the system in Fig. 14.1(d) as a result of the chemical reactions since the vessel is insulated and its volume is fixed.

14.2   Derivation of the equilibrium criterion

If we apply First law to the system in Fig. 14.1(d), we get

$$dE = dU = \cancel{\delta Q}^{\;0} - \cancel{P\,dV}^{\;0} \Rightarrow \quad dU = 0$$

The change in entropy of the system is given by

$$dS = \frac{\cancel{\delta Q}^{\;0}}{T} + \delta\sigma_{int}$$

As the system moves from the non-equilibrium state in Fig.   14.1(d) to the equilibrium state Fig. 14.1(e), the entropy keeps *increasing* (since $\delta\sigma_{int} > 0$ and so $dS > 0$). Once the equilibrium state is attained, pressure, temperature and composition do not change any longer. It follows then that the entropy attains a maximum and does not change any more *i.e.*, $dS = 0$ at the equilibrium state. The criterion for equilibrium in this case is thus,

$$dS|_{U,\,V} = 0 \tag{14.1}$$

where we have explicitly indicated that U and V are constant since $dU = 0$ and $dV = 0$ throughout.

---

†This corresponds to the situation when the pressure and/or the temperature difference across the piston is very small to begin with, in the above example. In this case, it is immaterial whether a constraint is present or not.

consisting of a gas (or mixture of gases) in chemical non-equilibrium

$$dE = dU = \delta Q - P\,dV$$
$$dU + P\,dV + V\,dP = \delta Q + V\,dP$$
$$dH = \delta Q + V\,dP$$

The δQ term may be written in terms of the change in entropy of the system and this gives

$$dH = T\,dS - \delta\sigma_{int} + V\,dP$$
$$dH - T\,dS - S\,dT = -S\,dT + V\,dP - \delta\sigma_{int}$$
$$d(H - TS) = dG = -S\,dT + V\,dP - \delta\sigma_{int}$$

Here, $G = H - TS$ is a new property known as the Gibbs function. Since we are seeking a criterion for which P and T are constant, the above expression may be simplified as

$$dG = -S\,d\!\!\!\!\!\overset{0}{T} + V\,d\!\!\!\!\!\overset{0}{P} - \delta\sigma_{int} = -\delta\sigma_{int}$$

Since $\delta\sigma_{int} > 0$, $dG < 0$ and so the Gibbs function keeps *decreasing* as the system evolves towards the equilibrium state (Fig. 14.1e) starting from a non-equilibrium state (Fig. 14.1d). When the equilibrium state is attained, G reaches a minimum and does not change any more *i.e.*, $dG = 0$. Thus,

$$dG|_{T,P} = 0 \qquad (14.2)$$

is the criterion for equilibrium. In other words, with P and T known and fixed, this criterion should allow us to determine the equilibrium composition. This is discussed next.

14.3 Equilibrium mixture composition for a given P and T

For a mixture of gases, $G = \sum_i n_i \bar{g}_i$, where $\bar{g}_i = \bar{h}_i - T\bar{s}_i$ is the molar specific Gibbs function of species i. If we use Eqn. 13.4, to rewrite $\bar{s}_i$, we get

$$\bar{g}_i = \bar{h}_i - T\left(\bar{s}_i^\circ - \mathcal{R}\ln\frac{y_i P}{P_{ref}}\right) = \bar{g}_i^\circ + \mathcal{R}T\ln\frac{y_i P}{P_{ref}} \qquad (14.3)$$

It is clear from this equation that, $\bar{g}_i$ is a function of T and P only. However, G is a function of T, P and n_i. Therefore,

$$dG = \frac{\partial G}{\partial T}\,dT + \frac{\partial G}{\partial P}\,dP + \sum_i \frac{\partial G}{\partial n_i}\,dn_i$$

The change in the number of moles of the various species is, of course, determined by the chemical reaction.

Let a system consist of certain amounts of species A, B, C, D and E initially at temperature T and pressure P. Let a chemical reaction of the form

$$\alpha\, A + \beta\, B \rightleftharpoons \gamma\, C + \delta\, D \tag{14.5}$$

take place now. As a result, the composition of the system changes towards an equilibrium value with T and P remaining constant. Note that species E is inert as it does not participate in the reaction and hence the number of moles of species E remains the same. We may write for the above reaction,

$$\frac{-dn_A}{\alpha} = \frac{-dn_B}{\beta} = \frac{dn_C}{\gamma} = \frac{dn_D}{\delta} = d\xi \tag{14.6}$$

where ξ is the extent of the reaction. The negative sign in front of dn_A and dn_B account for the fact that species A and B are consumed in the forward reaction. The equilibrium criterion, namely, Eqn. 14.4, may now be written for the reaction given in Eqn. 14.5 as

$$\left(-\alpha\,\bar{g}_A - \beta\,\bar{g}_B + \gamma\,\bar{g}_C + \delta\,\bar{g}_D\right)\,d\xi = 0$$

In the general case,

$$-\alpha\,\bar{g}_A - \beta\,\bar{g}_B + \gamma\,\bar{g}_C + \delta\,\bar{g}_D = 0$$

If we use Eqn. 14.3 to rewrite $\bar{g}$, we get, after rearranging

$$\frac{-\alpha\bar{g}_A^\circ - \beta\bar{g}_B^\circ + \gamma\bar{g}_C^\circ + \delta\bar{g}_D^\circ}{\mathcal{R}T} = \ln\left[\left(\frac{y_A P}{P_{ref}}\right)^\alpha \left(\frac{y_B P}{P_{ref}}\right)^\beta \left(\frac{y_C P}{P_{ref}}\right)^{-\gamma} \left(\frac{y_D P}{P_{ref}}\right)^{-\delta}\right]$$

This may be written as

$$e^{-\frac{\Delta G^\circ}{\mathcal{R}T}} = \frac{y_C^\gamma\, y_D^\delta}{y_A^\alpha\, y_B^\beta}\left(\frac{P}{P_{ref}}\right)^{\gamma+\delta-\alpha-\beta} \tag{14.7}$$

where $\Delta G^\circ = \gamma\bar{g}_C^\circ + \delta\bar{g}_D^\circ - \alpha\bar{g}_A^\circ - \beta\bar{g}_B^\circ$. For a given T and P, the above expression may be used to evaluate the equilibrium composition and hence this is the expression that we sought at the beginning of this section. It may seem at first sight that there are more unknowns (y_A, y_B, y_C and y_D) than equations (Eqn. 14.7). However, sufficient number of additional equations arising from atom balances in Eqn. 14.5 are also available.

in Eqn. 14.5 and is usually denoted K. Thus

$$K(T) = e^{-\frac{\Delta G^\circ}{RT}} \tag{14.8}$$

$$= \frac{y_C^\gamma \, y_D^\delta}{y_A^\alpha \, y_B^\beta} \left(\frac{P}{P_{ref}} \right)^{\gamma + \delta - \alpha - \beta} \tag{14.9}$$

$$= \frac{n_C^\gamma \, n_D^\delta}{n_A^\alpha \, n_B^\beta} \left(\frac{1}{n} \frac{P}{P_{ref}} \right)^{\gamma + \delta - \alpha - \beta} \tag{14.10}$$

$$= \frac{P_C^\gamma \, P_D^\delta}{P_A^\alpha \, P_B^\beta} \left(\frac{1}{P_{ref}} \right)^{\gamma + \delta - \alpha - \beta} \tag{14.11}$$

Here, $n = n_A + n_B + n_C + n_D + n_E$ is the total number of moles and $P = P_A + P_B + P_C + P_D + P_E$.

It is easy to see that if the reaction in Eqn. 14.5 had been written as

$$\gamma C + \delta D \rightleftharpoons \alpha A + \beta B$$

then the negative sign in front of dn_A and dn_B in Eqn. 14.6 would not be necessary but would be required in front of dn_C and dn_D. Consequently, the equilibrium constant for this reaction, denoted K_{-1}, would just be the reciprocal of that of the reaction in Eqn. 14.5. It may then be inferred that, at a given temperature and pressure, if the value for K is high, then the equilibrium mixture would contain more products than reactants and *vice versa*.

Values of equilibrium constants of a few reactions calculated using Eqn. 14.8 for temperatures ranging from 300 K to 4000 K are given in Table I.

▣ EXAMPLE 14.1

One kmol of CO_2 at 1 atm and 2700 K decomposes to form an equilibrium mixture of CO_2, CO and O_2. Determine the composition of the mixture.

Solution : At equilibrium, let there be χ kmol (< 1) of CO_2. Carbon and oxygen atom balance give, respectively, that $1 - \chi$ kmol of CO and $(1 - \chi) \div 2$ kmol of O_2 are present in the equilibrium mixture. The balanced chemical reaction is

$$CO_2 \rightarrow \chi CO_2 + (1 - \chi) CO + \frac{1 - \chi}{2} O_2$$

The mole fractions of the species in the equilibrium mixture are given in the table below:

CO	0	$1 - \chi$	$2(1 - \chi) \div (3 - \chi)$
O_2	0	$(1 - \chi) \div 2$	$(1 - \chi) \div (3 - \chi)$
Total		$(3 - \chi) \div 2$	1

The final composition is determined by the reaction

$$CO_2 \rightleftharpoons CO + \frac{1}{2} O_2$$

From Table I, we can retrieve $\log_{10} K = -1.0094$ at 2700 K, for this equilibrium reaction. With $P = 1$ atm $= P_{ref}$, Eqn. 14.9 becomes

$$10^{-1.0094} = \frac{y_{CO}\, y_{O_2}^{1/2}}{y_{CO_2}} = \frac{1 - \chi}{\chi} \sqrt{\frac{1 - \chi}{3 - \chi}}$$

This non-linear equation for χ may be solved to give $\chi = 0.7674$. Therefore, the equilibrium mixture contains 0.7674 kmol of CO_2, 0.2326 kmol of CO and 0.1163 kmol of O_2.

EXAMPLE 14.2

Carbon monoxide at 298 K enters a reactor that is operating at 1 atm. It is burnt with a stoichiometric amount of air that enters at 298 K. The products of combustion leave the reactor as an equilibrium mixture of CO_2, CO and O_2 at 2000 K. Determine the composition of the products and the rate of heat removal from the reactor. Assume steady state operation. The N_2 in the air may be assumed to be inert.

Solution : The balanced chemical reaction for this case with χ kmol of CO_2 in the equilibrium product mixture is given as

$$CO + \frac{1}{2} (O_2 + 3.76\, N_2) \rightarrow \chi\, CO_2 + (1 - \chi)\, CO + \frac{1 - \chi}{2} O_2 + 1.88\, N_2$$

The species mole fractions in the products may be evaluated as shown below:

	Initial	Equilibrium	
	(kmol)	(kmol)	Mole fraction
CO_2	0	χ	$2\chi \div (6.76 - \chi)$
CO	1	$1 - \chi$	$2(1 - \chi) \div (6.76 - \chi)$
O_2	0.5	$(1 - \chi) \div 2$	$(1 - \chi) \div (6.76 - \chi)$
N_2	1.88	1.88	$3.76 \div (6.76 - \chi)$
Total		$(6.76 - \chi) \div 2$	1

we can retrieve $\log_{10} K = -2.8814$ at 2000 K for this reaction from Table I. With $P = 1$ atm $= P_{ref}$, Eqn. 14.9 becomes

$$10^{-2.8814} = \frac{y_{CO}\, y_{O_2}^{1/2}}{y_{CO_2}} = \frac{1-\chi}{\chi}\sqrt{\frac{1-\chi}{6.76-\chi}}$$

This non-linear equation for χ may be solved to give $\chi = 0.9788$. Therefore, the product stream contains 0.9788 kmol of CO_2, 0.0212 kmol of CO, 0.0106 kmol of O_2 and 1.88 kmoles of N_2.

Application of Eqn. 13.2 to the combustor gives, after simplification,

$$-\frac{\dot{Q}}{\dot{n}_F} = \bar{h}^\circ_{f,CO} - \left[\chi\left(\bar{h}^\circ_f + \Delta\bar{h}\right)_{CO_2}\right.$$

$$\left. + (1-\chi)\left(\bar{h}^\circ_f + \Delta\bar{h}\right)_{CO} + \frac{1-\chi}{2}\Delta\bar{h}_{O_2} + 1.88\,\Delta\bar{h}_{N_2}\right]_{2000\,K}$$

The required values on the right hand side of the above expression may be retrieved from Table G. The rate at which heat is removed comes out to be 80.117 MJ/kmol of CO.

■ EXAMPLE 14.3

LPG is a mixture of 40 percent propane (C_3H_8) and 60 percent butane (C_4H_{10}) by volume. It undergoes partial combustion (cracking) in a reactor operating steadily at 1 atm with 75 percent theoretical air. Assuming the products to be an equilibrium mixture of CO_2, CO, H_2O, H_2 and N_2 (inert) at 1800 K, determine the composition and the rate at which heat is removed from the reactor. Assume that the reactants enter at 298 K. The enthalpy of formation of propane and butane may be taken to be -103847 kJ/kmol and -124733 kJ/kmol respectively.

Solution : Combustion of LPG with stoichiometric air is represented by the reaction

$$0.4\,C_3H_8 + 0.6\,C_4H_{10} + 5.9\,(O_2 + 3.76N_2) \rightarrow 3.6\,CO_2 + 4.6\,H_2O + 22.184\,N_2$$

With 75 percent theoretical air and χ kmol of CO_2 in the equilibrium product mixture, the chemical reaction may be written as

$$0.4\,C_3H_8 + 0.6\,C_4H_{10} + 0.75 \times 5.9\,(O_2 + 3.76N_2) \rightarrow 0.75 \times 22.184\,N_2$$
$$\chi\,CO_2 + (3.6 - \chi)\,CO + (5.25 - \chi)\,H_2O + (\chi - 0.65)\,H_2$$

Note that the moles of CO, H_2O and H_2 in the above reaction have been obtained using C, O and H atom balances. The species mole fractions in the equilibrium product mixture can now be evaluated as shown below:

CO	0	$3.6 - \chi$	$(3.6 - \chi) \div 25.33$
H_2O	0	$5.25 - \chi$	$(5.25 - \chi) \div 25.33$
H_2	0	$\chi - 0.65$	$(\chi - 0.65) \div 25.33$
N_2	17.13	17.13	$17.13 \div 25.33$
Total		25.33	1

Since the mole fraction of CO_2 alone is unknown, we have the freedom to select only a single equilibrium reaction. The appropriate equilibrium reaction for the specified final composition is

$$CO_2 + H_2 \rightleftharpoons CO + H_2O$$

and we can retrieve $\log_{10} K = 0.5758$ at 1800 K for this reaction from Table I. With $P = 1 \text{ atm} = P_{ref}$, Eqn. 14.9 becomes

$$10^{0.5758} = \frac{y_{CO} \, y_{H_2O}}{y_{CO_2} \, y_{H_2}} = \frac{(3.6 - \chi)(5.25 - \chi)}{\chi(\chi - 0.65)}$$

Solving for χ, we get $\chi = 1.70151$. Hence, the products contain 1.70151 kmoles of CO_2, 1.89849 kmoles of CO, 3.54849 kmoles of H_2O, 1.05151 kmoles of H_2 and 17.13 kmoles of N_2.

Application of Eqn. 13.2 to the combustor gives, after simplification,

$$
\begin{aligned}
-\frac{\dot{Q}}{\dot{n}_F} = \ & 0.4\,\bar{h}^\circ_{f,C_3H_8} + 0.6\,\bar{h}^\circ_{f,C_4H_{10}} \\
& - \left[1.70151\,(\bar{h}^\circ_f + \Delta\bar{h})_{CO_2} + 1.89849\,(\bar{h}^\circ_f + \Delta\bar{h})_{CO} \right. \\
& + \left. 3.54849\,(\bar{h}^\circ_f + \Delta\bar{h})_{H_2O} + 1.05151\,\Delta\bar{h}_{H_2} + 17.13\,\Delta\bar{h}_{N_2}\right]_{1800\,K} \\
= \ & 281940.43\ \frac{\text{kJ}}{\text{kmol of fuel}}
\end{aligned}
$$

◼ EXAMPLE 14.4

One kmol of CO, H_2 and O_2 each, at 5 atm, react to form an equilibrium mixture of CO, CO_2, H_2O, H_2 and O_2 at 2500 K. Determine the composition of the equilibrium mixture.

$$CO + H_2 + O_2 \rightarrow \chi\, CO + (1-\chi)\, CO_2 + \nu\, H_2 + (1-\nu)\, H_2O + \frac{\chi + \nu - 1}{2} O_2$$

The mole fractions of the species in the equilibrium mixture may be evaluated as shown below:

	Initial (kmol)	Equilibrium (kmol)	Equilibrium Mole fraction
CO_2	0	$1 - \chi$	$2(1-\chi) \div (\chi + \nu + 3)$
CO	1	χ	$2\chi \div (\chi + \nu + 3)$
O_2	1	$(\chi + \nu - 1) \div 2$	$(\chi + \nu - 1) \div (\chi + \nu + 3)$
H_2	1	ν	$2\nu \div (\chi + \nu + 3)$
H_2O	0	$1 - \nu$	$2(1-\nu) \div (\chi + \nu + 3)$
Total		$(\chi + \nu + 3) \div 2$	1

We now assume that the final composition is determined by the following two reactions proceeding independently:

$$CO_2 \;\rightleftharpoons\; CO + \frac{1}{2} O_2$$

$$H_2O \;\rightleftharpoons\; H_2 + \frac{1}{2} O_2$$

The equilibrium constant for these two reactions may be retrieved from Table I as $\log_{10} K_1 = -1.4338$ and $\log_{10} K_2 = -2.2253$ at 2500 K, respectively. Since the reactions are independent, Eqn. 14.9 may be applied to these reactions independently and this leads to (with $P = 5$ atm)[‡]

$$10^{-1.4338} = \frac{y_{CO}\, y_{O_2}^{1/2}}{y_{CO_2}} \times 5^{1/2} = \sqrt{5}\,\frac{\chi}{1-\chi}\sqrt{\frac{\chi + \nu - 1}{\chi + \nu + 3}}$$

$$10^{-2.2253} = \frac{y_{H_2}\, y_{O_2}^{1/2}}{y_{H_2O}} \times 5^{1/2} = \sqrt{5}\,\frac{\nu}{1-\nu}\sqrt{\frac{\chi + \nu - 1}{\chi + \nu + 3}}$$

[†] In reality, there are five unknowns – the number of moles of each species in the equilibrium mixture. Since there are three different atoms, namely, C, H and O, atom balance provides three equations, which reduces the number of unknowns to two.

[‡] In this case, Eqn. 14.6 may be written for each reaction with its own extent of reaction, namely, $d\xi_1$ and $d\xi_2$. Substitution of these expressions into Eqn. 14.4 and invoking the condition that $d\xi_1$ and $d\xi_2$ are independent, yields two expressions similar to Eqn. 14.7 – one for each reaction.

reader to determine the amount of heat removed from the combustor.

■ EXAMPLE 14.5

Consider Example 13.9 again. If the products form an equilibrium mixture of CO_2, CO, H_2O, H_2, O_2, N_2 and NO at 20 atm, determine the composition of the mixture, assuming that the same amount of air is supplied.

Solution : There are *seven* different species in the product mixture composed of *four* different atoms – C, H, O and N. The number of unknowns is thus 7 - 4 = 3. Accordingly, let the product mixture contain χ, μ and ν kmoles of CO_2, H_2O and N_2 respectively. The combustion is then represented by the following balanced chemical reaction:

$$C_{12}H_{26,(l)} \quad + \quad (1+x)(18.5)\,(O_2 + 3.76\,N_2) \rightarrow \chi\,CO_2 + (12-\chi)\,CO$$
$$+\mu\,H_2O + (13-\mu)\,H_2 + \nu\,N_2 + 2[69.56(1+x) - \nu]\,NO$$
$$+\frac{2\nu - \chi - \mu - 102.12\,x - 114.12}{2}\,O_2$$

The mole fractions of the species in the equilibrium mixture are given in the table below:

CO_2	$2\chi \div (214.12 + 176.12\,x - \chi - \mu)$
CO	$2(12-\chi) \div (214.12 + 176.12\,x - \chi - \mu)$
H_2O	$2\mu \div (214.12 + 176.12\,x - \chi - \mu)$
H_2	$2(13-\mu) \div (214.12 + 176.12\,x - \chi - \mu)$
N_2	$2\nu \div (214.12 + 176.12\,x - \chi - \mu)$
NO	$4[69.56(1+x) - \nu] \div (214.12 + 176.12\,x - \chi - \mu)$
O_2	$(2\nu - \chi - \mu - 102.12\,x - 114.12) \div (214.12 + 176.12\,x - \chi - \mu)$

We now assume that the final composition is determined by the following three reactions proceeding independently:

$$CO_2 \rightleftharpoons CO + \frac{1}{2}O_2$$

$$H_2O \rightleftharpoons H_2 + \frac{1}{2}O_2$$

$$\frac{1}{2}O_2 + \frac{1}{2}N_2 \rightleftharpoons NO$$

Since the reactions are independent, Eqn. 14.9 may be applied to these reactions

$$K_2 = \frac{y_{H_2} y_{O_2}^{1/2}}{y_{H_2O}} \times 20^{1/2} = \sqrt{20} \, \frac{13 - \mu}{\mu} \sqrt{\frac{2\nu - \chi - \mu - 102.12\,x - 114.12}{214.12 + 176.12\,x - \chi - \mu}}$$

$$K_3 = \frac{y_{NO}}{y_{O_2}^{1/2} y_{N_2}^{1/2}} = \frac{4[69.56(1 + x) - \nu]}{\sqrt{(2\nu - \chi - \mu - 102.12\,x - 114.12)} \times 2\nu}$$

The solution to this system of equations as well as the mole fractions of the remaining species are shown in the table below for 1700 K and 2200 K.

T	CO_2	H_2O	N_2	O_2	CO	H_2	NO
(K)	(kmol)	(kmol)	(kmol)	(kmol)	(kmol)	(kmol)	(kmol)
			Non-equilibrium (Example 13.9)				
1700	12	13	157.866	23.4858	0	0	0
2200	12	13	97.6066	7.4592	0	0	0
			Equilibrium				
1700	11.999	13	157.61	23.23	10^{-3}	0	0.51284
2200	11.931	12.986	97.135	7.0306	0.069	0.014	0.9432

It is clear that very little CO and no H_2 is formed at 1700 K (which is the peak temperature that gas turbine engines operate at today). However, some NO is formed, owing to the fact that the equilibrium constant for the NO formation reaction is two *orders of magnitude* higher than those of the CO_2 and H_2O decomposition reactions, even at 1700 K. As the temperature increases, understandably, the dissociation reactions also become quite active while the amount of NO that is formed keeps increasing. However, as the temperature of the products increase, the amount of dilution air and hence the amount of excess oxygen also decreases, thereby reducing the amount of NO formed. In summary, at 1700 K, more oxygen is available but the equilibrium constants are small; at 2200 K, the amount of oxygen available decreases but the equilibrium constants are higher.

14.5 Equilibrium flame temperature

It was mentioned in the previous chapter that the calculated values for the adiabatic flame temperature are unrealistically high. This is on account of the fact that decomposition reactions of product species such as CO_2 and H_2 are suppressed and also the fact that N_2 is not inert at such temperatures. If these artificial constraints are removed so that these additional reactions are allowed to take place then the products

equilibrium temperature is demonstrated below for a few examples.

■ EXAMPLE 14.6

Determine the equilibrium flame temperature for Example 13.12, assuming the products to an equilibrium mixture of H_2O, H_2, O_2 and N_2 (inert).

Solution : The number of moles of N_2 remains the same, since it is inert. Hence, there are three more species in the equilibrium product mixture composed of two atoms – H and O. Consequently, the number of kmoles of one species alone is unknown. Let there be χ kmoles of H_2O in the product mixture. The balanced chemical reaction for the combustion of H_2 with 150 percent theoretical air may now be written as (with products at equilibrium):

$$H_2+1.5\times\frac{1}{2}\,(O_2+3.76N_2) \rightarrow \chi\,H_2O+(1-\chi)\,H_2+\frac{1.5-\chi}{2}\,O_2+0.75\times3.76\,N_2$$

	Initial	Equilibrium	
	(kmol)	(kmol)	Mole fraction
H_2	1	$1-\chi$	$2(1-\chi)\div(9.14-\chi)$
O_2	0.75	$(1.5-\chi)\div2$	$(1.5-\chi)\div(9.14-\chi)$
H_2O	0	χ	$2\chi\div(9.14-\chi)$
N_2	2.82	2.82	$5.64\div(9.14-\chi)$
Total		$(9.14-\chi)\div2$	1

The mole fractions of the species in the equilibrium mixture may be evaluated as shown in the table. Assuming that the final composition is determined by the reaction

$$H_2O \rightleftharpoons \frac{1}{2}\,H_2 + \frac{1}{2}\,O_2$$

we can retrieve $\log_{10}K$ for this reaction from Table I for a given temperature. With $P = 1$ atm $=P_{\text{ref}}$, Eqn. 14.9 becomes

$$K = \frac{y_{H_2}\,y_{O_2}^{1/2}}{y_{H_2O}} = \frac{1-\chi}{\chi}\sqrt{\frac{1.5-\chi}{9.14-\chi}}$$

SFEE (Eqn. 13.2) applied to the combustor gives

$$0 = \frac{\overset{0}{\cancel{\dot{Q}}}}{\dot{n}_F}+\overset{0}{\bar{h}^{\circ}_{f,H_2}} +\Delta\bar{h}^{\overset{0}{\cancel{}}}_{H_2} +1.25\times\left[(\bar{h}^{\overset{0}{\cancel{}}}_{f,O_2} +\Delta\bar{h}^{\overset{0}{\cancel{}}}_{O_2}) +3.76\,(\bar{h}^{\overset{0}{\cancel{}}}_{f,N_2} +\Delta\bar{h}^{\overset{0}{\cancel{}}}_{N_2})\right]$$

$$+ \frac{1.5 - \chi}{2} \left(\bar{h}^{\circ}_{f,O_2} + \Delta \bar{h}_{O_2} \right) + 2.82 \left(\bar{h}^{\circ}_{f,N_2} + \Delta \bar{h}_{N_2} \right) \Bigg]$$

$$0 = \chi \left(-241830 + \Delta \bar{h}_{H_2O} \right) + (1 - \chi) \Delta \bar{h}_{H_2} + \frac{1.5 - \chi}{2} \Delta \bar{h}_{O_2} + 2.82 \Delta \bar{h}_{N_2}$$

We now have two equations for the two unknowns, namely, χ and T. The solution has to be obtained iteratively. An excellent starting guess for T is 1900 K and $\chi = 0.9$ (less than the corresponding values in Example 13.12). We get the equilibrium flame temperature to be 1973 K and $\chi = 0.99902$. These values are almost identical to the corresponding values in Example 13.12 on account of the fact that the equilibrium constant in this range of temperature is quite small.

▉ EXAMPLE 14.7

Determine the equilibrium flame temperature for Example 13.12, assuming the products to an equilibrium mixture of H_2O, H_2, O_2, N_2 and NO. Enthalpy values for NO may be taken from http://webbook.nist.gov.

Solution : In this example, there are *five* species in the equilibrium product mixture composed of *three* atoms – H, O and N. Consequently, the number of kmoles of two species are unknown. Let there be χ kmoles of H_2O and μ kmoles of N_2 in the product mixture. The balanced chemical reaction for the combustion of H_2 with 150 percent theoretical air may now be written as (with products at equilibrium):

$$H_2 + 1.5 \times \frac{1}{2}(O_2 + 3.76 N_2) \rightarrow \chi H_2O + (1 - \chi) H_2 + \mu N_2$$

$$+ 2(2.82 - \mu) NO + \frac{2\mu - \chi - 4.14}{2} O_2$$

The mole fractions of the species in the equilibrium mixture may be evaluated as shown in the table.

	Initial	Equilibrium	
	(kmol)	(kmol)	Mole fraction
H_2	1	$1 - \chi$	$2(1 - \chi) \div (9.14 - \chi)$
O_2	0.75	$(2\mu - \chi - 4.14) \div 2$	$(2\mu - \chi - 4.14) \div (9.14 - \chi)$
H_2O	0	χ	$2\chi \div (9.14 - \chi)$
N_2	2.82	μ	$2\mu \div (9.14 - \chi)$
NO	0	$2(2.82 - \mu)$	$4(2.82 - \mu) \div (9.14 - \chi)$
Total		$(9.14 - \chi) \div 2$	1

$$H_2O \rightleftharpoons H_2 + \frac{1}{2}O_2$$

$$\frac{1}{2}O_2 + \frac{1}{2}N_2 \rightleftharpoons NO$$

The equilibrium constant for these two reactions, denoted respectively as K_1 and K_2, may be retrieved from Table I for a given temperature. Since the reactions are independent, Eqn. 14.9 may be applied to these reactions independently and this leads to (with $P = 1$ atm)

$$K_1 = \frac{y_{H_2}\, y_{O_2}^{1/2}}{y_{H_2O}} = \frac{\chi}{1-\chi}\sqrt{\frac{2\mu - \chi - 4.14}{9.14 - \chi}}$$

$$K_2 = \frac{y_{NO}}{y_{O_2}^{1/2}\, y_{N_2}^{1/2}} = \frac{4(2.82 - \mu)}{\sqrt{2\mu(2\mu - \chi - 4.14)}}$$

SFEE (Eqn. 13.2) applied to the combustor gives, after simplification,

$$0 = \chi\,(-241830 + \Delta\bar{h}_{H_2O}) + (1 - \chi)\,\Delta\bar{h}_{H_2} + \frac{2\mu - \chi - 4.14}{2}\,\Delta\bar{h}_{O_2}$$
$$+\mu\,\Delta\bar{h}_{N_2} + 2(2.82 - \mu)\,(\bar{h}^\circ_{f,NO} + \Delta\bar{h}_{NO})$$

We now have three equations for the three unknowns, namely, χ, μ and T. Proceeding as before, we can get the equilibrium flame temperature to be 1964 K, $\chi = 0.9991$ and $\mu = 2.8125^\dagger$. The equilibrium product composition is H_2O – 0.9991 kmoles, $H_2 - 9\times10^{-4}$ kmoles, N_2 – 2.8125 kmoles, NO – 0.015 kmoles and O_2 – 0.24295 kmoles. Note that the equilibrium flame temperature is less than the adiabatic flame temperature calculated in Example 13.12 on account of the dissociation of H_2O and the formation of NO.

---

†The solution has been obtained using Cantera, an open source software for solving problems in Thermodynamics, Chemical Kinetics and Transport (https://cantera.org/). The relevant code fragment and the associated properties file are given in Appendix B.

COMPRESSIBLE FLOW IN NOZZLES

Compressible flows are encountered in many applications in Aerospace and Mechanical engineering. Some examples are flows in nozzles, compressors, turbines and diffusers. In aerospace engineering, in addition to these examples, compressible flows are seen in external aerodynamics, aircraft and rocket engines. In almost all of these applications, air (or some other gas or mixture of gases) is the working fluid. However, steam is the working substance in steam turbines and steam nozzles. Furthermore, gas dynamics of refrigerants is important in capillary tubes that are used for throttling in vapor compression refrigeration systems. Thus, the range of engineering applications in which compressible flow occurs is quite large and hence a clear understanding of the dynamics of compressible flow is essential for engineers. It is important to note that air may be treated as calorically perfect in most applications whereas steam and refrigerants may not be.

$$\tau = -\frac{1}{v}\frac{\partial v}{\partial P} \tag{15.1}$$

where v is the specific volume and P is the pressure. The change in specific volume corresponding to a given change in pressure, will, of course, depend upon the compression process. That is, for a given change in pressure, the change in specific volume will be different between an isothermal and an adiabatic compression process.

The definition of compressibility actually comes from thermodynamics. Since the specific volume $v = v(T, P)$, we can write

$$dv = \underbrace{\left(\frac{\partial v}{\partial P}\right)_T dP}_{compressibility} + \underbrace{\left(\frac{\partial v}{\partial T}\right)_P dT}_{volume\ expansion}$$

From the first term, we can define the isothermal compressibility as $-\frac{1}{v}\left(\frac{\partial v}{\partial P}\right)_T$ and, from the second term, we can define the coefficient of volume expansion as $\frac{1}{v}\left(\frac{\partial v}{\partial T}\right)_P$. The second term represents the change in specific volume (or equivalently density) due to a change in temperature. For example, when a gas is heated at constant pressure, the density decreases and the specific volume increases. This change can be large, as is the case in most combustion equipment, without necessarily having any implications on the compressibility of the fluid. It thus follows that compressibility effect is important only when the change in specific volume (or equivalently density) is due largely to a change in pressure.

If the above equation is written in terms of the density ρ, we get

$$\tau = -\frac{1}{\rho}\frac{\partial \rho}{\partial P} \tag{15.2}$$

The isothermal compressibility of water and air under standard atmospheric conditions are $5 \times 10^{-10} m^2/N$ and $10^{-5} m^2/N$. Thus, water (in liquid phase) can be treated as an incompressible fluid in all applications. On the contrary, it would seem that, air, with a compressibility that is five orders of magnitude higher, has to be treated as a compressible fluid in all applications. Fortunately, this is not true when flow is involved.

15.2 Compressible and Incompressible Flows

It is well known from high school physics that sound (pressure waves) propagates in any medium with a speed which depends on the bulk compressibility. The less

automobile travelling at 120 kph (about 33 m/s). This speed is 1/10th of the speed of sound. In other words, compared with 120 kph, sound waves travel 10 times faster. Since the speed of sound *appears* to be high compared with the highest velocity in the flow field, the medium behaves as though it were incompressible. As the flow velocity becomes comparable to the speed of sound, compressibility effects become more prominent. In reality, the speed of sound itself can vary from one point to another in the flow field and so the velocity at each point has to be compared with the speed of sound at that point. This ratio is called the Mach number, after Ernst Mach who made pioneering contributions in the study of the propagation of sound waves. Thus, the Mach number at a point in the flow can be written as

$$M = \frac{V}{a} \tag{15.3}$$

where V is the velocity magnitude at any point and a is the speed of sound at that point.

We can come up with a quantitative criterion to give us an idea about the importance of compressibility effects in the flow by using simple scaling arguments as follows. From Bernoulli's equation for steady flow, it follows that $\Delta P \sim \rho V_{ref}^2$, where V_{ref} is a characteristic speed. It will be shown in the next chapter that the speed of sound $a = \sqrt{\Delta P / \Delta \rho}$, wherein ΔP and $\Delta \rho$ correspond to an isentropic process. Thus,

$$\frac{\Delta \rho}{\rho} = \frac{1}{\rho} \frac{\Delta \rho}{\Delta P} \Delta P = \frac{V_{ref}^2}{a^2} = M^2 \tag{15.4}$$

On the other hand, upon rewriting Eqn. 15.2 for an isentropic process, we get

$$\frac{\Delta \rho}{\rho} = \tau_{isentropic} \Delta P$$

Comparison of these two equations shows clearly that, in the presence of a flow, density changes are proportional to the square of the Mach number[†]. It is customary to assume that the flow is essentially incompressible if the change in density is less than 10% of the mean value[‡]. It thus follows that compressibility effects are significant only when the Mach number exceeds 0.3.

[†] This is true for steady flows only. For unsteady flows, density changes are proportional to the Mach number.

[‡] Provided the change is predominantly due to a change in pressure.

15.3.1 Governing Equations

The governing equations for frictionless, adiabatic, steady, one dimensional compressible flow may be written in differential form as

$$d(\rho V) = 0 \tag{15.5}$$

$$dP + \rho V \, dV = 0 \tag{15.6}$$

and

$$dh + d\left(\frac{V^2}{2}\right) = 0 \tag{15.7}$$

These equations express mass, momentum and energy conservation respectively. In addition, changes in flow properties must also obey the second law of thermodynamics. Thus,

$$ds \geq \left(\frac{\delta q}{T}\right)_{rev} \tag{15.8}$$

where s is the entropy per unit mass and q is the heat interaction, also expressed on a per unit mass basis. The subscript refers to a reversible process. Equations 15.5, 15.6, 15.7 and 15.8 are applicable for both calorically perfect as well as real gases. For a calorically perfect gas, we have, from section 9.3,

$$ds = C_v \frac{dT}{T} + R\frac{dv}{v} = C_v \frac{dP}{P} + C_p \frac{dv}{v} = C_p \frac{dT}{T} - R\frac{dP}{P} \tag{15.9}$$

Note that Eqn. 15.6 is written in the so-called non-conservative form. By using Eqn. 15.5, we can rewrite Eqn. 15.6 in conservative form as follows.

$$dP + d\left(\rho V^2\right) = 0 \tag{15.10}$$

Equations 15.5,15.10, 15.7 and 15.8 can be integrated between any two points in the flow field to give

$$\rho_1 V_1 = \rho_2 V_2 \tag{15.11}$$

$$P_1 + \rho_1 V_1^2 = P_2 + \rho_2 V_2^2 \tag{15.12}$$

$$h_1 + \frac{V_1^2}{2} = h_2 + \frac{V_2^2}{2} \tag{15.13}$$

and

$$s_2 - s_1 = \int_1^2 \frac{\delta q}{T} + \sigma_{irr} \tag{15.14}$$

$$s_2 - s_1 \;=\; C_v \ln \frac{T_2}{T_1} + R \ln \frac{v_2}{v_1}$$

$$=\; C_v \ln \frac{P_2}{P_1} + C_p \ln \frac{v_2}{v_1} \qquad (15.15)$$

$$=\; C_p \ln \frac{T_2}{T_1} - R \ln \frac{P_2}{P_1}$$

The flow area does not appear in any of the above equations as they stand. When we discuss one dimensional flow in ducts and passages, this can be introduced quite easily. Also, it is important to keep in mind that, when points 1 and 2 are located across a wave (say, a sound wave or shock wave), the derivatives of the flow properties will be discontinuous.

15.3.2 Acoustic Wave Propagation Speed

Equations 15.11, 15.12, 15.13 and 15.15 admit different solutions, which we will see in the subsequent chapters. The most basic solution is the expression for the speed of sound, which we will derive in this section.

Consider an acoustic wave propagating into quiescent air as shown in Fig. 15.1. Although the wave front is spherical, at any point on the wave front, the flow is essentially one dimensional as the radius of curvature of the wave front is large when compared with the distance across which flow properties change. If we switch to a reference frame in which the wave appears stationary, then the flow approaches the wave with a velocity equal to the wave speed in the stationary frame of reference and moves away from the wave with a slightly different velocity. As a result of going through the acoustic wave, the flow properties change by an infinitesimal amount and the process is isentropic. Thus, we can take $V_2 = V_1 + dV_1$, $P_2 = P_1 + dP_1$, and $\rho_2 = \rho_1 + d\rho_1$. Substitution of these into Eqns. 15.11 and 15.12 gives

$$\rho_1 V_1 = (\rho_1 + d\rho_1)(V_1 + dV_1)$$

and

$$P_1 + \rho_1 V_1^2 = P_1 + dP_1 + (\rho_1 + d\rho_1)(V_1 + du_1)^2$$

If we neglect the product of differential terms, then we can write

$$\rho_1 dV_1 + V_1 d\rho_1 = 0$$

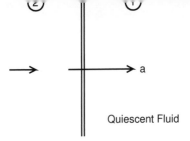

Quiescent Fluid

Observer Moving With Wave

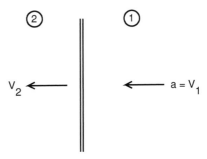

Figure 15.1: Propagation of a Sound Wave into a Quiescent Fluid

and

$$dP_1 + 2\rho_1 V_1 dV_1 + V_1^2 d\rho_1 = 0$$

Upon combining these two equations, we get

$$\frac{dP_1}{d\rho_1} = V_1^2$$

As mentioned earlier, V_1 is equal to the speed of sound a and so

where the subscript 1 has been dropped for convenience. Furthermore, we have also explicitly indicated that the process is isentropic. Since the process is isentropic, $ds = 0$, and so from Eqn. 15.9 for a calorically perfect gas,

$$C_v \frac{dP}{P} + C_p \frac{dv}{v} = 0$$

Since $\rho = 1/v$, $dv/v = -d\rho/\rho$ and so

$$\frac{dP}{d\rho} = \frac{C_p}{C_v} \frac{P}{\rho} = \gamma RT$$

Thus,

$$a = \sqrt{\gamma RT} \tag{15.17}$$

for a calorically perfect gas.

15.3.3 Mach Number

The Mach number has already been defined in Eqn. 15.3 and we are now in a position to take a closer look at it. Since it is defined as a ratio, changes in the Mach number are the outcome of either changes in velocity, speed of sound or both. Speed of sound itself varies from point to point and is proportional to the square root of the temperature as seen from Eqn. 15.17. Thus, any deductions of the velocity or temperature variation from a given variation of Mach number cannot be made in a straightforward manner. For example, the velocity at the entry to the combustor in an aircraft gas turbine engine may be as high as 200 m/s, but the Mach number is usually 0.3 or less due to the high static temperature of the fluid.

15.3.4 Reference States

In the study of compressible flows and indeed in fluid mechanics, it is conventional to define certain reference states. These allow the governing equations to be simplified and written in dimensionless form so that the important parameters can be identified. In the context of compressible flows, the solution procedure can also be made simpler and in addition, the important physics in the flow can be brought out clearly by the use of these reference states. Two such reference states are discussed next.

15.3.5 Sonic State

Since the speed of sound plays a crucial role in compressible flows, it is convenient to use the sonic state as a reference state. The sonic state is the state of the fluid at that

only at one or a few points in the flow field. For example, in the case of choked isentropic flow through a nozzle, the sonic state is achieved in the throat section. In some other cases, such as flow with heat addition or flow with friction, the sonic state may not even be attained anywhere in the actual flow field, but is still defined in a hypothetical sense and is useful for analysis. The importance of the sonic state lies in the fact that it separates subsonic ($M < 1$) and supersonic ($M > 1$) regions of the flow. Since information travels in a compressible medium through acoustic waves, the sonic state separates regions of flow that are fully accessible (subsonic) and those that are not (supersonic).

Note that the dimensionless velocity V/V^* at a point is **not** equal to the Mach number at that point since V^* is not the speed of sound at that point[‡].

15.3.6 Stagnation State

Let us consider a point in a one dimensional flow and assume that the state at this point is completely known. This means that the pressure, temperature and velocity at this point are known. We now carry out a thought experiment in which an isentropic, deceleration process takes the fluid from the present state to one with zero velocity. The resulting end state is called the stagnation state corresponding to the known initial state. Thus, the stagnation state at a point in the flow field is defined as the thermodynamic state that would be reached from the given state at that point, at the end of an isentropic, deceleration process to zero velocity. Note that the stagnation state is a local state contrary to the sonic state. Hence, the stagnation state can change from one point to the next in the flow field. Also, it is important to note that the stagnation process alone is isentropic, and the flow need not be isentropic. Properties at the stagnation state are usually indicated with a subscript 0 *viz.*, P_0, T_0, ρ_0 and so on. Here P_0 is the stagnation pressure, T_0 is the stagnation temperature and ρ_0 is the stagnation density. Hereafter, P and T will be referred to as the static pressure and static temperature and the corresponding state point will be called the static state.

To derive the relationship between the static and stagnation states, we start by integrating Eqn. 15.7 between these two states. This gives,

$$\int_1^0 dh + \int_1^0 d\left(\frac{V^2}{2}\right) = 0$$

If we integrate this equation and rearrange, we get

[‡]Except, of course, at the point where the sonic state occurs

after noting that the velocity is zero at the stagnation state. For a calorically perfect gas, $dh = C_p dT$ and so

$$T_{0,1} - T_1 = \frac{V_1^2}{2C_p}$$

This may be finally written as

$$\frac{T_0}{T} = 1 + \frac{\gamma - 1}{2} M^2 \qquad (15.19)$$

where the subscript for the static state has been dropped for convenience. Although the stagnation process is isentropic, this fact is not required for the calculation of stagnation temperature.

Since the stagnation process is isentropic, the static and stagnation states lie on the same isentrope. If we apply Eqn. 15.15 between the static and stagnation states and use the fact that $s_0 = s_1$, we get

$$\frac{P_{0,1}}{P_1} = \left(\frac{T_{0,1}}{T_1}\right)^{\frac{\gamma}{\gamma - 1}}$$

If we substitute from Eqn. 15.19, we get

$$\frac{P_0}{P} = \left(1 + \frac{\gamma - 1}{2} M^2\right)^{\frac{\gamma}{\gamma - 1}} \qquad (15.20)$$

where, the subscript denoting the static state has been dropped. This equation can be derived in an alternative way, in a manner similar to the one used for the derivation of the stagnation temperature. This is somewhat longer but gives some interesting insights into the stagnation process. We start by rewriting Eqn. 15.6 in the following form

$$\frac{dP}{\rho} + d\left(\frac{V^2}{2}\right) = 0$$

By substituting Eqn. 15.7, this can be simplified to read

$$\frac{dP}{\rho} - dh = 0$$

Integrating this between the static and stagnation states leads to

$$\int_1^0 \frac{dP}{\rho} - \int_1^0 dh = 0$$

isentropic, from Eqn. 15.9 we can show that

$$C_v \frac{dP}{P} + C_p \frac{dv}{v} = 0 \Rightarrow Pv^\gamma = constant = P_1 v_1^\gamma$$

Thus, the above equation reduces to

$$\int_1^0 \frac{P_1^{1/\gamma}}{\rho_1} \frac{dP}{P^{1/\gamma}} = C_p(T_{0,1} - T_1)$$

where we have invoked the calorically perfect gas assumption. With a little bit of algebra, this can be easily shown to lead to Eqn. 15.20.

The stagnation density can be evaluated by using the ideal equation of state $P_0 = \rho_0 R T_0$. Thus

$$\frac{\rho_0}{\rho} = \left(1 + \frac{\gamma - 1}{2} M^2 \right)^{\frac{1}{\gamma - 1}} \tag{15.21}$$

This derivation brings out the fact that unlike the stagnation temperature, the nature of the stagnation process has to be known in order to evaluate the stagnation pressure. This, in itself, arises from the fact that Eqn. 15.6 is not a perfect differential. It would appear that we could have circumvented this difficulty by integrating Eqn. 15.10 instead, which is a perfect differential. This would have led to the following expression

$$P_{0,1} = P_1 + \rho_1 V_1^2$$

If we divide through by P_1 and use the fact that $P_1 = \rho_1 RT$ and $a_1 = \sqrt{\gamma RT_1}$, we get

$$\frac{P_0}{P} = 1 + \gamma M^2$$

This expression for stagnation pressure is disconcertingly (and erroneously!) quite different from Eqn. 15.20. The inconsistency arises due to the use of the continuity equation while deriving Eqn. 15.10. Continuity equation 15.5 is not applicable during the stagnation process, as otherwise $\rho_0 \to \infty$ as $u \to 0$. Hence, Eqn. 15.10 is not applicable for the stagnation process.

Another important fact about stagnation quantities is that they depend on the frame of reference unlike static quantities which are frame independent. This is best illustrated through a numerical example.

moving frames of reference.

Solution : In the stationary frame of reference, $V_1 = 0$ and so, $T_{0,1} = T_1 = 300$ K and $P_{0,1} = P_1 = 100$ kPa.

In the moving frame of reference, $V_1 = a_1$ and so $M_1 = 1$. Substituting this into Eqns. 15.19 and 15.20, we get $T_{0,1} = 360$ K and $P_{0,1} = 189$ kPa.

The difference between the values evaluated in different frames becomes more pronounced at higher Mach numbers.

As already mentioned, stagnation temperature and pressure are local quantities and so they can change from one point to another in the flow field. Changes in stagnation temperature can be achieved by the addition or removal of heat or work[†]. Heat addition increases the stagnation temperature, while removal of heat results in a decrease in stagnation temperature. Changes in stagnation pressure are brought about by work interaction or irreversibilities. Across a compressor where work is done on the flow, stagnation pressure increases while across a turbine where work is extracted from the fluid, stagnation pressure decreases. It is for this reason, that any loss of stagnation pressure in the flow is undesirable as it is tantamount to a loss of work. To see the effect of irreversibilities, we start with the last equality in Eqn. 15.15 and substitute for T_2/T_1 and P_2/P_1 as follows:

$$\frac{T_2}{T_1} = \frac{T_2}{T_{0,2}} \frac{T_{0,2}}{T_{0,1}} \frac{T_{0,1}}{T_1}$$

and

$$\frac{P_2}{P_1} = \frac{P_2}{P_{0,2}} \frac{P_{0,2}}{P_{0,1}} \frac{P_{0,1}}{P_1}$$

[†] In such cases, the energy equation has to be modified suitably. For example Eqn. 15.7 will read as

$$dh + d\left(\frac{u^2}{2}\right) = \delta q - \delta w$$

and Eqn. 15.13 will read as

$$h_1 + \frac{V_1^2}{2} = h_2 + \frac{V_2^2}{2} - Q + W$$

where q (and Q) and w (and W) refer to the heat and work interaction per unit mass. We have also used the customary sign convention from thermodynamics *i.e.*, heat added to a system is positive and work done by a system is positive.

If we use Eqns. 15.19 and 15.20, we get

$$s_2 - s_1 = C_p \ln \frac{T_{0,2}}{T_{0,1}} - R \ln \frac{P_{0,2}}{P_{0,1}} \tag{15.22}$$

This equation shows that irreversibilities in an adiabatic flow lead to a loss of stagnation pressure, since, for such a flow, $s_2 > s_1$ and $T_{0,2} = T_{0,1}$ and so $P_{0,2} < P_{0,1}$.

15.3.7 T-s and P-v Diagrams in Compressible Flows

T-s and P-v diagrams are familiar to most of the readers from their basic thermodynamics course. These diagrams are extremely useful in illustrating states and processes graphically. Both of these diagrams display the same information, since the thermodynamic state is fully fixed by the specification of two properties, either P, v or T, s. Nevertheless, they are both useful as some processes can be depicted better in one than the other.

Let us review some basic concepts from thermodynamics in relation to T-s and P-v diagrams. Figure 15.2 shows thermodynamic states (filled circles) and contours of P, v (isobars and isochors) and contours of T, s (isotherms and isentropes) for a calorically perfect gas. From the first equality in Eqn. 15.9, we can write,

$$dv = \frac{v}{R} ds - C_v \frac{v}{RT} dT \tag{15.23}$$

From this equation, it is easy to see that, as we move along a $s = constant$ line in the direction of increasing temperature, v decreases, since, $dv = -(C_v/P)dT$, along such a line. Also, the change in v for a given change in T is higher at lower values of pressure than at higher values of pressure. This fact is of tremendous importance in compressible flows as we will see later.

Since $dv = 0$ along a $v = constant$ contour, from the above equation,

$$\frac{dT}{ds}\bigg|_v = \frac{T}{C_v} \tag{15.24}$$

This equation shows that the slope of the contours of v on a T-s diagram is always positive and is not a constant. Hence, the contours are not straight lines. Furthermore, the slope increases with increasing temperature and so the contours are shallow at

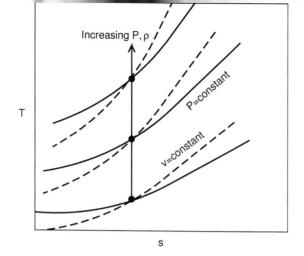

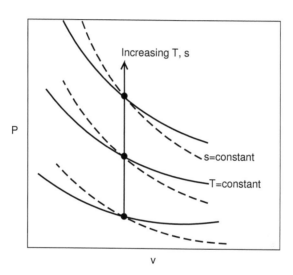

Figure 15.2: Constant pressure and constant volume lines on a T-s diagram; constant temperature and constant entropy lines on a P-v diagram for a calorically perfect gas.

$$\left.\frac{dT}{ds}\right|_P = \frac{T}{C_p} \tag{15.25}$$

for any isobar. The same observation made above regarding the slope of the contours of v on a T-s diagram are applicable to isobars as well. In addition, since $C_p > C_v$, at any state point, isochors are steeper than isobars on a T-s diagram and pressure increases along a $s = constant$ line in the direction of increasing temperature. These observations regarding isochors and isobars are shown in Fig. 15.2.

From the second equality in Eqn. 15.9, the equation for isentropes on a P-v diagram can be obtained after setting $ds = 0$. Thus

$$\left.\frac{dP}{dv}\right|_s = -\frac{C_p}{C_v}\frac{P}{v} \tag{15.26}$$

By equating the second and the last term in Eqn. 15.9, we get

$$C_v\frac{dT}{T} + R\frac{dv}{v} = C_p\frac{dT}{T} - R\frac{dP}{P}$$

This can be rearranged to give (after setting $dT = 0$)

$$\left.\frac{dP}{dv}\right|_T = -\frac{P}{v} \tag{15.27}$$

This equation shows that isotherms also have a negative slope on a P-v diagram and they are less steep than isentropes (Fig. 15.2). Furthermore, s and T increase with increasing pressure as we move along a $v = constant$ line.

Let us now look at using T-s and P-v diagrams for graphically illustrating states in 1D compressible flows. In this case, in addition to T, s (or P, v), velocity information also has to be displayed. Equation 15.7 tells us how this can be done. For a calorically perfect gas, this can be written as

$$d\left(T + \frac{V^2}{2C_p}\right) = 0$$

Hence, at each state point, the static temperature is depicted as usual, and the quantity $V^2/2C_p$ is added to the ordinate (in case of a T-s diagram). Note that this quantity has units of temperature and the sum $T + V^2/2C_p$ is equal to the stagnation temperature T_0 corresponding to this state. This is shown in Fig. 15.3 for the subsonic state point marked 1. Also shown in this figure is the sonic state

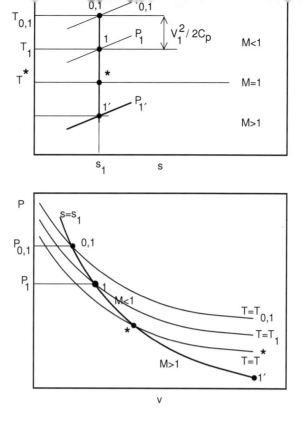

Figure 15.3: Illustration of states for a 1D compressible flow on T-s and P-v diagram

corresponding to this state. Once T_0 is known, T^* can be evaluated from Eqn. 15.19 by setting $M = 1$. Thus

$$\frac{T_0}{T^*} = \frac{\gamma + 1}{2}$$

Depicting the sonic state is useful since it tells at a glance whether the flow is subsonic or supersonic. All subsonic states will lie above the sonic state and all supersonic states will lie below. State point $1'$ shown in Fig. 15.3 is a supersonic state. This figure also shows that the stagnation process (1-0 or $1'$-0) is an isentropic process. All this information is shown in Fig. 15.3 on T-s as well as P-v diagram.

The discussion above is applicable to a calorically perfect gas. In the case of a real gas such as steam or a refrigerant, the quantity $V^2/2$ has the same unit as the static enthalpy h and so for such working substances, it is customary and better to use $h - s$ coordinates for depicting states and processes.

15.4 Normal Shock Waves

Normal shock waves are compression waves that are seen in nozzles, turbomachinery blade passages, supersonic intakes and shock tubes, to name a few. In the first three examples, normal shock usually occurs under off-design operating conditions or during start-up. The compression process across the shock wave is highly irreversible and so it is undesirable in such cases. In the last example, normal shock is designed to achieve extremely fast compression and heating of a gas with the aim of studying highly transient phenomena. Normal shocks are seen in external flows also. The term "normal" is used to denote the fact that the shock wave is normal (perpendicular) to the flow direction, before and after passage through the shock wave. This latter fact implies that there is no change in flow direction as a result of passing through the shock wave. In this chapter, we take a detailed look at the thermodynamic and flow aspects of normal shock waves in calorically perfect gases.

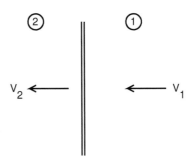

Figure 15.4: Illustration of a Normal shock Wave

15.4.1 Governing Equations

Figure 15.4 shows a stationary normal shock wave. This figure is almost identical to Fig. 15.1, where the propagation of an acoustic wave is shown. The main differences

The governing equations for the flow are Eqns. 15.11, 15.12, 15.13 and 15.15. It can be seen from the energy equation that, for a calorically perfect gas, the stagnation temperature is constant across the shock wave.

15.4.2 Mathematical Derivation of the Normal Shock Solution for a Calorically Perfect Gas

The continuity equation, Eqn. 15.11 can be written as

$$\frac{P_2}{P_1} = \sqrt{\frac{T_2}{T_1} \frac{M_1}{M_2}} \tag{15.28}$$

after using the fact that $V = M\sqrt{\gamma RT}$ and $\rho = P/RT$ from the ideal gas equation of state. Similarly, we can get from the momentum equation, Eqn. 15.12 ,

$$\frac{P_2}{P_1} = \frac{1 + \gamma M_1^2}{1 + \gamma M_2^2} \tag{15.29}$$

and

$$\frac{T_2}{T_1} = \left(1 + \frac{\gamma - 1}{2} M_1^2\right) / \left(1 + \frac{\gamma - 1}{2} M_2^2\right) \tag{15.30}$$

from the energy equation, Eqn. 15.13. Combining these three equations, we get

$$\frac{1 + \frac{\gamma - 1}{2} M_1^2}{1 + \frac{\gamma - 1}{2} M_2^2} = \frac{M_2^2}{M_1^2} \left(\frac{1 + \gamma M_1^2}{1 + \gamma M_2^2}\right)^2$$

This is a quadratic equation in M_2^2. Given M_1, we can solve this equation to get M_2. With M_2 known, all the other properties at state 2 can be evaluated. This equation has only one meaningful solution, namely,

$$M_2^2 = \frac{2 + (\gamma - 1)M_1^2}{2\gamma M_1^2 - (\gamma - 1)} \tag{15.31}$$

The other solutions are either trivial ($M_2 = M_1$) or imaginary. Note that, if we set $M_1 = 1$ in Eqn. 15.31 we get $M_2 = 1$, which is, of course, the solution corresponding to an acoustic wave. Also, a simple rearrangement of the expression in Eqn. 15.31 shows that

$$M_2^2 = 1 - \frac{\gamma + 1}{2\gamma} \frac{M_1^2 - 1}{M_1^2 - 1 + \frac{\gamma + 1}{2\gamma}} \tag{15.32}$$

across the shock wave. From Eqn. 15.15, the entropy change across the shock wave is given as

$$s_2 - s_1 = C_p \ln \frac{T_2}{T_1} - R \ln \frac{P_2}{P_1}$$

Upon substituting the relations obtained above for T_2/T_1 and P_2/P_1, we get

$$s_2 - s_1 = C_p \ln \frac{M_2^2}{M_1^2} \left(\frac{1 + \gamma M_1^2}{1 + \gamma M_2^2} \right)^2 - R \ln \frac{1 + \gamma M_1^2}{1 + \gamma M_2^2} .$$

This can be simplified to read

$$s_2 - s_1 = C_p \ln \frac{M_2^2}{M_1^2} + R \frac{\gamma + 1}{\gamma - 1} \ln \frac{1 + \gamma M_1^2}{1 + \gamma M_2^2} .$$

Substituting for M_2 from Eqn. 15.31, we get (after some tedious algebra!)

$$\frac{s_2 - s_1}{R} = \frac{1}{\gamma - 1} \ln \left[\frac{2\gamma M_1^2 - \gamma + 1}{\gamma + 1} \right] + \frac{\gamma}{\gamma - 1} \ln \left[\frac{2 + (\gamma - 1)M_1^2}{(\gamma + 1)M_1^2} \right]$$

With a slight rearrangement, this becomes

$$\begin{aligned}
\frac{s_2 - s_1}{R} &= \frac{1}{\gamma - 1} \ln \left[1 + \frac{2\gamma}{\gamma + 1}(M_1^2 - 1) \right] \\
&+ \frac{\gamma}{\gamma - 1} \ln \left[1 - \frac{2}{\gamma + 1} \left(1 - \frac{1}{M_1^2} \right) \right]
\end{aligned}$$

It is clear from this expression that entropy across the shock wave increases when $M_1 > 1$ and decreases when $M_1 < 1$. Thus, for a normal shock, M_1 is always greater than one and M_2 is always less than one.

The static pressure and temperature can both be seen to increase across the shock wave from Eqns. 15.29 and 15.30. However, the increase is such that

$$\frac{\rho_2}{\rho_1} = \left(\frac{P_2}{P_1} \right) / \left(\frac{T_2}{T_1} \right) > 1 \tag{15.33}$$

Of course, due to the irreversibility associated with the shock, there is a loss of stagnation pressure. From Eqn. 15.22, it is easy to show that

$$s_2 - s_1 = R \ln \frac{P_{0,1}}{P_{0,2}}$$

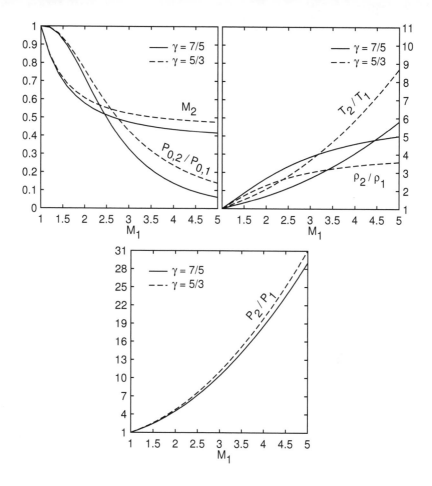

Figure 15.5: Variation of the downstream Mach number and property ratios across a normal shock wave for monatomic and diatomic gases

From Eqn. 15.32, we get

$$M_2^2 = 1 - \frac{6}{7} \frac{M_1^2 - 1}{M_1^2 - \frac{1}{7}}$$

[†] Strength of a shock is usually defined as $\frac{P_2}{P_1} - 1$.

for monatomic gases for which $\gamma = 5/3$. A comparison of these two expressions suggests that, for a given M_1, M_2 is higher for monatomic gases than diatomic gases. However, the strength of the shock as well as the temperature rise at a given M_1 is higher in the case of the former. This explains why monatomic gases are used extensively in shock tubes. Equations 15.31, 15.29, 15.30, 15.33 as well as the ratio $P_{0,2}/P_{0,1}$ are plotted in Fig. 15.5 for monatomic and diatomic gases.

In the limiting case when $M_1 = 1$, it is easy to see that the process is isentropic (as it should be, since it corresponds to the propagation of an acoustic wave). Also, $M_2 = 1$, $T_2/T_1 = 1$, $P_2/P_1 = 1$ and $\rho_2/\rho_1 = 1$ from Eqns. 15.32, 15.30, 15.29 and 15.33.

If we let $M_1 \to \infty$ in Eqn. 15.32, then we have

$$M_2 = \sqrt{\frac{\gamma - 1}{2\gamma}} \; ; \quad \frac{P_2}{P_1} \to \infty \; ; \quad \frac{T_2}{T_1} \to \infty \quad \text{and} \quad \frac{\rho_2}{\rho_1} = \frac{\gamma + 1}{\gamma - 1}$$

These trends can be clearly seen in Fig. 15.5.

15.4.3 Illustration of the Normal Shock Solution on T-s and P-v diagrams

In this section, we will try to draw some insight into the normal shock compression process through graphical illustrations on the T-s and P-v diagrams. Figure 15.6 shows the T-s and P-v diagram for the normal shock process. The static (P_1, T_1), stagnation $(P_{0,1}, T_0)$ and sonic state (*) corresponding to state 1 are shown in this figure. State point 2 lies to the right of state point 1 (owing to the increase in entropy across the shock) and above the sonic state (since the flow becomes subsonic after the shock). The corresponding stagnation state lies at the point of intersection of the isentrope (vertical line) through point 2 and the isotherm $T = T_0$. From the orientation of isobars in a T-s diagram (see Fig. 15.2), it is easy to see that the stagnation pressure corresponding to state point 2, $P_{0,2}$ is less than $P_{0,1}$. The normal shock itself is shown in this diagram as a heavy dashed line.

The same features are illustrated in a P-v diagram also in Fig. 15.6. Here, isotherms are shown as dashed lines and isentropes as solid lines. The stagnation state $(0, 1)$ lies at the point of intersection of the isentrope through state point 1 and the isotherm $T = T_0$. State point 2 lies to the left of and above point 1, since $v_2 < v_1$ and $P_2 > P_1$ and the increase in entropy causes it to lie on a higher isentrope. Since isentropes are steeper than isotherms, the isotherm $T = T_0$ intersects this isentrope at a lower

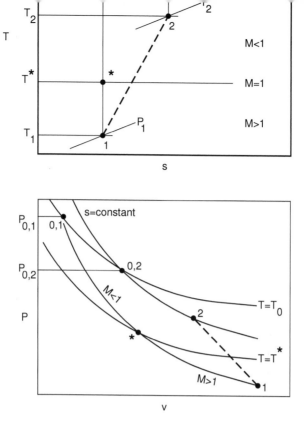

Figure 15.6: Illustration of Normal Shock in a T-s and P-v diagram

value of pressure and so $P_{0,2} < P_{0,1}$.

It can also be seen from this diagram that, for given values of v_2, v_1 and P_1, normal shock compression results in a higher value for P_2 than isentropic compression albeit with a loss of stagnation pressure. In other words, normal shock compression is more effective but less efficient than isentropic compression. The former attribute is of importance in intakes of supersonic vehicles, since it determines the length of the intake. However, the latter attribute is also important and so an optimal operating condition has to be determined. An inspection of Fig. 15.5 reveals that the loss of stagnation pressure is about 20% for $M_1 = 2$ and about 70% for $M_1 = 3$.

EXAMPLE 15.2

Air at 100 kPa and 300 K and moving at 696 m/s encounters a stationary normal shock. Determine the static and stagnation properties ahead of and behind the shock wave.

②   ①

T = 506 K   T = 300 K

P = 450 kPa   P = 100 kPa

←   ←
260.8 m/s   696 m/s

T_0 = 540 K T_0 = 540 K

P_0 = 564 kPa P_0 = 782.4 kPa

Solution : Since the shock wave is stationary,

$$P_1 = 100\,\text{kPa}, \quad T_1 = 300\,\text{K}, \quad V_1 = 696\,\text{m/s}$$
$$a_1 = \sqrt{\gamma R T_1} = \sqrt{1.4 \times 288 \times 300} = 348\,\text{m/s}$$
$$M_1 = 2, \quad T_{0,1} = 540\,\text{K}, \quad P_{0,1} = 782.4\,\text{kPa}$$

We can use Eqn. 15.31 to evaluate M_2 or use the gas tables. The latter choice allows us to look up pressure ratio, temperature ratio and other ratios, in one go. For $M_1 = 2$, from normal shock table, we get

$$\frac{P_2}{P_1} = 4.5 \Rightarrow P_2 = 450\,\text{kPa}$$
$$\frac{T_2}{T_1} = 1.687 \Rightarrow T_2 = 506\,\text{K}$$
$$\frac{P_{0,2}}{P_{0,1}} = 0.7209 \Rightarrow P_{0,2} = 564\,\text{kPa}$$
$$M_2 = 0.5774 \Rightarrow V_2 = 260.8\,\text{m/s}$$

These numbers are shown in the figure to illustrate them more clearly. Note that, stagnation temperature remains constant while stagnation pressure decreases.

static pressure downstream of the shock. Assuming an isentropic deceleration process from the same initial state, determine (a) the final static pressure for the same final speed, (b) the final speed for the same final static pressure and (c) the final static pressure and speed for the same final specific volume.

Solution : Given $P_1 = 100$ kPa, $T_1 = 295$ K and $V_1 = 766$ m/s, we can evaluate the following quantities:

$$M_1 = \frac{V_1}{\sqrt{\gamma R T_1}} = 2.22; \quad T_{0,1} = T_1 \left(1 + \frac{\gamma - 1}{2} M_1^2\right) = 586 \, K$$

From Table K, we can get $M_2 = 0.5444$ for $M_1 = 2.22$. It follows that $T_2 = 553$ K and $P_2 = 558.3$ kPa. Also,

$$V_2 = M_2 \sqrt{\gamma R T_2} = 257 \, \text{m/s}$$

(a) For adiabatic deceleration between the same speeds, the final static temperature remains the same *i.e.*, $T_{2s} = 553$ K, since the stagnation temperature remains the same. As the deceleration process is isentropic, the final pressure is given as

$$P_{2s} = P_1 \left(\frac{T_{2s}}{T_1}\right)^{\gamma/(\gamma-1)} = 901.9 \, \text{kPa}$$

(b) In this case, the final pressure at the end of the isentropic deceleration process $P_{2s} = 558.3$ kPa. Hence

$$T_{2s} = T_1 \left(\frac{P_{2s}}{P_1}\right)^{(\gamma-1)/\gamma} = 482 \, K$$

Since the stagnation temperature remains the same, we can get

$$V_{2s} = \sqrt{2C_p(T_{0,2s} - T_{2s})} = 458 \, \text{m/s}$$

(c) The specific volume after the normal shock compression process is given as

$$v_2 = \frac{R T_2}{P_2} = 0.2853 \, \text{m}^3/\text{kg}$$

Since it is given that the specific volume at the end of the isentropic deceleration process is the same, $v_{2s} = 0.2853$ m³/kg. Therefore,

$$P_{2s} = P_1 \left(\frac{v_1}{v_{2s}}\right)^{\gamma} = 460.8 \, \text{kPa}$$

$$V_{2s} = \sqrt{2C_p(T_{0,2s} - T_{2s})} = 512\,\mathrm{m/s}$$

15.5 Quasi One Dimensional Flows

In the previous chapters, 1D compressible flow solutions were presented, wherein the flow was either across a wave or through a constant area passage. A very important class of compressible flow is flow through a passage of finite but varying cross-sectional area such as flow through nozzles, diffusers and blade passages in turbomachines. The main difficulty that arises in this case is that the flow is no longer strictly one dimensional, since the variation in the cross-sectional area occurs in a direction normal to the main flow direction (see Fig. 15.7). This means that the velocity component in the normal direction is non-zero. However, it so happens that in most of the applications involving such flows, the magnitude of the normal component of velocity is *small* when compared with the axial component. Hence, as a first approximation, the former is usually neglected and only the axial component is considered. Thus, the flow is *approximately* one dimensional, or, as it is usually called, quasi one dimensional. In addition, the flow is assumed to be isentropic (except across normal shocks).

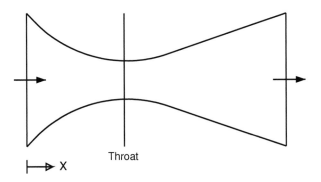

Figure 15.7: Flow Through a Varying Area Passage

15.5.1 Governing Equations

The equations governing the flow are almost the same as those for 1D flow and these are given below:

$$\left(P_1 + \rho_1 V_1^2\right) A_1 + \int_1^2 (P dA)_x = \left(P_2 + \rho_2 V_2^2\right) A_2 \qquad (15.35)$$

and Eqn. 15.13:

$$h_1 + \frac{V_1^2}{2} = h_2 + \frac{V_2^2}{2}$$

Since the flow is isentropic, entropy remains the same, $s_2 = s_1$. Note that the cross-sectional area appears in the continuity and momentum equation.

15.5.2 Area Velocity Relation

The general objective in quasi 1D flows is to determine the Mach number at any axial location, given the inlet conditions and the area at that location. Once the Mach number is known, all the other properties at that location can be determined from the inlet properties using the fact that the flow is isentropic. Before we do this, let us first explore the nature of the flow in detail. The continuity equation for this flow can be written in differential form as

$$d(\rho V A) = 0 \qquad (15.36)$$

which simply says that the mass flow rate at any section $\rho V A$ is a constant. Momentum and energy equation in differential form are the same as Eqns. 15.6 [†] and 15.7.

If we compute the derivative in Eqn. 15.36 using product rule and then divide by $\rho V A$, we get

$$\frac{d\rho}{\rho} + \frac{dV}{V} + \frac{dA}{A} = 0 \qquad (15.37)$$

We can write

$$\frac{d\rho}{\rho} = \frac{d\rho}{dP} \frac{dP}{\rho}$$

[†]For a differential fluid element of width dx, the forces on the fluid element in the x-direction are: PA (acting the positive x-direction), $(P + dP)(A + dA) = PA + AdP + PdA$ acting in the negative x-direction and $P\frac{dA}{\sin \theta} \sin \theta = PdA$ in the positive x-direction. Here θ is the inclination of the wall to the horizontal at the given axial location and the last term is the force exerted by the wall *on* the fluid element. Therefore, the net force on the fluid element in the x-direction is $PA - PA - AdP - PdA + PdA = -AdP$.

If we substitute this into Eqn. 15.37, we get

$$\frac{dA}{A} = (M^2 - 1)\frac{dV}{V} \qquad (15.39)$$

which is called the area-velocity relationship. The change in velocity for a given change in the area predicted by this equation for subsonic and supersonic flow is given in Table 15.1. It can be seen that a subsonic flow decelerates in a diverging

Table 15.1: Changes in velocity for a given change in area

	$A\uparrow$	$A\downarrow$
$M < 1$	$V\downarrow$	$V\uparrow$
$M > 1$	$V\uparrow$	$V\downarrow$

passage and accelerates in a converging passage. In contrast, a supersonic flow accelerates in a diverging passage and decelerates in a converging passage. This conclusion can be reached through a slightly different argument as follows.

It was mentioned earlier that the change in v for a given change in T (with $s = constant$) is higher at lower values of temperature. This qualitative statement is made more precise in Eqn. 15.38. Changes in density for a given change in velocity are higher when the flow is supersonic than when the flow is subsonic. Let us say that the velocity at a point in a subsonic flow increases by dV. From Eqn. 15.38, $d\rho$ is negative and since $M < 1$, $d\rho/\rho$ is less than dV/V. It follows from Eqn. 15.37 that dA has to be negative to make the left hand side zero. Hence, if a subsonic flow accelerates, it can do so only in a converging passage. Similarly, let us say that the velocity at a point in a supersonic flow increases by dV. From Eqn. 15.38, $d\rho$ is negative and since $M > 1$, $d\rho/\rho$ is greater than dV/V in magnitude. It follows from Eqn. 15.37 that dA has to be positive to make the left hand side zero, leading to the conclusion that a supersonic flow can accelerate only in a diverging passage. It is easy to demonstrate the remaining observations in Table 15.1 using similar arguments.

15.5.3 Geometric Choking

It is easy to see from Eqn. 15.39 that, as $M \to 1$ on the right hand side, then $dA \to 0$ on the left hand side for the velocity to remain finite. In fact, $dA = 0$ where $M = 1$. Expressed another way, in an isentropic flow in a passage of varying cross-section, the sonic state can be attained only in a location where $dA = 0$. This location can be

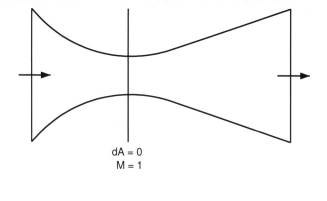

dA = 0
M = 1

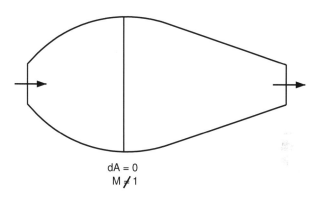

dA = 0
M ≠ 1

Figure 15.8: Illustration of Geometric Choking

Let us consider the geometry shown on the top in Fig. 15.8. If we assume the flow at the inlet to be subsonic, this will accelerate in the converging passage and can attain $M = 1$ at the throat. If, on the other hand, the flow is supersonic at the inlet, then this will decelerate in the converging passage and can possibly reach $M = 1$ at the throat. On the contrary, for the geometry shown in the bottom in Fig. 15.8, if we start with a subsonic flow at the inlet, it will decelerate in the diverging passage and so cannot attain $M = 1$ at the location where $dA = 0$. Similarly, if we start with a supersonic flow at the inlet, then since the supersonic flow accelerates in a diverging passage, it is not possible to reach $M = 1$ at the location where $dA = 0$. Thus, it is clear that an isentropic flow in a varying area passage can attain $M = 1$ only at a throat (minimum area of cross-section).

itself is finite. Of these two possibilities, the one realized in practice depends upon the prevailing operating condition. Hence, for the geometry on the top in Fig. 15.8, a flow, starting from a subsonic Mach number at the inlet can accelerate to a higher Mach number (but less than one) at the throat and decelerate afterwards. Similarly, the flow can start from a supersonic Mach number, decelerate to a value above 1 at the throat and accelerate again. The same argument is applicable to the geometry in the bottom in Fig. 15.8. Two things should thus become clear from the arguments given so far:

- Valid compressible flows are possible in both the geometries shown in Fig. 15.8. But, choking, if it occurs, can occur only in a geometric throat. This is a consequence of the fact that, in Eqn. 15.39, when $M \rightarrow 1$, $dA \rightarrow 0$ for the velocity to be finite.

- Flow need not always choke at a geometric throat. This follows from the fact that, in Eqn. 15.39, when $dA \rightarrow 0$ at the throat, $dV \rightarrow 0$, without any restriction on the value of M.

When the Mach number does become equal to 1 at the throat, the flow is said to be choked and this choking is a consequence of the area variation and is called geometric choking[†].

15.5.4 Area Mach number Relation for Choked Flow of a Calorically Perfect Gas

The state at any point in the nozzle can be conveniently related to the sonic state. We proceed to derive a relationship between the Mach number and cross-sectional area at any axial location to the area at the location where the sonic state occurs. We start by equating the mass flow rates at these two locations,

$$\dot{m} = \rho V A = \rho^* V^* A^*$$

We can write

$$\frac{A}{A^*} = \frac{\rho^*}{\rho} \frac{V^*}{V} = \frac{\rho^*}{\rho_0} \frac{\rho_0}{\rho} \frac{V^*}{V}$$

where ρ_0 is the stagnation density. We know from Eqn. 15.21 that

---

[†]Mathematically, it is possible to have $M = 1$ at a location where $dA/dx = 0$ and $d^2A/dx^2 > 0$. These conditions are satisfied at locations where the area reaches a local minimum. Thus, we can have multiple locations where these conditions are satisfied. However, the Mach number will be equal to one at one or more of these locations depending upon the actual flow conditions.

Setting $M = 1$ in this expression gives

$$\frac{\rho_0}{\rho^*} = \left(\frac{\gamma+1}{2}\right)^{\frac{1}{\gamma-1}}$$

Also, $V = Ma = M\sqrt{\gamma RT}$ and $V^* = a = \sqrt{\gamma RT^*}$. Substituting these expressions into the equation above for $\dot{m}$, we get

$$\frac{A}{A^*} = \left(1 + \frac{\gamma-1}{2}M^2\right)^{\frac{1}{\gamma-1}} \left(\frac{2}{\gamma+1}\right)^{\frac{1}{\gamma-1}} \frac{1}{M}\sqrt{\frac{T^*}{T}}$$

We can write T^*/T in terms of Mach number as follows.

$$\frac{T^*}{T} = \frac{T^*}{T_0}\frac{T_0}{T} = \frac{2}{\gamma+1}\left(1 + \frac{\gamma-1}{2}M^2\right)$$

If we substitute this into the equation above and simplify, we can finally write

$$\left(\frac{A}{A^*}\right)^2 = \frac{1}{M^2}\left[\frac{2}{\gamma+1}\left(1 + \frac{\gamma-1}{2}M^2\right)\right]^{\frac{\gamma+1}{\gamma-1}} \qquad (15.40)$$

This relationship is called the Area Mach number relationship for an isentropic flow in a varying area passage. Given A, the area of cross-section at a location and A^*, we can determine the Mach number at that location using this relation. Actually, this equation yields two solutions for a given A/A^*, one subsonic and the other supersonic, and the appropriate solution has to be chosen based on other details of the flow field. Also, it should be kept in mind that A^* is equal to the throat area, only when the flow is choked.

15.5.5 Mass Flow Rate for Choked Flow of a Calorically Perfect Gas

The mass flow rate at any section is given as

$$\dot{m} = \rho V A = \frac{\rho}{\rho_0}\rho_0 M\sqrt{\gamma R\frac{T}{T_0}T_0}\frac{A}{A^*}A^*$$

where we have used the same technique as in the previous section to rewrite the right hand side. This can be rearranged to give

$$\dot{m} = \frac{P_0}{RT_0}\sqrt{\gamma RT_0}\,A_{throat}\frac{\rho}{\rho_0}M\sqrt{\frac{T}{T_0}}\frac{A}{A^*}$$

$$\dot{m} = \frac{P_0 A_{throat}}{\sqrt{T_0}} \sqrt{\frac{\gamma}{R}\left(\frac{2}{\gamma+1}\right)^{(\gamma+1)/(\gamma-1)}} \qquad (15.41)$$

This equation is of tremendous importance in the design of intakes, nozzles and wind tunnels. The most striking feature of this expression is that it does not involve any downstream quantity. The quantity under the big square root depends only upon the nature of the gas, such as whether it is monatomic or diatomic and the molecular weight. For a given working substance such as air, Eqn. 15.41 shows that, once the flow is choked, the mass flow rate that can be realized through the passage is dependent only on the upstream stagnation pressure, temperature and the throat area. This means that the mass flow rate cannot be controlled anymore from downstream, *i.e.,* by adjusting the exit conditions (provided the exit pressure is not increased so much that the nozzle unchokes). In other words, this is the maximum mass flow rate that can be achieved by adjusting the back pressure. It is clear from this expression that the mass flow rate can be changed at will, by an adjustment of the upstream stagnation conditions or the throat area. These are active control measures which can be utilized in practical devices when they operate under off-design conditions. Although Eqn. 15.41 has been derived for a calorically perfect gas, the inferences drawn from it are equally applicable when the working substance is steam or a refrigerant, which are not calorically perfect.

15.5.6 Flow Through A Convergent Nozzle

Flow through a convergent nozzle can be established in one of two ways:

- By *pulling* the flow - lowering the back pressure or the pressure of the ambient environment into which the nozzle exhausts, while maintaining the inlet stagnation conditions

- By *pushing* the flow - increasing the inlet stagnation pressure while maintaining the back pressure

The first scenario is illustrated using T-s coordinates in Fig. 15.9. Initially, when the back pressure is less than P^* corresponding to the given inlet stagnation pressure P_0, the flow accelerates in the nozzle but the exit Mach number is less than one (Figs. 15.9(a)). The exit pressure of the fluid as it leaves the nozzle is the same as the ambient (back) pressure. It should be recalled that, for a calorically perfect gas,

$$\frac{P_0}{P^*} = \left(\frac{\gamma+1}{2}\right)^{\frac{\gamma}{\gamma-1}} = 1.8929$$

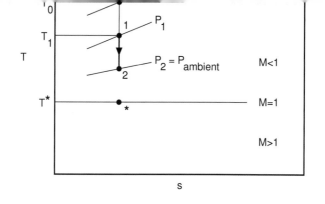

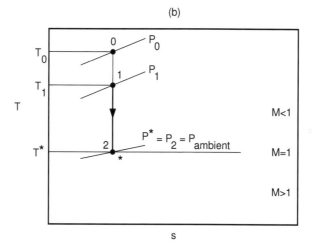

Figure 15.9: Illustration of flow through a convergent nozzle with inlet stagnation conditions fixed and varying back (ambient) pressure : T-s diagram

and

$$\frac{T_0}{T^*} = \frac{\gamma+1}{2} = 1.2$$

where we have set $\gamma = 1.4$. When the back pressure is lowered to a value equal to P^*, the flow accelerates and reaches a Mach number of one at the exit and the flow becomes choked. In this case also, the exit pressure of the fluid is the same as the ambient (back) pressure (Figs. 15.9(b)). Consequently, the diameter of the jet that issues out of the nozzle is exactly equal to the nozzle exit diameter. The mass flow

Physically, this is because the fluid is already travelling at the speed of sound at the exit and the changed back pressure condition is propagating *upstream* also at the speed of sound and so the flow becomes "aware" of the new back pressure value only after it reaches the exit. The static pressure of the fluid at the exit is still P^* but no longer equal to $P_{ambient}$. Since $P^* > P_{ambient}$ now, the fluid is "under-expanded" and it expands further outside the nozzle and equilibrates with the ambient conditions a few nozzle diameters downstream of the exit. The expansion is accomplished across an expansion fan (discussed in Chapter 8) centered at the nozzle lip. The jet swells initially as it comes out of the nozzle and expands, but shrinks afterwards due to entrainment of the ambient air and equalization of static pressure.

The second scenario, *i.e.*, pushing the flow is illustrated in Fig. 15.10 using T-s coordinates. When the stagnation pressure is not high enough, that is, $P_0/P_{ambient} < 1.8929$, then the flow is not choked at the exit. This is shown in Fig. 15.10(a). As the stagnation pressure is increased (keeping the stagnation temperature constant), the exit Mach number and the mass flow rate both increase. The exit state point slides down along the $P = P_{ambient}$ isobar. When $P_0/P_{ambient} = 1.8929$, the flow becomes choked (Fig. 15.10(b)). Contrary to what happened in the previous scenario, if the stagnation pressure is increased further, then the mass flow rate also increases. However, the exit Mach number remains at 1. The exit static pressure is not equal to $P_{ambient}$ any more but is equal to $P_0/1.8929$. Hence, the fluid is under-expanded and expands further outside the nozzle.

■ EXAMPLE 15.4

Air flows in a frictionless, adiabatic duct at $M = 0.6$ and $P_0 = 200$ kPa. The cross-sectional area of the duct is 6.5 cm^2 and the mass flow rate is 0.3 kg/s. A convergent nozzle is now attached at the exit of the duct. What is the minimum throat area possible without altering the flow properties in the duct? If the throat area is reduced to 3/4th of this value, determine the change (if any) in the mass flow rate as well as the static and stagnation conditions in the duct.

Solution : Since $P_0 = 200$ kPa and M $= 0.6$, we can get $P = 200 \div 1.2755 = 156.8$ kPa, using Table J. The continuity equation applied in the duct gives,

$$\dot{m} = \rho AV = \frac{P}{RT} AM \sqrt{\gamma RT} = \sqrt{\frac{\gamma}{RT}} AMP$$

Substitution of the known quantities into the above expression yields $T = 201$ K. Hence, the velocity in the duct, $V = 171$ m/s and $T_0 = 215$ K.

For the given stagnation condition, the nozzle can pass this mass flow rate as long as

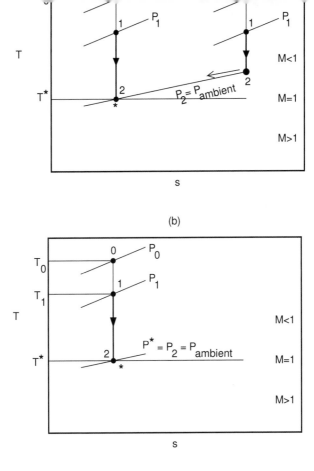

(b)

Figure 15.10: Illustration of flow through a convergent nozzle with varying inlet stagnation conditions and fixed back (ambient) pressure : T-s diagram

it is less than or equal to the maximum (choked) mass flow rate. Hence, the minimum throat area may be evaluated using Eqn. 15.41 as $A_t = 5.4576 \text{ cm}^2$.

If the throat area is reduced further, then the mass flow rate through the nozzle decreases, while the nozzle remains choked. As discussed earlier, the stagnation condition may be changed only by the addition or removal of work and/or heat, neither of which is applicable here. Hence, the flow in the duct adjusts to the reduced mass flow rate through a change in the static condition (Mach number). This may be

$$\dot{m} = \sqrt{\frac{\gamma}{R}} A M \frac{\dot{}}{\sqrt{T}} = \sqrt{\frac{\gamma}{R}} A M \frac{\dot{}}{\left(1 + \frac{\gamma-1}{2} M^2\right)^{\frac{\gamma}{\gamma-1}}} \sqrt{\frac{\dot{}}{T_0}}$$

$$\Rightarrow \dot{m} = \sqrt{\frac{\gamma}{R}} \frac{A P_0}{\sqrt{T_0}} \frac{M}{\left(1 + \frac{\gamma-1}{2} M^2\right)^3}$$

Everything except M is known in the above expression. With $\dot{m} = (3/4) \times 0.3 = 0.225$ kg/s, M can be obtained as 0.443.

15.5.7 Flow Through A Convergent Divergent Nozzle

Convergent divergent nozzles are used in supersonic wind tunnels, turbomachinery and in propulsion applications such as aircraft engines and rockets. In propulsion applications, convergent nozzles can be used without severe penalty on the thrust up to $P_0/P_{ambient} < 3$. Beyond this value, convergent divergent nozzles have to be used to utilize the momentum thrust fully.

Flow through a convergent divergent nozzle also can be established in one of the two ways mentioned above. We look at the sequence of events during the start-up of a convergent divergent nozzle with fixed inlet stagnation conditions and varying back pressure conditions, next. This sequence is illustrated in T-s coordinates using Fig. 15.11. The inlet and exit sections are denoted as before by 1 and 2. The corresponding variation of the static pressure along the length of the nozzle is shown in Fig. 15.12. Starting with Fig. 15.11(a), we can see that, when the back pressure is high, the flow accelerates in the converging portion and decelerates in the diverging portion, but remains subsonic throughout (curve labelled (a) in Fig. 15.12). When the back pressure is reduced, the Mach number at the throat becomes 1 as shown in Fig. 15.11(b) and the flow becomes choked. The flow field from the inlet state to the throat as well as the mass flow rate through the nozzle does not change anymore (curve labelled (b) in Fig. 15.12).

When the back pressure is reduced some more, the flow accelerates beyond the throat and becomes supersonic (Fig. 15.11(c)). However, the back pressure is too high, and this triggers a normal shock in the divergent portion of the nozzle. The state point just before and after the shock are denoted by x and y respectively in Fig. 15.11. The flow becomes subsonic after the normal shock and it decelerates in the rest of the divergent portion with the attendant increase in static pressure to the specified back pressure (curve labelled (c) in Fig. 15.12). The location of the normal shock is dictated by the exit area, throat area and the back pressure. As the back pressure is lowered further, the normal shock moves further downstream (Figs. 15.11(d),(e) and curves labelled (d) and (e) in Fig. 15.12). The situation shown in Fig. 15.11(e) where the normal shock stands just at the exit represents a threshold situation. If the

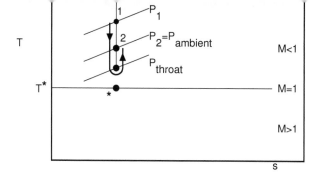

(b)

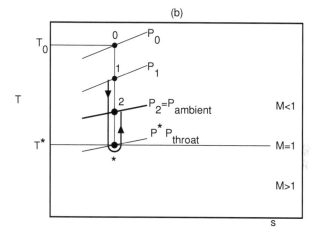

Figure 15.11: Illustration of flow through a convergent divergent nozzle with inlet stagnation conditions fixed and varying back (ambient) pressure : T-s diagram

back pressure were to be lowered further, then the normal shock moves out of the nozzle[†] and the flow inside the nozzle becomes shock free as shown in Fig. 15.11(f) and the curve labelled (f) in Fig. 15.12.

Two things should be noted in Figs. 15.11(c)-(e). Firstly, the Mach number before the shock keeps increasing as the shock moves downstream. Consequently, the loss

---

[†]The normal shock actually becomes an oblique shock that is anchored to the nozzle lip.

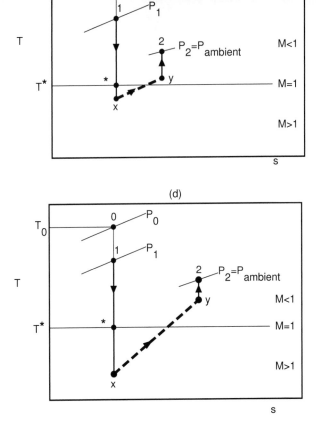

Figure 15.11: (cont'd) Illustration of flow through a convergent divergent nozzle with inlet stagnation conditions fixed and varying back (ambient) pressure : T-s diagram

of stagnation pressure across the shock wave also keeps increasing. Secondly, the flow field upstream of the shock wave does not change as the back pressure is lowered, since the flow is supersonic ahead of the shock wave. Finally, when the back pressure is decreased to the design value, the flow through the nozzle becomes shock free (isentropic) and the flow is supersonic throughout the divergent portion (Fig. 15.11(f) and curve labelled (f) in Fig. 15.12). Since the exit area and the throat area are known, and $M = 1$ at the throat, the exit Mach number can be calculated from Eqn. 15.40.

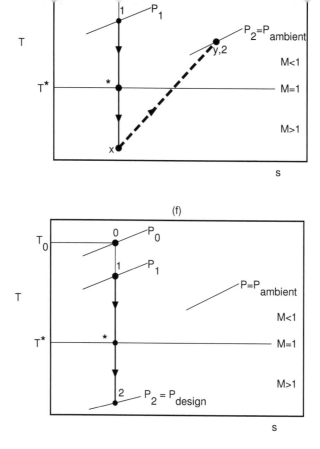

(f)

Figure 15.11: (cont'd) Illustration of flow through a convergent divergent nozzle with inlet stagnation conditions fixed and varying back (ambient) pressure : T-s diagram

If the back pressure is lowered below the design value, the nozzle exit pressure does not change. The jet is now said to be "under-expanded" (as in the case of the convergent nozzle) and further expansion takes place outside the nozzle.

In contrast to a convergent nozzle, it is possible to operate a convergent divergent nozzle in an "over-expanded" mode. This happens when the back pressure is higher than the design value, but lower than the value for which a normal shock would stand just at the exit (Fig. 15.12). Since the static pressure of the jet as it comes

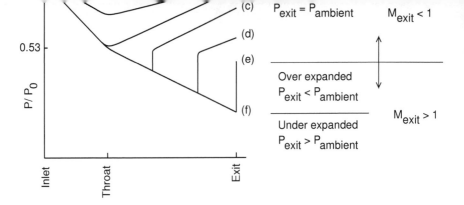

Figure 15.12: Variation of static pressure in a convergent divergent nozzle with inlet stagnation conditions fixed and varying back (ambient) pressure. For condition (f), $P_{exit} = P_{design}$ and $M_{exit} = M_{design}$.

out of the nozzle is less than the ambient value, it undergoes compression through oblique shocks outside the nozzle.

The start-up sequence described above remains the same when the back pressure is fixed and the inlet stagnation pressure is varied. This means that the normal shock that occurs in the divergent portion is inevitable and cannot be avoided. This is undesirable since the loss of stagnation pressure across the normal shock can be quite high. It is for these reasons, that convergent divergent nozzles are not used in propulsion applications unless the pressure ratio $P_0/P_{ambient}$ is high enough.

◤ EXAMPLE 15.5

Air at 1 MPa, 300 K in a reservoir expands through a convergent-divergent nozzle into the ambient. The exit to throat area ratio is 2. Determine (i) the ambient pressure(s) and the corresponding exit Mach number(s) for isentropic, choked flow (ii) the range of ambient pressure for which the nozzle will not choke (iii) the range of ambient pressure for which a normal shock would stand in the divergent part of the nozzle and (iv) the range of ambient pressure for which the nozzle exit conditions are independent of the ambient condition.

(ii) The nozzle does not choke for 1 MPa $\leq P_e \leq$ 937.2 kPa (see Fig. 15.12).

(iii) When a normal shock stands at the exit of the nozzle, we have $M_x = 2.1967$ and $P_x = 94$ kPa. From Table K, we can get $M_y = 0.5475$ and $P_y \div P_x = 5.46307$ (exit state corresponding to (e) in Fig. 15.12) . Hence $P_e = 5.46307 \times 94 = 513.53$ kPa. A normal shock will thus stand in the divergent part of the nozzle for 937.2 kPa $< P_e \leq 513.53$ kPa.

(iv) The nozzle exit condition becomes independent of the ambient pressure when $P_e < 513.53$ kPa (see Fig. 15.12).

▣ EXAMPLE 15.6

Air at 1100 kPa, 400 K in a reservoir expands through a convergent-divergent nozzle into the ambient. The exit to throat area ratio is 2.5. The ambient pressure is 100 kPa initially and is then gradually increased. Determine the Mach number and static and stagnation pressure at the exit as well as the ambient pressure at each of the following instants: (i) initial (ii) when a normal shock first appears (iii) when a normal shock occurs at a location where $A/A_{throat} = 1.5$ and (iv) when the normal shock disappears.

Solution : Similar to the previous example, let us determine the exit condition corresponding to (b), (f) and (e) for this problem also.

From Table J, corresponding to an area ratio of 2.5, the values of P_0/P_e for curves (b) and (f) in Fig. 15.12 are 1.04073 and 15.6346. The exit static pressures are 1056.95 kPa and 70.36 kPa. The corresponding exit Mach numbers are 0.2395 and 2.443 respectively.

For curve (f) in Fig. 15.12, $M_x = 2.443$ and $P_x = 70.36$ kPa. From Table K, we can get $M_y = 0.5186$, $P_y = 6.79629 \times 70.36 = 478.19$ kPa and $P_{0,y} = 0.522196 \times 1100 = 574.42$ kPa.

When a normal shock stands at $A/A_{throat} = 1.5$, we have, from Table J, $M_x = 1.8545$ and $P_{0,x} \div P_x = 6.24675$. From Table K, we can get $M_y = 0.60477$, $P_y \div P_x = 3.8457$ and $P_{0,y} \div P_{0,x} = 0.7882$. Hence, $P_{0,y} = 0.7882 \times 1100 = 867.02$ kPa. Since the entropy changes across the normal shock, so does the stagnation pressure

$$\Rightarrow \frac{A_y^*}{A_{throat}} = \frac{P_{0,x}}{P_{0,y}}$$

where we have used the fact that the stagnation temperature remains the same across the normal shock and $A_x^* = A_{throat}$. Therefore

$$\frac{A_e}{A_y^*} = \frac{A_e}{A_y} \frac{A_y}{A_y^*} = \frac{A_e}{A_{throat}} \frac{A_{throat}}{A_y} \frac{A_x}{A_{throat}} \frac{A_{throat}}{A_y^*}$$

$$= 2.5 \times \frac{1}{1.5} \times 1.5 \times 0.7882 = 1.9705$$

where we have set $A_y = A_x$ since the thickness of the normal shock is negligible. From Table J, for this value of A/A^*, we can retrieve $M_e = 0.311$ and $P_{0,e} \div P_e = 1.06938$. Hence, $P_e = 867.02 \div 1.06938 = 810.77$ kPa.

	M_e	P_e (kPa)	$P_{0,e}$ (kPa)	$P_{ambient}$ (kPa)
(i)	2.443	70.36	1100	100
(ii)	0.5186	478.19	574.42	478.19
(iii)	0.311	810.77	867.02	810.77
(iv)	0.2395	1056.95	1100	1056.95

■ EXAMPLE 15.7

A converging diverging nozzle with an exit to throat ratio of 3.5, operates with inlet stagnation conditions 1 MPa and 500 K. Determine the exit conditions when the back pressure is (a) 20 kPa (b) 500 kPa. Assume air to be the working fluid.

Solution : Given $P_0 = 1$ MPa, $T_0 = 500$ K and $A_e/A_{throat} = 3.5$. The exit conditions for this area ratio are:

Supersonic exit : From Table J, we can get $P_0 \div P_e = 27.14$ and $M_e = 2.8$. Therefore, $P_e = 36.85$ kPa.

Subsonic exit : From Table J, we can get $P_0 \div P_e = 1.02$ and $M_e = 0.1682$. Therefore, $P_e = 980.4$ kPa.

Normal shock at exit : From Table K, for $M_x = 2.8$, we get $M_y = 0.4882$ and $P_y \div P_x = 8.98$. Therefore, $P_e = 36.85 \times 8.98 = 330.91$ kPa.

(b) Since the given back pressure of 500 kPa lies in between 980.4 kPa and 330.91 kPa, there is a normal shock in the divergent part of the nozzle.

The mass flow rate through the nozzle is the same at the throat and the exit sections. Thus,

$$\dot{m} = \rho^* V^* A_{throat} \quad = \quad \rho_{exit} V_e A_e$$

$$\Rightarrow \quad \frac{P^*}{RT^*} \sqrt{\gamma R T^*} A_{throat} \quad = \quad \frac{P_e}{RT_e} M_e \sqrt{\gamma R T_e} A_e$$

$$\frac{P^*}{P_{0,1}} \frac{P_{0,1}}{P_e} \sqrt{\frac{T_0}{T^*}} \frac{A_{throat}}{A_e} \quad = \quad M_e \sqrt{\frac{T_0}{T_e}}$$

$$\frac{P^*}{P_{0,1}} \frac{P_{0,1}}{P_e} \sqrt{\frac{\gamma+1}{2}} \frac{A_{throat}}{A_e} \quad = \quad M_e \sqrt{1 + \frac{\gamma-1}{2} M_e^2}$$

$$\frac{1}{1.893} \frac{1000}{500} \sqrt{1.2} \frac{1}{3.5} \quad = \quad M_e \sqrt{1 + 0.2 M_e^2}$$

$$0.330675 \quad = \quad M_e \sqrt{1 + 0.2 M_e^2}$$

Therefore, $M_e = 0.327$ and

$$P_{0,e} = \frac{P_0}{P_e} P_e = 1.07687 \times 500 = 538.435 \, kPa.$$

For $P_{0,y} \div P_{0,x} = 538.435 \div 1000 = 0.538435$, from Table K, we get $M_x = 2.405$. From Table J, A/A^* corresponding to this value of Mach number is 2.414. Thus, the normal shock stands at a location in the divergent portion where A/A_{throat} is 2.414.

Alternatively, the locations of the shock may be obtained iteratively using the procedure from the previous example as follows.

$\frac{A_x}{A_{throat}}$	M_x	$\frac{P_{0,y}}{P_{0,x}}$	$\frac{A_e}{A_y^*}$	$\frac{P_{0,e}}{P_e}$	P_e
(Guess)	From Table J	From Table K		From Table J	(kPa)
2	2.1967	0.62964	2.20374	1.05383	597
2.5	2.443	0.522196	1.827686	1.08278	482
2.42	2.408	0.536776	1.8678716	1.07867	497

Since $1 < A_x/A_{throat} < A_e/A_{throat}$, a good initial guess for A_x/A_{throat} is $(1 + A_e/A_{throat})/2$ (this has not been done in the present example, as it would have been

15.5.8 Flow of Steam through Nozzles

In this subsection, we will study the dynamics of the flow of steam through nozzles. Historically, the theory of the flow of steam through nozzles was developed first in the late 1800s and early 1900s. The convergent divergent nozzle to accelerate steam to high speeds for use in impulse steam turbines was designed by de Laval in 1888. The importance of studying the flow of steam arises from the fact that steam turbines, even today, are used extensively in power generation. A clear understanding of the dynamics of the flow of steam through the blade passages of the steam turbines as well as through the nozzles which precede the blades is thus very important. The theory developed in the previous subsection is general, although the calorically perfect assumption has been invoked a few times. Hence, it can be carried over for steam with a few modifications, which will be mentioned shortly.

Isentropic expansion of steam It may be recalled that the calorically perfect model assumes that

- the gas obeys the ideal gas equation of state, $Pv = RT$ and

- the internal energy is a linear function of temperature.

In the case of steam (as well as a two phase saturated mixture of liquid and vapor), both these assumptions are unrealistic and must be abandoned. The T-v diagram shown in Fig. 5.6 illustrates how well superheated steam obeys the ideal gas equation of state. In an actual application, there is no guarantee that the initial state of the steam (before expansion in the nozzle) will lie within the shaded region in this figure. More importantly, the state at the end of an isentropic expansion process in a nozzle (process 1-2 in Fig. 9.6) will most likely lie in the two phase region.

The internal energy of water in the two-phase region as well as the superheated region is a function of temperature and pressure, *i.e.,* $u = u(T, P)$. Hence, even the thermally perfect assumption is invalid let alone the calorically perfect assumption.

In view of the arguments presented above, actual calculations involving expansion of steam in nozzles and blade passages have to be carried out using tabulated property data or the Mollier ($h - s$) diagram. However, it would be convenient, to the extent possible, to be able to use closed form expressions as was done for ideal gases. This is explored next.

When steam undergoes an isentropic expansion process in a nozzle, if the process line crosses the saturated vapor line as shown in Fig. 9.6, condensation takes place and the steam becomes a two phase mixture of liquid and vapor. The nucleation

the divergent portion of a convergent divergent nozzle. At this point, the steam would have undergone a considerable amount of expansion and is likely to be quite wet. As a result of the increase in pressure, temperature and entropy across the normal shock, the steam will become dry and saturated or even superheated. The gas dynamics of a two phase mixture is complex and beyond the scope of this book. However, if the dryness fraction does not become too low *i.e.,* the steam does not become too wet, then the aforementioned complications can be ignored. We assume this to be the case in the rest of this chapter. In addition, we will assume that the flow is shock free and hence isentropic.

We know that isentropic expansion of an ideal gas obeys $Pv^\gamma = \text{constant}$. Isentropic expansion of steam can be represented using a similar expression of the form

$$Pv^n = \text{constant} \tag{15.42}$$

where the exponent n has to be determined from experimental data through a curve fit. This is given as

$$n = \begin{cases} 1.3 & \text{if superheated} \\ 1.035 + 0.1\,x_1 & \text{saturated mixture} \end{cases} \tag{15.43}$$

where x_1 is the *initial* dryness fraction. The exponent for superheated steam is due to Callendar and the expression for the saturated mixture is due to Zeuner. It will be demonstrated later through numerical examples that Eqn. 15.42 with the exponent given by Eqn. 15.43 is an excellent description of the isentropic expansion process of steam in nozzles.

Since the propagation of an acoustic wave through a medium causes changes in the properties of the medium that are governed by an isentropic process, the speed of sound a, in superheated steam or a saturated mixture can be calculated using Eqns. 15.16 and 15.43 as follows:

$$a = \sqrt{\left.\frac{dP}{d\rho}\right|_s}$$

$$= \sqrt{nPv} \tag{15.44}$$

Flow of steam through nozzles The theory developed in the previous subsections for the flow of a calorically perfect gas is applicable for the flow of steam as well, subject to the constraints mentioned above. In view of this, this material will not be repeated here. However, the material will be developed along the lines customarily

expansion between states 1 and 2,

$$\frac{V_2^2 - V_1^2}{2} = -\int_1^2 v\, dP$$

Since $Pv^n = $ constant, the integral on the right hand can be evaluated. This leads to

$$\frac{V_2^2 - V_1^2}{2} = \frac{n}{n-1}\left(P_1 v_1 - P_2 v_2\right)$$

This can be rewritten as

$$V_2 = \sqrt{V_1^2 + \frac{2n}{n-1} P_1 v_1 \left[1 - \frac{P_2 v_2}{P_1 v_1}\right]}$$

$$= \sqrt{V_1^2 + \frac{2n}{n-1} P_1 v_1 \left[1 - \left(\frac{P_2}{P_1}\right)^{(n-1)/n}\right]} \qquad (15.45)$$

where we have used $P_1 v_1^n = P_2 v_2^n$.
If we integrate Eqn. 15.7 between states 1 and 2, we get

$$h_1 - h_2 = \frac{V_2^2 - V_1^2}{2}$$

Upon substituting for V_2 from Eqn. 15.45, we get

$$h_1 - h_2 = \frac{n}{n-1} P_1 v_1 \left[1 - \left(\frac{P_2}{P_1}\right)^{(n-1)/n}\right] \qquad (15.46)$$

Without any loss of generality, if we simply assume that the steam is expanded from a steam chest, where stagnation conditions prevail, to a final pressure of P, then the above expression may be modified to give the velocity at the end of the expansion process to be

$$V = \sqrt{\frac{2n}{n-1} P_0 v_0 \left[1 - \left(\frac{P}{P_0}\right)^{(n-1)/n}\right]} \qquad (15.47)$$

In contrast to nozzle flows involving perfect gases where the Area-Mach number

The mass flow rate at any section in a nozzle is given as

$$\dot{m} = \frac{AV}{v}$$

If we use the fact that $Pv^n = P_0 v_0^n$ in the above expression, we get

$$\dot{m} = \frac{AV}{v_0} \left(\frac{P}{P_0}\right)^{1/n}$$

Upon substituting for V from Eqn. 15.47, we are led to

$$\dot{m} = A \sqrt{\frac{2n}{n-1} \frac{P_0}{v_0} \left[\left(\frac{P}{P_0}\right)^{2/n} - \left(\frac{P}{P_0}\right)^{(n+1)/n}\right]}$$

For a given value of A, P_0 and v_0, it is easy to show (by differentiating the above expression with respect to P and setting the derivative to zero) that the mass flow rate is a maximum when the steam is expanded to a pressure P given by

$$\frac{P}{P_0} = \left(\frac{2}{n+1}\right)^{n/(n-1)} \tag{15.48}$$

We can rewrite Eqn. 15.47 as

$$V = \sqrt{\frac{2n}{n-1} Pv \left[\left(\frac{P_0}{P}\right)^{n/(n-1)} - 1\right]}$$

If we substitute for P from Eqn. 15.48, then, at the section where the pressure is given by this equation, the velocity of the steam is given as

$$V = \sqrt{nPv}$$

It follows from Eqn. 15.44 that this velocity is equal to the local speed of sound. Hence, Eqn. 15.48 may be rewritten as

$$\frac{P^*}{P_0} = \left(\frac{2}{n+1}\right)^{n/(n-1)} \tag{15.49}$$

It is worthwhile emphasizing that Eqns. 15.47, 15.48 and 15.49 have been derived

in the value of n between the inlet and the throat (which will be true only when the degree of superheat is sufficiently high).

■ EXAMPLE 15.8

Steam at a pressure of 600 kPa and dryness fraction of 0.9 in a steam chest expands through a convergent nozzle having a throat area of $0.002m^2$ into a region maintained at (a) 400 kPa and (b) 300 kPa. In each case, determine the exit pressure, exit velocity and the mass flow rate.

Solution : Given $P_0 = 600$ kPa, we can get $v_0 = 0.2845$ m³/kg and $s_0 = 6.2771$ kJ/kg.K using Table B. Since $x_1 = 0.9$, the index of expansion $n = 1.125$ from Eqn. 15.43. From Eqn. 15.49, we have $P^* = 0.5795\,P_0 = 347.7$ kPa.

(a) Since the ambient pressure is greater than P^*, the flow is not choked and the exit pressure $P_e = 400$ kPa. We have from Eqn. 15.46

$$h_0 - h_e = \frac{n}{n-1} P_0 v_0 \left[1 - \left(\frac{P_e}{P_0} \right)^{(n-1)/n} \right]$$

where the subscript e denotes the exit. Upon substituting numerical values, we get

$$h_0 - h_e = 67.68 \,\text{kJ/kg}$$

Therefore

$$V_e = \sqrt{2(h_0 - h_e)} = 368 \,\text{m/s}$$

From Eqn. 15.42, we can get

$$v_e = v_0 \left(\frac{P_e}{P_0} \right)^{-1/n} = 0.408 \,\text{m}^3/\text{kg}$$

The mass flow rate may be calculated as

$$\dot{m} = \frac{A_e V_e}{v_e} = 1.804 \,\text{kg/s}$$

(b) In this case, since the ambient pressure is less than the critical value, the nozzle is choked and the exit pressure $P_e = P^* = 347.7$ kPa. The enthalpy drop comes out to be 90.366 kJ/kg and hence the exit velocity is 425 m/s. The specific volume at the exit is $v_e = 0.4621$ m³/kg. The mass flow rate in this case comes out to be 1.84 kg/s. Since the nozzle is choked in this case, the exit velocity may also be calculated as $V_e = \sqrt{n P_e v_e} = 425$ m/s.

and 0.025 m respectively, determine the velocity at the inlet and exit and the stagnation pressure.

Solution : Given $P_1 = 500$ kPa and $x_1 = 1$, we can get $v_1 = 0.3748$ m³/kg, $h_1 = 2748.49$ kJ/kg and $s_1 = 6.8215$ kJ/kg.K using Table B. Since the steam is initially dry and saturated, $n = 1.135$ from Eqn. 15.43. From Eqn. 15.42, we can get

$$v_e = v_1 \left(\frac{P_e}{P_1}\right)^{-1/n} = 0.5878 \, \text{m}^3/\text{kg}$$

and from Eqn. 15.46

$$h_1 - h_e = \frac{n}{n-1} P_1 v_1 \left[1 - \left(\frac{P_e}{P_1}\right)^{(n-1)/n}\right] = 92.88 \, \text{kJ/kg}$$

$$= \frac{V_e^2 - V_1^2}{2}$$

Also, since the mass flow rates at the inlet and outlet are the same, we have

$$V_1 = \frac{A_e}{A_1} \frac{v_1}{v_e} V_e$$

(a) We can thus obtain the exit velocity $V_e = 436.6$ m/s. It follows that the inlet velocity $V_1 = 69.6$ m/s. Note that the speed of sound at the exit can be calculated from Eqn. 15.44 to be 447.4 m/s. Hence, the flow is not choked.

(b) The stagnation enthalpy can be evaluated using

$$h_0 = h_1 + \frac{V_1^2}{2} = 2751 \, \text{kJ/kg}$$

Since $s_0 = s_1 = 6.8215$ kJ/kg.K, we can get the stagnation pressure $P_0 = 507$ kPa from the steam table. For this value of P_0, we can obtain $P^* = 294$ kPa from Eqn. 15.49. The given exit pressure is higher than this value confirming our earlier observation that the flow is not choked.

■ EXAMPLE 15.10

Dry, saturated steam at 1 MPa in a steam chest expands through a nozzle to a final pressure of 100 kPa. Determine (a) if the nozzle is convergent or

Solution : We have $P_0 = 1$ MPa and we can get $v_0 = 0.19436$ m³/kg using Table B.

(a) Since the steam is initially dry and saturated, $n = 1.135$ from Eqn. 15.43. From Eqn. 15.49, we have $P^* = 0.58\,P_0 = 580$ kPa. Since the flow is expanded to a final pressure of 100 kPa in the nozzle, it is clear that the nozzle is convergent-divergent.

(b) We have from Eqn. 15.46

$$h_0 - h_e = \frac{n}{n-1} P_0 v_0 \left[1 - \left(\frac{P_e}{P_0} \right)^{(n-1)/n} \right]$$

where the subscript e denotes the exit. Upon substituting numerical values, we get

$$h_0 - h_e = 391.48\,\text{kJ/kg}$$

Therefore

$$V_e = \sqrt{2(h_0 - h_e)} = 885\,\text{m/s}$$

(c) From Eqn. 15.42, we can get

$$v_e = v_0 \left(\frac{P_e}{P_0} \right)^{-1/n} = 1.478\,\text{m}^3/\text{kg}$$

Using steam tables, the dryness fraction at the exit can now be calculated as 0.87.

(d) From Eqn. 15.42, we can get

$$v^* = v_0 \left(\frac{P^*}{P_0} \right)^{-1/n} = 0.3141\,\text{m}^3/\text{kg}$$

and

$$V^* = \sqrt{nP^*v^*} = 454.7\,\text{m/s}$$

Since

$$\dot{m} = \frac{A^* V^*}{v^*} = \frac{A_e V_e}{v_e}$$

we can obtain $A_e/A^* = 2.42$.

An alternative solution method is to use the Mollier diagram or the steam table. From Table B, for dry saturated vapor at 1 MPa, we can get $h_0 = 2778.1$ kJ/kg and $s_0 = 6.5865$ kJ/kg.K. Since the expansion process is isentropic, we have at the exit, $s_e = 6.5865$ kJ/kg.K and $P_e = 100$ kPa. From Table B, we can get $x_e = 0.872$ and

At the throat, since $P^* = 580$ kPa and $s^* = s_0 = 6.5865$ kJ/kg.K, from the tables, we can get $v^* = 0.3135$ m³/kg, by interpolation. Hence, $V^* = \sqrt{nP^*v^*} = 454$ m/s. The exit to throat area ratio may be evaluated to be 2.42.

◼ EXAMPLE 15.11

Steam at 700 kPa, 250°C in a steam chest expands through a nozzle to a final pressure of 100 kPa. The mass flow rate is 0.076 kg/s. Determine (a) if the nozzle is convergent or convergent-divergent (b) the throat diameter (c) the exit diameter and (d) the dryness fraction at the exit. Assume the expansion process to be isentropic and in equilibrium throughout.

Solution : From the steam table, it is easy to establish that for the given steam chest conditions of $P_0 = 700$ kPa and $T_0 = 250°C$ the steam is initially superheated. Therefore, $v_0 = 0.336343$ m³/kg and $s_0 = 7.1062$ kJ/kg.K from Table C.

(a) Since the steam is initially superheated, $n = 1.3$ from Eqn. 15.43. From Eqn. 15.49, we have $P^* = 0.545\, P_0 = 380$ kPa (assuming that the steam is superheated at the throat). Since the flow is expanded to a final pressure of 100 kPa in the nozzle, it is clear that the nozzle is convergent-divergent.

(b) From Eqn. 15.42, we can get

$$ v^* = v_0 \left(\frac{P^*}{P_0} \right)^{-1/n} = 0.53648 \, \text{m}^3/\text{kg} $$

and

$$ V^* = \sqrt{nP^*v^*} = 514.8 \, \text{m/s} $$

Since

$$ A^* = \frac{\dot{m}v^*}{V^*} = 7.92 \times 10^{-5} \, \text{m}^2 $$

the throat diameter $D^* = 10$ mm.

(c) Since the expansion process is isentropic, the process line crosses the saturated vapor line at $P_g = 212$ kPa. With $n = 1.3$ (as the flow is superheated until it crosses the saturated vapor line), we can get

$$ h_0 - h_g = \frac{n}{n-1} P_0 v_0 \left[1 - \left(\frac{P_g}{P_0} \right)^{(n-1)/n} \right] = 245.8 \, \text{kJ/kg} $$

It follows that $V_g = \sqrt{2(h_0 - h_g)} = 701.14$ m/s. Note that the value for P_g calculated above also confirms that the steam is superheated at the throat, sine $P^* > P_g$. Once the flow crosses the saturated vapor line, $n = 1.135$ (this implies that the flow continues to be in equilibrium after crossing the saturated vapor line). Hence

$$h_g - h_e = \frac{n}{n-1} P_g v_g \left[1 - \left(\frac{P_e}{P_g} \right)^{(n-1)/n} \right] = 127.77 \,\text{kJ/kg}$$

and

$$v_e = v_g \left(\frac{P_e}{P_g} \right)^{-1/n} = 1.6344 \,\text{m}^3/\text{kg}$$

Therefore

$$V_e = \sqrt{V_g^2 + 2(h_g - h_e)} = 864 \,\text{m/s}$$

For the given mass flow rate, the exit area can be calculated as

$$A_e = \frac{\dot{m} v_e}{V_e} = 1.44 \times 10^{-4} \,\text{m}^2$$

and the exit diameter can be calculated as $D_e = 13.5$ mm.

(d) From the given value of the exit pressure P_e and the calculated value of v_e, we can get the dryness fraction at the exit to be 0.96 using steam tables.

Alternatively, we can use the steam table (or Mollier diagram) to solve the problem. For the given steam chest condition, we can get $h_0 = 2954$ kJ/kg and $s_0 = 7.1062$ kJ/kg.K from Table C.

The pressure at the throat $P^* = 0.545 P_0 = 380$ kPa. Since the expansion is isentropic, $s^* = s_0 = 7.1062$ kJ/kg.K. Therefore, the fluid is superheated at the throat with $v^* = 0.537$ m³/kg and $h^* = 2820$ kJ/kg. It follows that

$$V^* = \sqrt{2(h_0 - h^*)} = 518 \,\text{m/s}$$

and

$$A^* = \frac{\dot{m} v^*}{V^*} = 7.86 \times 10^{-5} \,\text{m}^2$$

Thus, the throat diameter $D^* = 10$ mm.

$$V_e = \sqrt{2(h_0 - h_e)} = 864 \, \text{m/s}$$

and

$$A_e = \frac{\dot{m} v_e}{V_e} = 1.43 \times 10^{-4} \, \text{m}^2$$

Thus, the exit diameter $D_e = 13.5$ mm.

SUGGESTED READING

Four Laws that Drive the Universe, P. W. Atkins, Oxford University Press, 2007.

The Second Law, P. W. Atkins, Scientific American Books, W. H. Freeman and Co, New York, 1994.

Thermodynamics: An Engineering Approach, Y. A. Cengel and M. A. Boles, 8th edition, McGraw Hill, 2015.

Thermodynamics, S. Klein and G. Nellis, Cambridge University Press, 2012.

Fundamentals of Engineering Thermodynamics, M. J. Moran, H. N. Shapiro, D. D. Boettner and M. B. Bailey, 8th edition, Wiley, New York, 2014.

Fundamentals of Thermodynamics, R. E. Sonntag, C. Borgnakke and G. J. Van Wylen, 5th edition, Wiley, New York, 2008.

Engineering Thermodynamics, D. B. Spalding and E. H. Cole, Edward Arnold Publishers Ltd., London, 1976.

Thermodynamics: Concepts and Applications, S. Turns, Cambridge University Press, 2006.

Table A: Thermodynamic properties of saturated steam, temperature table

T °C	P bar	$v_f \times 10^3$ m^3/kg	v_g m^3/kg	u_f kJ/kg	u_g kJ/kg	h_f kJ/kg	h_g kJ/kg	s_f kJ/kg.K	s_g kJ/kg.K
0.01	0.00611	1.0002	206.136	0.00	2375.3	0.01	2501.3	0.0000	9.1562
1	0.00657	1.0002	192.439	4.18	2375.9	4.183	2502.4	0.0153	9.1277
2	0.00706	1.0001	179.762	8.40	2377.3	8.401	2504.2	0.0306	9.1013
3	0.00758	1.0001	168.016	12.61	2378.7	12.61	2506.0	0.0459	9.0752
4	0.00814	1.0001	157.126	16.82	2380.0	16.82	2507.9	0.0611	9.0492
5	0.00873	1.0001	147.024	21.02	2381.4	21.02	2509.7	0.0763	9.0236
6	0.00935	1.0001	137.647	25.22	2382.8	25.22	2511.5	0.0913	8.9981
7	0.01002	1.0001	128.939	29.41	2384.2	29.42	2513.4	0.1063	8.9729
8	0.01073	1.0002	120.847	33.61	2385.6	33.61	2515.2	0.1213	8.9479
9	0.01148	1.0002	113.323	37.80	2386.9	37.80	2517.1	0.1361	8.9232
10	0.01228	1.0003	106.323	41.99	2388.3	41.99	2518.9	0.1510	8.8986
11	0.01313	1.0004	99.808	46.17	2389.7	46.18	2520.7	0.1657	8.8743
12	0.01403	1.0005	93.740	50.36	2391.1	50.36	2522.6	0.1804	8.8502
13	0.01498	1.0006	88.086	54.55	2392.4	54.55	2524.4	0.1951	8.8263
14	0.01599	1.0008	82.814	58.73	2393.8	58.73	2526.2	0.2097	8.8027
15	0.01706	1.0009	77.897	62.92	2395.2	62.92	2528.0	0.2242	8.7792
16	0.01819	1.0011	73.308	67.10	2396.6	67.10	2529.9	0.2387	8.7560
17	0.01938	1.0012	69.023	71.28	2397.9	71.28	2531.7	0.2532	8.7330
18	0.02064	1.0014	65.019	75.47	2399.3	75.47	2533.5	0.2676	8.7101
19	0.02198	1.0016	61.277	79.65	2400.7	79.65	2535.3	0.2819	8.6875
20	0.02339	1.0018	57.778	83.83	2402.0	83.84	2537.2	0.2962	8.6651
21	0.02488	1.0020	54.503	88.02	2403.4	88.02	2539.0	0.3104	8.6428
22	0.02645	1.0023	51.438	92.20	2404.8	92.20	2540.8	0.3246	8.6208
23	0.02810	1.0025	48.568	96.38	2406.1	96.39	2542.6	0.3388	8.5990
24	0.02985	1.0027	45.878	100.57	2407.5	100.57	2544.5	0.3529	8.5773
25	0.03169	1.0030	43.357	104.75	2408.9	104.75	2546.3	0.3670	8.5558
26	0.03363	1.0033	40.992	108.93	2410.2	108.94	2548.1	0.3810	8.5346
27	0.03567	1.0035	38.773	113.12	2411.6	113.12	2549.9	0.3949	8.5135
28	0.03782	1.0038	36.690	117.30	2413.0	117.30	2551.7	0.4088	8.4926
29	0.04008	1.0041	34.734	121.48	2414.3	121.49	2553.5	0.4227	8.4718
30	0.04246	1.0044	32.896	125.67	2415.7	125.67	2555.3	0.4365	8.4513

Table A: Thermodynamic properties of saturated steam, temperature table

T °C	P bar	$v_f \times 10^3$ m³/kg	v_g m³/kg	u_f kJ/kg	u_g kJ/kg	h_f kJ/kg	h_g kJ/kg	s_f kJ/kg.K	s_g kJ/kg.K
31	0.04495	1.0047	31.168	129.85	2417.0	129.85	2557.1	0.4503	8.4309
32	0.04758	1.0050	29.543	134.03	2418.4	134.04	2559.0	0.4640	8.4107
33	0.05033	1.0054	28.014	138.22	2419.8	138.22	2560.8	0.4777	8.3906
34	0.05323	1.0057	26.575	142.40	2421.1	142.41	2562.6	0.4914	8.3708
35	0.05627	1.0060	25.220	146.58	2422.5	146.59	2564.4	0.5050	8.3511
40	0.07381	1.0079	19.528	167.50	2429.2	167.50	2573.4	0.5723	8.2550
45	0.09593	1.0099	15.263	188.41	2435.9	188.42	2582.3	0.6385	8.1629
50	0.12345	1.0122	12.037	209.31	2442.6	209.33	2591.2	0.7037	8.0745
55	0.1575	1.0146	9.5726	230.22	2449.2	230.24	2600.0	0.7679	7.9896
60	0.1993	1.0171	7.6743	251.13	2455.8	251.15	2608.8	0.8312	7.9080
65	0.2502	1.0199	6.1996	272.05	2462.4	272.08	2617.5	0.8935	7.8295
70	0.3118	1.0228	5.0446	292.98	2468.8	293.01	2626.1	0.9549	7.7540
75	0.3856	1.0258	4.1333	313.92	2475.2	313.96	2634.6	1.0155	7.6813
80	0.4737	1.0290	3.4088	334.88	2481.6	334.93	2643.1	1.0753	7.6112
85	0.5781	1.0324	2.8289	355.86	2487.9	355.92	2651.4	1.1343	7.5436
90	0.7012	1.0359	2.3617	376.86	2494.0	376.93	2659.6	1.1925	7.4784
95	0.8453	1.0396	1.9828	397.89	2500.1	397.98	2667.7	1.2501	7.4154
100	1.013	1.0434	1.6736	418.96	2506.1	419.06	2675.7	1.3069	7.3545
105	1.208	1.0474	1.4200	440.05	2512.1	440.18	2683.6	1.3630	7.2956
110	1.432	1.0515	1.2106	461.19	2517.9	461.34	2691.3	1.4186	7.2386
115	1.690	1.0558	1.0370	482.36	2523.5	482.54	2698.8	1.4735	7.1833
120	1.985	1.0603	0.8922	503.57	2529.1	503.78	2706.2	1.5278	7.1297
125	2.320	1.0649	0.7709	524.82	2534.5	525.07	2713.4	1.5815	7.0777
130	2.700	1.0697	0.6687	546.12	2539.8	546.41	2720.4	1.6346	7.0272
135	3.130	1.0746	0.5824	567.46	2545.0	567.80	2727.2	1.6873	6.9780
140	3.612	1.0797	0.5090	588.85	2550.0	589.24	2733.8	1.7394	6.9302
145	4.153	1.0850	0.4464	610.30	2554.8	610.75	2740.2	1.7910	6.8836
150	4.757	1.0904	0.3929	631.80	2559.5	632.32	2746.4	1.8421	6.8381
160	6.177	1.1019	0.3071	674.97	2568.3	675.65	2758.0	1.9429	6.7503
170	7.915	1.1142	0.2428	718.40	2576.3	719.28	2768.5	2.0421	6.6662
180	10.02	1.1273	0.1940	762.12	2583.4	763.25	2777.8	2.1397	6.5853

Table A: Thermodynamic properties of saturated steam, temperature table

| T | P | $v_f \times 10^3$ | v_g | u_f | u_g | h_f | h_g | s_f | s_g |
°C	bar	m³/kg	m³/kg	kJ/kg	kJ/kg	kJ/kg	kJ/kg	kJ/kg.K	kJ/kg.K
190	12.54	1.1414	0.1565	806.17	2589.6	807.60	2785.8	2.2358	6.5071
200	15.55	1.1564	0.1273	850.58	2594.7	852.38	2792.5	2.3308	6.4312
210	19.07	1.1726	0.1044	895.43	2598.7	897.66	2797.7	2.4246	6.3572
220	23.18	1.1900	0.0862	940.75	2601.6	943.51	2801.3	2.5175	6.2847
230	27.95	1.2088	0.0716	986.62	2603.1	990.00	2803.1	2.6097	6.2131
240	33.45	1.2292	0.0597	1033.1	2603.1	1037.2	2803.0	2.7013	6.1423
250	39.74	1.2515	0.0501	1080.4	2601.6	1085.3	2800.7	2.7926	6.0717
260	46.89	1.2758	0.0422	1128.4	2598.4	1134.4	2796.2	2.8838	6.0009
270	55.00	1.3026	0.0356	1177.4	2593.2	1184.6	2789.1	2.9751	5.9293
280	64.13	1.3324	0.0302	1227.5	2585.7	1236.1	2779.2	3.0669	5.8565
290	74.38	1.3658	0.0256	1279.0	2575.7	1289.1	2765.9	3.1595	5.7818
300	85.84	1.4037	0.0217	1332.0	2562.8	1344.1	2748.7	3.2534	5.7042
310	98.61	1.4473	0.0183	1387.0	2546.2	1401.2	2727.0	3.3491	5.6226
320	112.8	1.4984	0.0155	1444.4	2525.2	1461.3	2699.7	3.4476	5.5356
340	145.9	1.6373	0.0108	1569.9	2463.9	1593.8	2621.3	3.6587	5.3345
360	186.6	1.8936	0.0070	1725.6	2352.2	1761.0	2482.0	3.9153	5.0542
374.12	220.9	3.1550	0.0031	2029.6	2029.6	2099.3	2099.3	4.4298	4.4298

Table B: Thermodynamic properties of saturated steam, pressure table

P bar	T °C	$v_f \times 10^3$ m³/kg	v_g m³/kg	u_f kJ/kg	u_g kJ/kg	h_f kJ/kg	h_g kJ/kg	s_f kJ/kg.K	s_g kJ/kg.K
0.06	36.17	1.0065	23.737	151.47	2424.0	151.47	2566.5	0.5208	8.3283
0.08	41.49	1.0085	18.128	173.73	2431.0	173.74	2576.0	0.5921	8.2272
0.10	45.79	1.0103	14.693	191.71	2436.8	191.72	2583.7	0.6489	8.1487
0.12	49.40	1.0119	12.377	206.82	2441.6	206.83	2590.1	0.6960	8.0849
0.16	55.30	1.0147	9.4447	231.47	2449.4	231.49	2600.5	0.7718	7.9846
0.20	60.05	1.0171	7.6591	251.32	2455.7	251.34	2608.9	0.8318	7.9072
0.25	64.95	1.0198	6.2120	271.85	2462.1	271.88	2617.4	0.8929	7.8302
0.30	69.09	1.0222	5.2357	289.15	2467.5	289.18	2624.5	0.9438	7.7676
0.40	75.85	1.0263	3.9983	317.48	2476.1	317.52	2636.1	1.0257	7.6692
0.50	81.31	1.0299	3.2442	340.38	2483.1	340.43	2645.3	1.0908	7.5932
0.60	85.92	1.0330	2.7351	359.73	2488.8	359.79	2652.9	1.1451	7.5314
0.70	89.93	1.0359	2.3676	376.56	2493.8	376.64	2659.5	1.1917	7.4793
0.80	93.48	1.0385	2.0895	391.51	2498.1	391.60	2665.3	1.2327	7.4342
0.90	96.69	1.0409	1.8715	405.00	2502.0	405.09	2670.5	1.2693	7.3946
1.00	99.61	1.0431	1.6958	417.30	2505.5	417.40	2675.1	1.3024	7.3592
2.00	120.2	1.0605	0.8865	504.49	2529.2	504.70	2706.5	1.5301	7.1275
2.50	127.4	1.0672	0.7193	535.12	2537.0	535.39	2716.8	1.6073	7.0531
3.00	133.5	1.0731	0.6063	561.19	2543.4	561.51	2725.2	1.6719	6.9923
3.50	138.9	1.0785	0.5246	584.01	2548.8	584.38	2732.4	1.7276	6.9409
4.00	143.6	1.0835	0.4627	604.38	2553.4	604.81	2738.5	1.7768	6.8963
5.00	151.8	1.0925	0.3751	639.74	2561.1	640.29	2748.6	1.8608	6.8216
6.00	158.8	1.1006	0.3158	669.96	2567.2	670.62	2756.7	1.9313	6.7602
7.00	165.0	1.1079	0.2729	696.49	2572.2	697.27	2763.3	1.9923	6.7081
8.00	170.4	1.1147	0.2405	720.25	2576.5	721.14	2768.9	2.0462	6.6627
9.00	175.4	1.1211	0.2150	741.84	2580.1	742.85	2773.6	2.0947	6.6224
10.00	179.9	1.1272	0.1945	761.67	2583.2	762.80	2777.7	2.1387	6.5861
11.00	184.1	1.1329	0.1775	780.06	2585.9	781.31	2781.2	2.1791	6.5531
12.00	188.0	1.1384	0.1633	797.23	2588.3	798.60	2784.3	2.2165	6.5227
13.00	191.6	1.1437	0.1513	813.37	2590.4	814.85	2787.0	2.2514	6.4946
14.00	195.1	1.1488	0.1408	828.60	2592.2	830.21	2789.4	2.2840	6.4684
15.00	198.3	1.1538	0.1318	843.05	2593.9	844.78	2791.5	2.3148	6.4439
16.00	201.4	1.1586	0.1238	856.81	2595.3	858.66	2793.3	2.3439	6.4208
17.00	204.3	1.1633	0.1167	869.95	2596.5	871.93	2795.0	2.3715	6.3990

Table B: Thermodynamic properties of saturated steam, pressure table

P bar	T °C	$v_f \times 10^3$ m³/kg	v_g m³/kg	u_f kJ/kg	u_g kJ/kg	h_f kJ/kg	h_g kJ/kg	s_f kJ/kg.K	s_g kJ/kg.K
18.00	207.1	1.1678	0.1104	882.54	2597.7	884.64	2796.4	2.3978	6.3782
19.00	209.8	1.1723	0.1047	894.63	2598.6	896.86	2797.6	2.4230	6.3585
20.00	212.4	1.1766	0.0996	906.27	2599.5	908.62	2798.7	2.4470	6.3397
25.00	224.0	1.1973	0.0800	958.92	2602.3	961.92	2802.2	2.5543	6.2561
30.00	233.9	1.2165	0.0667	1004.59	2603.2	1008.2	2803.3	2.6453	6.1856
35.00	242.6	1.2348	0.0571	1045.26	2602.9	1049.6	2802.6	2.7250	6.1240
40.00	250.4	1.2523	0.0498	1082.18	2601.5	1087.2	2800.6	2.7961	6.0690
45.00	257.5	1.2694	0.0441	1116.14	2599.3	1121.9	2797.6	2.8607	6.0188
50.00	264.0	1.2861	0.0394	1147.74	2596.5	1154.2	2793.7	2.9201	5.9726
55.00	270.0	1.3026	0.0356	1177.39	2593.1	1184.6	2789.1	2.9751	5.9294
60.00	275.6	1.3190	0.0324	1205.42	2589.3	1213.3	2783.9	3.0266	5.8886
65.00	280.9	1.3352	0.0297	1232.06	2584.9	1240.7	2778.1	3.0751	5.8500
70.00	285.9	1.3515	0.0274	1257.52	2580.2	1267.0	2771.8	3.1211	5.8130
75.00	290.6	1.3678	0.0253	1281.96	2575.1	1292.2	2765.0	3.1648	5.7774
80.00	295.0	1.3843	0.0235	1305.51	2569.6	1316.6	2757.8	3.2066	5.7431
85.00	299.3	1.4009	0.0219	1328.27	2563.8	1340.2	2750.1	3.2468	5.7097
90.00	303.4	1.4177	0.0205	1350.36	2557.6	1363.1	2742.0	3.2855	5.6771
95.00	307.3	1.4348	0.0192	1371.84	2551.1	1385.5	2733.4	3.3229	5.6452
100.0	311.0	1.4522	0.0180	1392.79	2544.3	1407.3	2724.5	3.3592	5.6139
105.0	314.6	1.4699	0.0170	1413.27	2537.1	1428.7	2715.1	3.3944	5.5830
110.0	318.1	1.4881	0.0160	1433.34	2529.5	1449.7	2705.4	3.4288	5.5525
115.0	321.5	1.5068	0.0151	1453.06	2521.6	1470.4	2695.1	3.4624	5.5221
120.0	324.7	1.5260	0.0143	1472.47	2513.4	1490.8	2684.5	3.4953	5.4920
125.0	327.9	1.5458	0.0135	1491.61	2504.7	1510.9	2673.4	3.5277	5.4619
130.0	330.9	1.5663	0.0128	1510.55	2495.7	1530.9	2661.8	3.5595	5.4317
135.0	333.8	1.5875	0.0121	1529.31	2486.2	1550.7	2649.7	3.5910	5.4015
140.0	336.7	1.6097	0.0115	1547.94	2476.3	1570.5	2637.1	3.6221	5.3710
145.0	339.5	1.6328	0.0109	1566.49	2465.9	1590.2	2623.9	3.6530	5.3403
150.0	342.2	1.6572	0.0103	1585.01	2455.0	1609.9	2610.0	3.6838	5.3091
155.0	344.8	1.6828	0.0098	1603.55	2443.4	1629.6	2595.5	3.7145	5.2774
160.0	347.4	1.7099	0.0093	1622.17	2431.3	1649.5	2580.2	3.7452	5.2450
165.0	349.9	1.7388	0.0088	1640.92	2418.4	1669.6	2564.1	3.7761	5.2119
170.0	352.3	1.7699	0.0084	1659.89	2404.8	1690.0	2547.1	3.8073	5.1777

Table B: Thermodynamic properties of saturated steam, pressure table

P bar	T °C	$v_f \times 10^3$ m³/kg	v_g m³/kg	u_f kJ/kg	u_g kJ/kg	h_f kJ/kg	h_g kJ/kg	s_f kJ/kg.K	s_g kJ/kg.K
175.0	354.7	1.8033	0.0079	1679.18	2390.2	1710.7	2529.0	3.8390	5.1423
180.0	357.0	1.8399	0.0075	1698.88	2374.6	1732.0	2509.7	3.8714	5.1054
185.0	359.3	1.8801	0.0071	1719.17	2357.7	1754.0	2488.9	3.9047	5.0667
190.0	361.5	1.9251	0.0067	1740.22	2339.3	1776.8	2466.3	3.9393	5.0256
195.0	363.7	1.9762	0.0063	1762.34	2319.0	1800.9	2441.4	3.9756	4.9815
200.0	365.8	2.0357	0.0059	1785.94	2296.2	1826.7	2413.7	4.0144	4.9331
205.0	367.9	2.1076	0.0055	1811.76	2269.7	1855.0	2381.6	4.0571	4.8787
210.0	369.9	2.1999	0.0050	1841.25	2237.5	1887.5	2343.0	4.1060	4.8144
215.0	371.8	2.3362	0.0045	1878.57	2193.9	1928.8	2291.0	4.1684	4.7299
220.9	374.1	3.1550	0.0316	2029.60	2029.6	2099.3	2099.3	4.4298	4.4298

Table C: Thermodynamic properties of superheated steam

T °C	v m³/kg	u kJ/kg	h kJ/kg	s kJ/kg.K	T °C	v m³/kg	u kJ/kg	h kJ/kg	s kJ/kg.K
		$P = 0.06$ bar					$P = 0.35$ bar		
36.17	23.739	2424.0	2566.5	8.3283	72.67	4.531	2472.1	2630.7	7.7148
80	27.133	2486.7	2649.5	8.5794	80	4.625	2483.1	2645.0	7.7553
120	30.220	2544.1	2725.5	8.7831	120	5.163	2542.0	2722.7	7.9637
160	33.303	2602.2	2802.0	8.9684	160	5.697	2600.7	2800.1	8.1512
200	36.384	2660.9	2879.2	9.1390	200	6.228	2659.9	2877.9	8.3229
240	39.463	2720.6	2957.4	9.2975	240	6.758	2719.8	2956.3	8.4821
280	42.541	2781.2	3036.4	9.4458	280	7.287	2780.6	3035.6	8.6308
320	45.620	2842.7	3116.4	9.5855	320	7.816	2842.2	3115.8	8.7707
360	48.697	2905.2	3197.4	9.7176	360	8.344	2904.8	3196.9	8.9031
400	51.775	2968.8	3279.5	9.8433	400	8.872	2968.5	3279.0	9.0288
440	54.852	3033.4	3362.6	9.9632	440	9.400	3033.2	3362.2	9.1488
500	59.468	3132.4	3489.2	10.134	500	10.192	3132.2	3488.9	9.3194
		$P = 0.7$ bar					$P = 1$ bar		
89.93	2.368	2493.8	2659.5	7.4793	99.61	1.6958	2505.5	2675.1	7.3592
120	2.5709	2539.3	2719.3	7.6370	120	1.7931	2537.0	2716.3	7.4665
160	2.8407	2599.0	2797.8	7.8272	160	1.9838	2597.5	2795.8	7.6591
200	3.1082	2658.7	2876.2	8.0004	200	2.1723	2657.6	2874.8	7.8335
240	3.3744	2718.9	2955.1	8.1603	240	2.3594	2718.1	2954.0	7.9942
280	3.6399	2779.8	3034.6	8.3096	280	2.5458	2779.2	3033.8	8.1438
320	3.9049	2841.6	3115.0	8.4498	320	2.7317	2841.1	3114.3	8.2844
360	4.1697	2904.4	3196.2	8.5824	360	2.9173	2904.0	3195.7	8.4171
400	4.4342	2968.1	3278.5	8.7083	400	3.1027	2967.7	3278.0	8.5432
440	4.6985	3032.8	3361.7	8.8285	440	3.2879	3032.5	3361.3	8.6634
480	4.9627	3098.6	3446.0	8.9434	480	3.4730	3098.3	3445.6	8.7785
520	5.2269	3165.4	3531.3	9.0538	520	3.6581	3165.2	3531.0	8.8889

Table C: Thermodynamic properties of superheated steam

T	v	u	h	s	T	v	u	h	s
°C	m³/kg	kJ/kg	kJ/kg	kJ/kg.K	°C	m³/kg	kJ/kg	kJ/kg	kJ/kg.K

		$P = 1.5$ bar					$P = 3$ bar		
111.37	1.1600	2519.3	2693.3	7.2234	133.55	0.6060	2543.4	2725.2	6.9923
160	1.3174	2594.9	2792.5	7.4660	160	0.6506	2586.9	2782.1	7.1274
200	1.4443	2655.8	2872.4	7.6425	200	0.7163	2650.2	2865.1	7.3108
240	1.5699	2716.7	2952.2	7.8044	240	0.7804	2712.6	2946.7	7.4765
280	1.6948	2778.2	3032.4	7.9548	280	0.8438	2775.0	3028.1	7.6292
320	1.8192	2840.3	3113.2	8.0958	320	0.9067	2837.8	3109.8	7.7716
360	1.9433	2903.3	3194.8	8.2289	360	0.9692	2901.2	3191.9	7.9057
400	2.0671	2967.2	3277.2	8.3552	400	1.0315	2965.4	3274.9	8.0327
440	2.1908	3032.0	3360.6	8.4756	440	1.0937	3030.5	3358.7	8.1536
480	2.3144	3097.9	3445.1	8.5908	480	1.1557	3096.6	3443.4	8.2692
520	2.4379	3164.8	3530.5	8.7013	520	1.2177	3163.7	3529.0	8.3800
560	2.5613	3232.9	3617.0	8.8077	560	1.2796	3231.9	3615.7	8.4867

		$P = 5$ bar					$P = 7$ bar		
151.86	0.3751	2561.1	2748.6	6.8216	164.97	0.2729	2572.2	2763.3	6.7081
180	0.4045	2609.5	2811.7	6.9652	180	0.2846	2599.6	2798.8	6.7876
220	0.4449	2674.9	2897.4	7.1463	220	0.3146	2668.1	2888.4	6.9771
260	0.4840	2738.9	2980.9	7.3092	260	0.3434	2733.9	2974.2	7.1445
300	0.5225	2802.5	3063.7	7.4591	300	0.3714	2798.6	3058.5	7.2970
340	0.5606	2866.3	3146.6	7.5989	340	0.3989	2863.2	3142.4	7.4385
380	0.5985	2930.7	3229.9	7.7304	380	0.4262	2928.1	3226.4	7.5712
420	0.6361	2995.7	3313.8	7.8550	420	0.4533	2993.6	3310.9	7.6966
460	0.6736	3061.6	3398.4	7.9738	460	0.4802	3059.8	3395.9	7.8160
500	0.7109	3128.5	3483.9	8.0873	500	0.5070	3126.9	3481.8	7.9300
540	0.7482	3196.3	3570.4	8.1964	540	0.5338	3194.9	3568.5	8.0393
560	0.7669	3230.6	3614.0	8.2493	560	0.5741	3229.2	3612.2	8.09243

Table C: Thermodynamic properties of superheated steam

T °C	v m³/kg	u kJ/kg	h kJ/kg	s kJ/kg.K	T °C	v m³/kg	u kJ/kg	h kJ/kg	s kJ/kg.K
		$P = 10$ bar					$P = 15$ bar		
179.91	0.1945	2583.2	2777.7	6.5861	198.32	0.1318	2593.9	2791.5	6.4439
220	0.2169	2657.5	2874.3	6.7904	220	0.1405	2638.1	2849.0	6.5630
260	0.2378	2726.1	2963.9	6.9652	260	0.1556	2712.6	2945.9	6.7521
300	0.2579	2792.7	3050.6	7.1219	300	0.1696	2782.5	3036.9	6.9168
340	0.2776	2858.5	3136.1	7.2661	340	0.1832	2850.4	3125.2	7.0657
380	0.2970	2924.2	3221.2	7.4006	380	0.1965	2917.6	3212.3	7.2033
420	0.3161	2990.3	3306.5	7.5273	420	0.2095	2984.8	3299.1	7.3322
460	0.3352	3057.0	3392.2	7.6475	460	0.2224	3052.3	3385.9	7.4540
500	0.3541	3124.5	3478.6	7.7622	500	0.2351	3120.4	3473.1	7.5699
540	0.3729	3192.8	3565.7	7.8721	540	0.2478	3189.2	3561.0	7.6806
580	0.3917	3262.0	3653.7	7.9778	580	0.2605	3258.8	3649.6	7.7870
620	0.4105	3332.2	3742.7	8.0797	620	0.2730	3329.4	3739.0	7.8894
		$P = 20$ bar					$P = 30$ bar		
212.42	0.0996	2599.5	2798.7	6.3397	233.90	0.06667	2603.2	2803.3	6.1856
240	0.1084	2658.8	2875.6	6.4937	240	0.06818	2618.9	2823.5	6.2251
280	0.1200	2735.6	2975.6	6.6814	280	0.07710	2709.0	2940.3	6.4445
320	0.1308	2807.3	3068.8	6.8441	320	0.08498	2787.6	3042.6	6.6232
360	0.1411	2876.7	3158.9	6.9911	360	0.09232	2861.3	3138.3	6.7794
400	0.1512	2945.1	3247.5	7.1269	400	0.09935	2932.7	3230.7	6.9210
440	0.1611	3013.4	3335.6	7.2539	440	0.10618	3003.0	3321.5	7.0521
480	0.1708	3081.9	3423.6	7.3740	480	0.11287	3073.0	3411.6	7.1750
520	0.1805	3150.9	3511.9	7.4882	520	0.11946	3143.2	3501.6	7.2913
560	0.1901	3220.6	3600.7	7.5975	560	0.12597	3213.8	3591.7	7.4022
600	0.1996	3291.0	3690.2	7.7024	600	0.13243	3285.0	3682.3	7.5084
640	0.2091	3362.4	3780.5	7.8036	640	0.13884	3357.0	3773.5	7.6105

Table C: Thermodynamic properties of superheated steam

T	v	u	h	s	T	v	u	h	s
°C	m³/kg	kJ/kg	kJ/kg	kJ/kg.K	°C	m³/kg	kJ/kg	kJ/kg	kJ/kg.K

			$P = 40$ bar					$P = 60$ bar	
250.38	0.04978	2601.5	2800.6	6.0690	276.62	0.03244	2589.3	2783.9	5.8886
280	0.05544	2679.0	2900.8	6.2552	280	0.03317	2604.7	2803.7	5.9245
320	0.06198	2766.6	3014.5	6.4538	320	0.03874	2719.0	2951.5	6.1830
360	0.06787	2845.3	3116.7	6.6207	360	0.04330	2810.6	3070.4	6.3771
400	0.07340	2919.8	3213.4	6.7688	400	0.04739	2892.7	3177.0	6.5404
440	0.07872	2992.3	3307.2	6.9041	440	0.05121	2970.2	3277.4	6.6854
480	0.08388	3064.0	3399.5	7.0301	480	0.05487	3045.3	3374.5	6.8179
520	0.08894	3135.4	3491.1	7.1486	520	0.05840	3119.4	3469.8	6.9411
560	0.09392	3206.9	3582.6	7.2612	560	0.06186	3193.0	3564.1	7.0571
600	0.09884	3278.9	3674.3	7.3687	600	0.06525	3266.0	3658.1	7.1673
640	0.10372	3351.5	3766.4	7.4718	640	0.06859	3340.5	3752.1	7.2725
680	0.10855	3424.9	3859.1	7.5711	680	0.07189	3414.9	3846.3	7.3736

			$P = 80$ bar					$P = 100$ bar	
295.04	0.02352	2569.6	2757.8	5.7431	311.04	0.01802	2544.3	2724.5	5.6139
320	0.02681	2661.7	2876.2	5.9473	320	0.01925	2588.2	2780.6	5.7093
360	0.03088	2771.9	3018.9	6.1805	360	0.02330	2728.0	2961.0	6.0043
400	0.03431	2863.5	3138.0	6.3630	400	0.02641	2832.0	3096.1	6.2114
440	0.03742	2946.8	3246.2	6.5192	440	0.02911	2922.3	3213.4	6.3807
480	0.04034	3026.0	3348.6	6.6589	480	0.03160	3005.8	3321.8	6.5287
520	0.04312	3102.9	3447.8	6.7873	520	0.03394	3085.9	3425.3	6.6625
560	0.04582	3178.6	3545.2	6.9070	560	0.03619	3164.0	3525.8	6.7862
600	0.04845	3254.0	3641.5	7.0200	600	0.03836	3241.1	3624.7	6.9022
640	0.05102	3329.3	3737.5	7.1274	640	0.04048	3317.9	3722.7	7.0119
680	0.05356	3404.9	3833.4	7.2302	680	0.04256	3394.6	3820.3	7.1165
720	0.05607	3480.9	3929.4	7.3289	720	0.04461	3471.6	3917.7	7.2167

Table C: Thermodynamic properties of superheated steam

T °C	v m³/kg	u kJ/kg	h kJ/kg	s kJ/kg.K	T °C	v m³/kg	u kJ/kg	h kJ/kg	s kJ/kg.K
		$P = 120$ bar					$P = 140$ bar		
324.75	0.01426	2513.4	2684.5	5.4920	336.75	0.01148	2476.3	2637.1	5.3710
360	0.01810	2677.1	2894.4	5.8341	360	0.01421	2616.0	2815.0	5.6579
400	0.02108	2797.8	3050.7	6.0739	400	0.01722	2760.2	3001.3	5.9438
440	0.02355	2896.3	3178.9	6.2589	440	0.01955	2868.8	3142.5	6.1477
480	0.02576	2984.9	3294.0	6.4161	480	0.02157	2963.1	3265.2	6.3152
520	0.02781	3068.4	3402.1	6.5559	520	0.02343	3050.3	3378.3	6.4616
560	0.02976	3149.0	3506.1	6.6839	560	0.02517	3133.6	3485.9	6.5940
600	0.03163	3228.0	3607.6	6.8029	600	0.02683	3214.7	3590.3	6.7163
640	0.03345	3306.3	3707.7	6.9150	640	0.02843	3294.5	3692.5	6.8309
680	0.03523	3384.3	3807.0	7.0214	680	0.02999	3373.8	3793.6	6.9392
720	0.03697	3462.3	3906.0	7.1231	720	0.03152	3452.8	3894.1	7.0425
760	0.03869	3540.6	4004.8	7.2207	760	0.03301	3532.0	3994.2	7.1413
		$P = 160$ bar					$P = 180$ bar		
347.44	0.00931	2431.3	2580.2	5.2450	357.06	0.00750	2374.6	2509.7	5.1054
360	0.01105	2537.5	2714.3	5.4591	360	0.00810	2418.3	2564.1	5.1916
400	0.01427	2718.5	2946.8	5.8162	400	0.01191	2671.7	2886.0	5.6872
440	0.01652	2839.6	3104.0	6.0433	440	0.01415	2808.5	3063.2	5.9432
480	0.01842	2940.5	3235.3	6.2226	480	0.01596	2916.9	3204.2	6.1358
520	0.02013	3031.8	3353.9	6.3761	520	0.01756	3012.7	3328.8	6.2971
560	0.02172	3117.9	3465.4	6.5133	560	0.01903	3101.9	3444.5	6.4394
600	0.02322	3201.1	3572.6	6.6390	600	0.02041	3187.3	3554.8	6.5687
640	0.02466	3282.6	3677.2	6.7561	640	0.02173	3270.5	3661.7	6.6885
680	0.02606	3363.1	3780.1	6.8664	680	0.02301	3352.4	3766.5	6.8008
720	0.02742	3443.3	3882.1	6.9712	720	0.02424	3433.7	3870.0	6.9072
760	0.02876	3523.4	3983.5	7.0714	760	0.02545	3514.7	3972.8	7.0086

Table C: Thermodynamic properties of superheated steam

T	v	u	h	s	T	v	u	h	s
°C	m^3/kg	kJ/kg	kJ/kg	kJ/kg.K	°C	m^3/kg	kJ/kg	kJ/kg	kJ/kg.K

		P = 200 bar					P = 240 bar		
365.81	0.00588	2296.2	2413.7	4.9331					
400	0.00995	2617.9	2816.9	5.5521	400	0.00673	2476.0	2637.5	5.2365
440	0.01223	2775.2	3019.8	5.8455	440	0.00929	2700.9	2923.9	5.6511
480	0.01399	2892.3	3172.0	6.0534	480	0.01100	2839.9	3103.9	5.8971
520	0.01551	2993.1	3303.2	6.2232	520	0.01241	2952.1	3250.0	6.0861
560	0.01688	3085.5	3423.2	6.3708	560	0.01366	3051.8	3379.5	6.2456
600	0.01817	3173.3	3536.7	6.5039	600	0.01480	3144.6	3499.8	6.3866
640	0.01939	3258.2	3646.0	6.6264	640	0.01587	3233.3	3614.3	6.5148
680	0.02056	3341.6	3752.8	6.7408	680	0.01690	3319.6	3725.1	6.6336
720	0.02170	3424.0	3857.9	6.8488	720	0.01788	3404.3	3833.4	6.7450
760	0.02280	3505.9	3961.9	6.9515	760	0.01883	3488.2	3940.2	6.8504
800	0.02388	3587.8	4065.4	7.0498	800	0.01976	3571.7	4046.0	6.9508

		P = 280 bar					P = 320 bar		
400	0.00383	2221.7	2328.8	4.7465	400	0.00237	1981.0	2056.8	4.3252
440	0.00712	2613.5	2812.9	5.4497	440	0.00543	2509.0	2682.9	5.2325
480	0.00885	2782.7	3030.5	5.7472	480	0.00722	2720.5	2951.5	5.5998
520	0.01019	2908.9	3194.3	5.9592	520	0.00853	2863.4	3136.2	5.8390
560	0.01135	3016.8	3334.6	6.1319	560	0.00962	2980.6	3288.4	6.0263
600	0.01239	3115.1	3462.1	6.2815	600	0.01059	3084.9	3423.8	6.1851
640	0.01336	3207.9	3582.0	6.4158	640	0.01148	3182.0	3549.4	6.3258
680	0.01428	3297.2	3697.0	6.5390	680	0.01232	3274.6	3668.8	6.4538
720	0.01516	3384.4	3808.8	6.6539	720	0.01312	3364.3	3784.0	6.5722
760	0.01600	3470.3	3918.4	6.7621	760	0.01388	3452.3	3896.4	6.6832
800	0.01682	3555.5	4026.5	6.8647	800	0.01462	3539.2	4006.9	6.7881

Table D: Thermodynamic properties of saturated R134a, temperature table

T °C	P kPa	$v_f \times 10^3$ m³/kg	v_g m³/kg	u_f kJ/kg	u_g kJ/kg	h_f kJ/kg	h_g kJ/kg	s_f kJ/kg.K	s_g kJ/kg.K
-40	51.25	0.7053	0.36064	0.00	207.38	0.000	225.860	0.0000	0.9687
-35	66.19	0.7126	0.28373	6.25	210.25	6.290	229.030	0.0267	0.9619
-30	84.43	0.7201	0.22577	12.58	213.12	12.640	232.190	0.0530	0.9559
-25	106.50	0.728	0.18152	18.95	215.99	19.030	235.320	0.0789	0.9505
-20	132.80	0.7361	0.14735	25.37	218.86	25.470	238.430	0.1046	0.9457
-15	164.00	0.7445	0.12066	31.85	221.72	31.970	241.510	0.1299	0.9415
-10	200.70	0.7533	0.09960	38.38	224.56	38.530	244.550	0.1550	0.9378
-5	243.50	0.7625	0.08282	44.96	227.38	45.150	247.550	0.1798	0.9345
0	293.00	0.7722	0.06934	51.61	230.18	51.830	250.500	0.2043	0.9316
5	349.90	0.7823	0.05840	58.31	232.96	58.590	253.390	0.2287	0.9290
10	414.90	0.7929	0.04947	65.09	235.69	65.420	256.220	0.2528	0.9266
15	488.70	0.8041	0.04211	71.93	238.39	72.320	258.970	0.2768	0.9245
20	572.10	0.816	0.03601	78.85	241.04	79.320	261.640	0.3006	0.9225
25	665.80	0.8286	0.03092	85.85	243.64	86.400	264.230	0.3243	0.9207
30	770.60	0.8421	0.02665	92.93	246.17	93.580	266.710	0.3479	0.9190
35	887.50	0.8565	0.02304	100.11	248.63	100.870	269.080	0.3714	0.9173
40	1017.00	0.872	0.01997	107.39	251.00	108.280	271.310	0.3949	0.9155
45	1161.00	0.8889	0.01734	114.79	253.27	115.820	273.400	0.4184	0.9137
50	1319.00	0.9072	0.01509	122.30	255.42	123.500	275.320	0.4419	0.9117
55	1492.00	0.9274	0.01314	129.96	257.43	131.350	277.030	0.4655	0.9095
60	1688.00	0.9498	0.01144	137.79	259.25	139.380	278.510	0.4893	0.9069
65	1891.00	0.9751	0.00996	145.80	260.86	147.640	279.690	0.5133	0.9038
70	2118.00	1.0038	0.00865	154.04	262.20	156.160	280.520	0.5377	0.9000
75	2366.00	1.0372	0.00749	162.54	263.17	165.000	280.880	0.5625	0.8953
80	2635.00	1.0774	0.00644	171.43	263.66	174.270	280.630	0.5881	0.8893
85	2928.00	1.1273	0.00548	180.81	263.45	184.110	279.510	0.6149	0.8812
90	3247.00	1.1938	0.00459	190.94	262.13	194.820	277.040	0.6435	0.8699
95	3594.00	1.2945	0.00371	202.49	258.73	207.140	272.080	0.6760	0.8524
100	3975.00	1.5269	0.00266	218.73	248.46	224.800	259.020	0.7222	0.8139
101.03	4059.00	1.9685	0.00197	232.95	233.90	241.880	241.880	0.7678	0.7678

Table E: Thermodynamic properties of saturated R134a, pressure table

P kPa	T °C	$v_f \times 10^3$ m³/kg	v_g m³/kg	u_f kJ/kg	u_g kJ/kg	h_f kJ/kg	h_g kJ/kg	s_f kJ/kg.K	s_g kJ/kg.K
40	-44.61	0.699	0.45483	-5.79	204.74	-5.760	222.940	-0.0249	0.9757
60	-36.95	0.71	0.31108	3.79	209.13	3.840	227.800	0.0163	0.9644
80	-31.13	0.718	0.23749	11.14	212.48	11.200	231.470	0.0471	0.9572
100	-26.37	0.726	0.19255	17.19	215.21	17.270	234.460	0.0718	0.9519
200	-10.09	0.753	0.09995	38.26	224.51	38.410	244.500	0.1545	0.9379
300	0.65	0.773	0.06778	52.48	230.55	52.710	250.880	0.2075	0.9312
400	8.91	0.791	0.05127	63.61	235.10	63.920	255.610	0.2476	0.9271
500	15.71	0.806	0.04117	72.92	238.77	73.320	259.360	0.2802	0.9242
600	21.55	0.82	0.03433	81.01	241.86	81.500	262.460	0.3080	0.9220
700	26.69	0.833	0.02939	88.24	244.51	88.820	265.080	0.3323	0.9201
800	31.31	0.846	0.02565	94.80	246.82	95.480	267.340	0.3541	0.9185
900	35.51	0.858	0.02270	100.84	248.88	101.620	269.310	0.3738	0.9171
1000	39.37	0.87	0.02033	106.47	250.71	107.340	271.040	0.3920	0.9157
1200	46.29	0.893	0.01673	116.72	253.84	117.790	273.920	0.4245	0.9132
1400	52.40	0.917	0.01412	125.96	256.40	127.250	276.170	0.4532	0.9107
1600	57.88	0.94	0.01213	134.45	258.50	135.960	277.920	0.4792	0.9080
1800	62.87	0.964	0.01057	142.36	260.21	144.090	279.230	0.5030	0.9052
2000	67.45	0.989	0.00930	149.81	261.56	151.780	280.150	0.5252	0.9020
2200	71.70	1.015	0.00824	156.90	262.57	159.130	280.700	0.5460	0.8985
2400	75.66	1.042	0.00734	163.70	263.27	166.200	280.890	0.5658	0.8946
2600	79.37	1.072	0.00657	170.29	263.63	173.080	280.700	0.5848	0.8901
2800	82.86	1.104	0.00588	176.73	263.64	179.820	280.110	0.6033	0.8849
3000	86.16	1.141	0.00527	183.09	263.26	186.510	279.080	0.6213	0.8789
3200	89.29	1.182	0.00472	189.41	262.41	193.190	277.500	0.6392	0.8718
3400	92.26	1.233	0.00420	195.91	260.96	200.100	275.230	0.6575	0.8631
3600	95.08	1.297	0.00370	202.66	258.65	207.320	271.970	0.6765	0.8521
3800	97.76	1.387	0.00319	210.26	254.87	215.540	266.990	0.6980	0.8367
4000	100.31	1.562	0.00256	220.43	246.82	226.680	257.050	0.7272	0.8085
4059	101.03	1.9685	0.00197	232.95	233.90	241.880	241.880	0.7678	0.7678

Table F: Thermodynamic properties of superheated R134a

T °C	v m³/kg	u kJ/kg	h kJ/kg	s kJ/kg.K	T °C	v m³/kg	u kJ/kg	h kJ/kg	s kJ/kg.K
		P = 80 kPa					*P* = 100 kPa		
-30	0.2388	213.2	232.4	0.9608					
-20	0.2501	220.2	240.2	0.9922	-20	0.1984	219.7	239.5	0.9721
-10	0.2611	227.2	248.1	1.0230	-10	0.2074	226.8	247.5	1.0030
0	0.2720	234.3	256.1	1.0530	0	0.2163	234.0	255.6	1.0330
10	0.2828	241.6	264.3	1.0820	10	0.2251	241.3	263.8	1.0630
20	0.2935	249.1	272.6	1.1110	20	0.2337	248.8	272.2	1.0920
30	0.3041	256.7	281.0	1.1390	30	0.2423	256.5	280.7	1.1200
40	0.3147	264.5	289.7	1.1670	40	0.2509	264.3	289.4	1.1490
50	0.3252	272.4	298.5	1.1950	50	0.2594	272.2	298.2	1.1760
60	0.3357	280.6	307.4	1.2220	60	0.2678	280.4	307.1	1.2040
70	0.3462	288.8	316.5	1.2490	70	0.2763	288.7	316.3	1.2310
		P = 120 kPa					*P* = 140 kPa		
-20	0.1639	219.2	238.9	0.9553					
-10	0.1716	226.4	246.9	0.9866	-10	0.1461	225.9	246.4	0.9724
0	0.1792	233.6	255.1	1.0170	0	0.1526	233.2	254.6	1.0030
10	0.1866	241.0	263.4	1.0470	10	0.1591	240.7	262.9	1.0330
20	0.1939	248.5	271.8	1.0760	20	0.1654	248.2	271.4	1.0620
30	0.2011	256.2	280.3	1.1050	30	0.1717	255.9	280.0	1.0910
40	0.2083	264.0	289.0	1.1330	40	0.1779	263.8	288.7	1.1200
50	0.2155	272.0	297.9	1.1610	50	0.1841	271.8	297.6	1.1470
60	0.2226	280.2	306.9	1.1880	60	0.1903	280.0	306.6	1.1750
70	0.2296	288.5	316.0	1.2150	70	0.1964	288.3	315.8	1.2020

Table F: Thermodynamic properties of superheated R134a

T	v	u	h	s	T	v	u	h	s
°C	m³/kg	kJ/kg	kJ/kg	kJ/kg.K	°C	m³/kg	kJ/kg	kJ/kg	kJ/kg.K

		$P = 160$ kPa					$P = 180$ kPa		
-10	0.1268	225.5	245.8	0.9599	-10	0.1119	225.0	245.2	0.9485
0	0.1327	232.9	254.1	0.9909	0	0.1172	232.5	253.6	0.9799
10	0.1385	240.4	262.5	1.0210	10	0.1224	240.0	262.1	1.0100
20	0.1441	248.0	271.0	1.0510	20	0.1275	247.7	270.6	1.0400
30	0.1496	255.7	279.6	1.0800	30	0.1325	255.4	279.3	1.0690
40	0.1551	263.6	288.4	1.1080	40	0.1374	263.3	288.1	1.0980
50	0.1606	271.6	297.3	1.1360	50	0.1423	271.4	297.0	1.1260
60	0.1660	279.8	306.3	1.1640	60	0.1471	279.6	306.1	1.1530
70	0.1714	288.1	315.5	1.1910	70	0.1520	287.9	315.3	1.1810

		$P = 200$ kPa					$P = 300$ kPa		
-10	0.0999	224.6	244.6	0.9381					
0	0.1048	232.1	253.1	0.9699					
10	0.1096	239.7	261.6	1.0010	10	0.0709	237.9	259.2	0.9611
20	0.1142	247.4	270.2	1.0300	20	0.0742	245.8	268.1	0.9920
30	0.1187	255.2	278.9	1.0600	30	0.0775	253.8	277.0	1.0220
40	0.1232	263.1	287.7	1.0880	40	0.0806	261.9	286.1	1.0510
50	0.1277	271.2	296.7	1.1160	50	0.0837	270.1	295.2	1.0800
60	0.1321	279.4	305.8	1.1440	60	0.0868	278.4	304.4	1.1080
70	0.1364	287.7	315.0	1.1710	70	0.0898	286.8	313.8	1.1360

Table F: Thermodynamic properties of superheated R134a

T	v	u	h	s	T	v	u	h	s
°C	m³/kg	kJ/kg	kJ/kg	kJ/kg.K	°C	m³/kg	kJ/kg	kJ/kg	kJ/kg.K

		$P = 400$ kPa							
10	0.0515	236.0	256.6	0.9306					
20	0.0542	244.2	265.9	0.9628					
30	0.0568	252.4	275.1	0.9937					
40	0.0593	260.6	284.3	1.0240					
50	0.0617	268.9	293.6	1.0530					
60	0.0641	277.3	303.0	1.0810					
70	0.0664	285.9	312.5	1.1090					
		$P = 500$ kPa					$P = 600$ kPa		
20	0.0421	242.4	263.5	0.9384					
30	0.0443	250.9	273.0	0.9704	30	0.0360	249.2	270.8	0.9500
40	0.0465	259.3	282.5	1.0010	40	0.0379	257.9	280.6	0.9817
50	0.0485	267.7	292.0	1.0310	50	0.0397	266.5	290.3	1.0120
60	0.0505	276.3	301.5	1.0600	60	0.0414	275.2	300.0	1.0420
70	0.0524	284.9	311.1	1.0880	70	0.0431	283.9	309.8	1.0710
80	0.0543	293.7	320.8	1.1160	80	0.0447	292.7	319.6	1.0990
900	0.0562	302.5	330.6	1.1440	900	0.0463	301.7	329.5	1.1260
100	0.0580	311.5	340.5	1.1710	100	0.0479	310.7	339.5	1.1540
110	0.0599	320.6	350.6	1.1970	110	0.0494	319.9	349.6	1.1800
120	0.0617	329.9	360.8	1.2230	120	0.0510	329.2	359.8	1.2070

Table F: Thermodynamic properties of superheated R134a

T	v	u	h	s	T	v	u	h	s
°C	m³/kg	kJ/kg	kJ/kg	kJ/kg.K	°C	m³/kg	kJ/kg	kJ/kg	kJ/kg.K

		P = 700 kPa					P = 800 kPa		
30	0.0300	247.5	268.5	0.9314					
40	0.0317	256.4	278.6	0.9642	40	0.0270	254.8	276.5	0.9481
50	0.0333	265.2	288.5	0.9955	50	0.0286	263.9	286.7	0.9803
60	0.0349	274.0	298.4	1.0260	60	0.0300	272.8	296.8	1.0110
70	0.0364	282.9	308.3	1.0550	70	0.0313	281.8	306.9	1.0410
80	0.0378	291.8	318.3	1.0840	80	0.0327	290.9	317.0	1.0700
900	0.0393	300.8	328.3	1.1120	900	0.0339	300.0	327.1	1.0980
100	0.0406	310.0	338.4	1.1390	100	0.0352	309.2	337.3	1.1260
110	0.0420	319.2	348.6	1.1660	110	0.0364	318.5	347.6	1.1530
120	0.0434	328.6	358.9	1.1920	120	0.0376	327.9	358.0	1.1800

		P =900 kPa					P = 1000 kPa		
40	0.0234	253.2	274.2	0.9328	40	0.0204	251.3	271.7	0.9180
50	0.0248	262.5	284.8	0.9661	50	0.0218	261.0	282.8	0.9526
60	0.0262	271.6	295.1	0.9977	60	0.0231	270.3	293.4	0.9851
70	0.0274	280.7	305.4	1.0280	70	0.0243	279.6	303.9	1.0160
80	0.0286	289.9	315.6	1.0570	80	0.0254	288.9	314.3	1.0460
900	0.0298	299.1	325.9	1.0860	900	0.0265	298.2	324.7	1.0750
100	0.0310	308.4	336.2	1.1140	100	0.0276	307.5	335.1	1.1030
110	0.0321	317.7	346.6	1.1410	110	0.0286	317.0	345.5	1.1310
120	0.0332	327.2	357.0	1.1680	120	0.0296	326.5	356.1	1.1580

Table F: Thermodynamic properties of superheated R134a

T °C	v m³/kg	u kJ/kg	h kJ/kg	s kJ/kg.K	T °C	v m³/kg	u kJ/kg	h kJ/kg	s kJ/kg.K
		$P = 1100$ kPa					$P = 1200$ kPa		
50	0.0193	259.4	280.6	0.9396	50	0.0172	257.6	278.3	0.9268
60	0.0205	269.0	291.6	0.9730	60	0.0184	267.6	289.7	0.9615
70	0.0217	278.4	302.3	1.0050	70	0.0195	277.2	300.6	0.9939
80	0.0228	287.8	312.9	1.0350	80	0.0205	286.8	311.4	1.0250
900	0.0238	297.2	323.4	1.0650	900	0.0215	296.3	322.1	1.0550
100	0.0248	306.7	333.9	1.0930	100	0.0224	305.8	332.7	1.0840
110	0.0257	316.2	344.5	1.1210	110	0.0234	315.4	343.4	1.1120
120	0.0267	325.8	355.1	1.1480	120	0.0242	325.1	354.1	1.1390
		$P = 1300$ kPa							
50	0.0154	255.8	275.8	0.914					
60	0.0166	266.1	287.6	0.9501					
70	0.0177	276	298.9	0.9835					
80	0.0187	285.7	309.9	1.015					
90	0.0196	295.3	320.8	1.045					
100	0.0205	304.9	331.5	1.075					
110	0.0213	314.6	342.3	1.103					
120	0.0222	324.3	353.1	1.131					

Table G: Sensible enthalpies and enthalpy of formation of some gases

T (K)	O_2 (MJ/kmol)	N_2 (MJ/kmol)	CO_2 (MJ/kmol)	H_2O (MJ/kmol)	CO (MJ/kmol)	H_2 (MJ/kmol)
			$\Delta\bar{h}(T)$			
	$\bar{h}_f^\circ=0$	$\bar{h}_f^\circ=0$	$\bar{h}_f^\circ=-393.52$	$\bar{h}_f^\circ=-241.83$	$\bar{h}_f^\circ=-110.53$	$\bar{h}_f^\circ=0$
300	0.03	0.05	0.07	0.062	0.06	0.05
400	3.02	2.97	4.00	3.458	2.97	2.96
500	6.13	5.91	8.31	6.92	5.93	5.88
600	9.32	8.89	12.91	10.50	8.94	8.81
700	12.59	11.94	17.75	14.19	12.03	11.75
800	15.91	15.05	22.81	18	15.18	14.70
900	19.29	18.22	28.03	21.94	18.40	17.68
1000	22.71	21.46	33.40	26.00	21.69	20.68
1100	26.18	24.76	38.89	30.19	25.03	23.72
1200	29.70	28.11	44.47	34.51	28.43	26.80
1300	33.25	31.50	50.15	38.94	31.87	29.92
1400	36.84	34.94	55.89	43.49	35.34	33.08
1500	40.46	38.41	61.71	48.15	38.85	36.29
1600	44.12	41.90	67.57	52.91	42.39	39.54
1700	47.81	45.43	73.48	57.76	45.95	42.83
1800	51.53	48.98	79.44	62.69	49.53	46.17
1900	55.28	52.55	85.42	67.70	53.12	49.54
2000	59.06	56.14	91.44	72.79	56.74	52.95
2100	62.86	59.74	97.49	77.94	60.37	56.40
2200	66.69	63.36	103.6	83.16	64.01	59.88
2300	70.54	66.99	109.70	88.42	67.66	63.39
2400	74.41	70.64	115.80	93.74	71.33	66.93
2500	78.31	74.30	121.90	99.11	75.00	70.50
2600	82.22	77.96	128.10	104.50	78.69	74.10
2700	86.16	81.64	134.20	110.00	82.38	77.72
2800	90.11	85.32	140.40	115.50	86.08	81.37
2900	94.08	89.01	146.60	121.00	89.79	85.04
3000	98.07	92.71	152.80	126.50	93.51	88.74
3100	102.0	96.42	159.10	132.10	97.23	92.46
3200	106.0	100.1	165.30	137.80	101.00	96.20
3300	110.1	103.9	171.60	143.40	104.70	99.97
3400	114.1	107.6	177.80	149.10	108.40	103.80
3500	118.2	111.3	184.10	154.80	112.2	107.60
3600	122.2	115.0	190.40	160.50	115.90	111.40
3700	126.3	118.8	196.70	166.20	119.70	115.20
3800	130.4	122.5	203.00	172.0	123.50	119.10
3900	134.6	126.3	209.30	177.80	127.20	123.00
4000	138.7	130.0	215.60	183.60	131.00	126.90

Source: http://webbook.nist.gov

Table H: Absolute entropy of some gases

T	O_2	N_2	CO_2	H_2O	CO	H_2
(K)	(kJ/kmol.K)	(kJ/kmol.K)	(kJ/kmol.K)	(kJ/kmol.K)	(kJ/kmol.K)	(kJ/kmol.K)
298	205.15	191.61	213.79	188.84	197.66	130.68
300	205.30	191.80	214.00	189.04	197.80	130.90
400	213.90	200.20	225.30	198.55	206.20	139.20
500	220.70	206.70	234.90	206.50	212.80	145.70
600	226.50	212.20	243.30	213.10	218.30	151.10
700	231.50	216.90	250.80	218.70	223.10	155.60
800	235.90	221.00	257.50	223.80	227.30	159.50
900	239.90	224.80	263.60	228.50	231.10	163.10
1000	243.60	228.20	269.30	232.70	234.50	166.20
1100	246.90	231.30	274.50	236.70	237.70	169.10
1200	250.00	234.20	279.40	240.50	240.70	171.80
1300	252.90	236.90	283.90	244.00	243.40	174.30
1400	255.60	239.50	288.20	247.40	246.00	176.60
1500	258.10	241.90	292.20	250.60	248.40	178.80
1600	260.40	244.10	296.00	253.70	250.70	180.90
1700	262.70	246.30	299.60	256.60	252.90	182.90
1800	264.80	248.30	303.00	259.40	254.90	184.80
1900	266.80	250.20	306.20	262.20	256.90	186.70
2000	268.70	252.10	309.30	264.80	258.70	188.40
2100	270.60	253.80	312.20	267.30	260.50	190.10
2200	272.40	255.50	315.10	269.70	262.20	191.70
2300	274.10	257.10	317.80	272.00	263.80	193.30
2400	275.70	258.70	320.40	274.30	265.40	194.80
2500	277.30	260.20	322.90	276.50	266.90	196.20
2600	278.80	261.60	325.30	278.60	268.30	197.70
2700	280.30	263.00	327.60	280.70	269.70	199.00
2800	281.70	264.30	329.90	282.70	271.00	200.30
2900	283.10	265.60	332.10	284.60	272.30	201.60
3000	284.50	266.90	334.20	286.50	273.60	202.90
3100	285.80	268.10	336.20	288.30	274.80	204.10
3200	287.10	269.30	338.20	290.10	276.00	205.30
3300	288.30	270.40	340.10	291.90	277.20	206.50
3400	289.50	271.50	342.00	293.60	278.30	207.60
3500	290.70	272.60	343.80	295.20	279.40	208.70
3600	291.80	273.70	345.60	296.80	280.40	209.80
3700	292.90	274.70	347.30	298.40	281.50	210.80
3800	294.0	275.70	349.00	299.90	282.50	211.90
3900	295.10	276.70	350.60	301.40	283.40	212.90
4000	296.20	277.60	352.20	302.90	284.40	213.80

Source: http://webbook.nist.gov

Table I: Equilibrium constants of a few reactions

T (K)	$CO_2 \rightleftharpoons$ $CO + \frac{1}{2}O_2$	$H_2O \rightleftharpoons$ $H_2 + \frac{1}{2}O_2$	$CO_2 + H_2 \rightleftharpoons$ $CO + H_2O$	$\frac{1}{2}N_2 + \frac{1}{2}O_2 \rightleftharpoons$ NO
298	-45.0898	-40.0693	-5.0205	-15.1738
300	-44.7597	-39.7853	-4.9744	-14.8204
400	-32.4297	-29.2264	-3.2033	-10.9547
500	-25.0265	-22.8848	-2.1417	-8.6274
600	-20.0876	-18.6363	-1.4513	-7.0789
700	-16.5612	-15.5844	-0.9768	-5.9677
800	-13.9155	-13.2880	-0.6275	-5.1339
900	-11.8577	-11.5019	-0.3558	-4.4941
1000	-10.2193	-10.0600	-0.1593	-3.9781
1100	-8.8757	-8.8828	0.0071	-3.5523
1200	-7.7589	-7.8951	0.1363	-3.2043
1300	-6.8149	-7.0576	0.2427	-2.9058
1400	-6.0079	-6.3430	0.3351	-2.6519
1500	-5.3102	-5.7176	0.4074	-2.4311
1600	-4.7023	-5.1799	0.4776	-2.2356
1700	-4.1623	-4.6954	0.5331	-2.0656
1800	-3.6886	-4.2644	0.5758	-1.9119
1900	-3.2590	-3.8861	0.6271	-1.7760
2000	-2.8814	-3.5435	0.6621	-1.6530
2100	-2.5307	-3.2287	0.6980	-1.5432
2200	-2.2197	-2.9398	0.7202	-1.4422
2300	-1.9351	-2.6789	0.7438	-1.3511
2400	-1.6734	-2.4424	0.7690	-1.2662
2500	-1.4338	-2.2253	0.7915	-1.1916
2600	-1.2145	-2.0179	0.8034	-1.1187
2700	-1.0094	-1.8351	0.8257	-1.0518
2800	-0.8258	-1.6590	0.8332	-0.9904
2900	-0.6511	-1.4959	0.8448	-0.9340
3000	-0.4838	-1.3445	0.8607	-0.8818
3100	-0.3281	-1.2510	0.9229	-0.8347
3200	-0.1856	-1.1660	0.9804	-0.7845
3300	-0.0477	-1.0948	1.0471	-0.7379
3400	0.0757	-1.0240	1.0997	-0.7004
3500	0.1960	-0.9560	1.1520	-0.6599
3600	0.3037	-0.8966	1.2003	-0.6247
3700	0.4132	-0.8440	1.2572	-0.5893
3800	0.5115	-0.7906	1.3021	-0.5607
3900	0.6051	-0.7430	1.3481	-0.5291
4000	0.6981	-0.6997	1.3978	-1.3756

Source: http://webbook.nist.gov

Table J: Isentropic table for $\gamma = 1.4$

M	$\frac{T_0}{T}$	$\frac{P_0}{P}$	$\frac{\rho_0}{\rho}$	$\frac{A}{A^*}$
0.00	1.00000E+00	1.00000E+00	1.00000E+00	∞
0.10	1.00200E+00	1.00702E+00	1.00501E+00	5.82183
0.20	1.00800E+00	1.02828E+00	1.02012E+00	2.96352
0.30	1.01800E+00	1.06443E+00	1.04561E+00	2.03507
0.40	1.03200E+00	1.11655E+00	1.08193E+00	1.59014
0.50	1.05000E+00	1.18621E+00	1.12973E+00	1.33984
0.60	1.07200E+00	1.27550E+00	1.18984E+00	1.18820
0.70	1.09800E+00	1.38710E+00	1.26330E+00	1.09437
0.80	1.12800E+00	1.52434E+00	1.35137E+00	1.03823
1.00	1.20000E+00	1.89293E+00	1.57744E+00	1.00000
1.10	1.24200E+00	2.13514E+00	1.71911E+00	1.00793
1.20	1.28800E+00	2.42497E+00	1.88274E+00	1.03044
1.30	1.33800E+00	2.77074E+00	2.07081E+00	1.06630
1.40	1.39200E+00	3.18227E+00	2.28612E+00	1.11493
1.50	1.45000E+00	3.67103E+00	2.53175E+00	1.17617
1.60	1.51200E+00	4.25041E+00	2.81112E+00	1.25024
1.70	1.57800E+00	4.93599E+00	3.12801E+00	1.33761
1.80	1.64800E+00	5.74580E+00	3.48653E+00	1.43898
1.90	1.72200E+00	6.70064E+00	3.89119E+00	1.55526
2.00	1.80000E+00	7.82445E+00	4.34692E+00	1.68750
2.10	1.88200E+00	9.14468E+00	4.85902E+00	1.83694
2.20	1.96800E+00	1.06927E+01	5.43329E+00	2.00497
2.30	2.05800E+00	1.25043E+01	6.07594E+00	2.19313
2.40	2.15200E+00	1.46200E+01	6.79369E+00	2.40310
2.50	2.25000E+00	1.70859E+01	7.59375E+00	2.63672
2.60	2.35200E+00	1.99540E+01	8.48386E+00	2.89598
2.70	2.45800E+00	2.32829E+01	9.47228E+00	3.18301
2.80	2.56800E+00	2.71383E+01	1.05679E+01	3.50012
2.90	2.68200E+00	3.15941E+01	1.17800E+01	3.84977
3.00	2.80000E+00	3.67327E+01	1.31188E+01	4.23457
3.10	2.92200E+00	4.26462E+01	1.45949E+01	4.65731
3.20	3.04800E+00	4.94370E+01	1.62195E+01	5.12096
3.30	3.17800E+00	5.72188E+01	1.80047E+01	5.62865
3.40	3.31200E+00	6.61175E+01	1.99630E+01	6.18370
3.50	3.45000E+00	7.62723E+01	2.21079E+01	6.78962

Table K: Normal shock properties for $\gamma = 1.4$

M_1	M_2	$\frac{P_2}{P_1}$	$\frac{T_2}{T_1}$	$\frac{\rho_2}{\rho_1}$	$\frac{P_{0,2}}{P_{0,1}}$
1.00	1.00000E+00	1.00000E+00	1.00000E+00	1.00000E+00	1.00000E+00
1.10	9.11770E-01	1.24500E+00	1.06494E+00	1.16908E+00	9.98928E-01
1.20	8.42170E-01	1.51333E+00	1.12799E+00	1.34161E+00	9.92798E-01
1.30	7.85957E-01	1.80500E+00	1.19087E+00	1.51570E+00	9.79374E-01
1.40	7.39709E-01	2.12000E+00	1.25469E+00	1.68966E+00	9.58194E-01
1.50	7.01089E-01	2.45833E+00	1.32022E+00	1.86207E+00	9.29787E-01
1.60	6.68437E-01	2.82000E+00	1.38797E+00	2.03175E+00	8.95200E-01
1.70	6.40544E-01	3.20500E+00	1.45833E+00	2.19772E+00	8.55721E-01
1.80	6.16501E-01	3.61333E+00	1.53158E+00	2.35922E+00	8.12684E-01
1.90	5.95616E-01	4.04500E+00	1.60792E+00	2.51568E+00	7.67357E-01
2.00	5.77350E-01	4.50000E+00	1.68750E+00	2.66667E+00	7.20874E-01
2.10	5.61277E-01	4.97833E+00	1.77045E+00	2.81190E+00	6.74203E-01
2.20	5.47056E-01	5.48000E+00	1.85686E+00	2.95122E+00	6.28136E-01
2.30	5.34411E-01	6.00500E+00	1.94680E+00	3.08455E+00	5.83295E-01
2.40	5.23118E-01	6.55333E+00	2.04033E+00	3.21190E+00	5.40144E-01
2.50	5.12989E-01	7.12500E+00	2.13750E+00	3.33333E+00	4.99015E-01
2.60	5.03871E-01	7.72000E+00	2.23834E+00	3.44898E+00	4.60123E-01
2.70	4.95634E-01	8.33833E+00	2.34289E+00	3.55899E+00	4.23590E-01
2.80	4.88167E-01	8.98000E+00	2.45117E+00	3.66355E+00	3.89464E-01
2.90	4.81380E-01	9.64500E+00	2.56321E+00	3.76286E+00	3.57733E-01
3.00	4.75191E-01	1.03333E+01	2.67901E+00	3.85714E+00	3.28344E-01
3.10	4.69534E-01	1.10450E+01	2.79860E+00	3.94661E+00	3.01211E-01
3.20	4.64349E-01	1.17800E+01	2.92199E+00	4.03150E+00	2.76229E-01
3.30	4.59586E-01	1.25383E+01	3.04919E+00	4.11202E+00	2.53276E-01
3.40	4.55200E-01	1.33200E+01	3.18021E+00	4.18841E+00	2.32226E-01
3.50	4.51154E-01	1.41250E+01	3.31505E+00	4.26087E+00	2.12948E-01

Table L: Ideal gas properties of air

T	h	u	s°	P_r	v_r
(K)	(kJ/kg)	(kJ/kg)	(kJ/kg.K)		
298	298.18	212.64	1.69528	1.3543	631.9
300	300.19	214.07	1.70203	1.386	621.2
310	310.24	221.25	1.73498	1.5546	572.3
320	320.29	228.42	1.7669	1.7375	528.6
340	340.42	242.82	1.8279	2.149	454.1
350	350.49	250.02	1.85708	2.379	422.2
360	360.58	257.24	1.88543	2.626	393.4
370	370.67	264.46	1.91313	2.892	367.2
380	380.77	271.69	1.94001	3.176	343.4
390	390.88	278.93	1.96633	3.481	321.5
400	400.98	286.16	1.99194	3.806	301.6
410	411.12	293.43	2.01699	4.153	283.3
420	421.26	300.69	2.04142	4.522	266.6
430	431.43	307.99	2.06533	4.915	251.1
440	441.61	315.3	2.0887	5.332	236.8
450	451.8	322.62	2.11161	5.775	223.6
460	462.02	329.97	2.13407	6.245	211.4
470	472.24	337.32	2.15604	6.742	200.1
480	482.49	344.7	2.1776	7.268	189.5
490	492.74	352.08	2.19876	7.824	179.7
500	503.02	359.49	2.21952	8.411	170.6
510	513.32	366.92	2.23993	9.031	162.1
520	523.63	374.36	2.25997	9.684	154.1
530	533.98	381.84	2.27967	10.37	146.7
540	544.35	389.34	2.29906	11.1	139.7
550	555.74	396.86	2.31809	11.86	133.1
560	565.17	404.42	2.33685	12.66	127
570	575.59	411.97	2.35531	13.5	121.2
580	586.04	419.55	2.37348	14.38	115.7
590	596.52	427.15	2.3914	15.31	110.6
600	607.02	434.78	2.40902	16.28	105.8
610	617.53	442.42	2.42644	17.3	101.2
620	628.07	450.09	2.44356	18.36	96.92
630	638.63	457.78	2.46048	19.84	92.84
640	649.22	465.5	2.47716	20.64	88.99
650	659.84	473.25	2.49364	21.86	85.34
660	670.47	481.01	2.50985	23.13	81.89
670	681.14	488.81	2.52589	24.46	78.61
680	691.82	496.62	2.54175	25.85	75.5
690	702.52	504.45	2.55731	27.29	72.56
700	713.27	512.33	2.57277	28.8	69.76

Table L: Ideal gas properties of air

T (K)	h (kJ/kg)	u (kJ/kg)	$s°$ (kJ/kg.K)	P_r	v_r
710	724.04	520.23	2.5881	30.38	67.07
720	734.82	528.14	2.60319	32.02	64.53
730	745.62	536.07	2.61803	33.72	62.13
740	756.44	544.02	2.6328	35.5	59.82
750	767.29	551.99	2.64737	37.35	57.63
760	778.18	560.01	2.66176	39.27	55.54
770	789.11	568.07	2.67595	41.31	53.39
780	800.03	576.12	2.69013	43.35	51.64
790	810.99	584.21	2.704	45.55	49.86
800	821.95	592.3	2.71787	47.75	48.08
820	843.98	608.59	2.74504	52.59	44.84
840	866.08	624.95	2.7717	57.6	41.85
860	888.27	641.4	2.79783	63.09	39.12
880	910.56	657.95	2.82344	68.98	36.61
900	932.93	674.58	2.84856	75.29	34.31
920	955.38	691.28	2.87324	82.05	32.18
940	977.92	708.08	2.89748	89.28	30.22
960	1000.55	725.02	2.92128	97	28.4
980	1023.25	741.98	2.94468	105.2	26.73
1000	1046.04	758.94	2.9677	114	25.17
1020	1068.89	776.1	2.99034	123.4	23.72
1040	1091.85	793.36	3.0126	133.3	23.29
1060	1114.86	810.62	3.03449	143.9	21.14
1080	1137.89	827.88	3.05608	155.2	19.98
1100	1161.07	845.33	3.07732	167.1	18.896
1120	1184.28	862.79	3.09825	179.7	17.886
1140	1207.57	880.35	3.11883	193.1	16.946
1160	1230.92	897.91	3.13916	207.2	16.064
1180	1254.34	915.57	3.15916	222.2	15.241
1200	1277.79	933.33	3.17888	238	14.47
1220	1301.31	951.09	3.19834	254.7	13.747
1240	1324.93	968.95	3.21751	272.3	13.069
1260	1348.55	986.9	3.23638	290.8	12.435
1280	1372.24	1004.76	3.2551	310.4	11.835
1300	1395.97	1022.82	3.27345	330.9	11.275
1320	1419.76	1040.88	3.2916	352.5	10.747
1340	1443.6	1058.94	3.30959	375.3	10.247
1360	1467.49	1077.1	3.32724	399.1	9.78
1380	1491.44	1095.26	3.34474	424.2	9.337
1400	1515.42	1113.52	3.362	450.5	8.919

Table L: Ideal gas properties of air

T	h	u	$s°$	P_r	v_r
(K)	(kJ/kg)	(kJ/kg)	(kJ/kg.K)		
1420	1539.44	1131.77	3.37901	478	8.526
1440	1563.51	1150.13	3.39586	506.9	8.153
1460	1587.63	1168.49	3.41247	537.1	7.801
1480	1611.79	1186.95	3.42892	568.8	7.468
1500	1635.97	1205.41	3.44516	601.9	7.152
1520	1660.23	1223.87	3.4612	636.5	6.854
1540	1684.51	1242.43	3.47712	672.8	6.569
1560	1708.82	1260.99	3.49276	710.5	6.301
1580	1733.17	1279.65	3.50829	750	6.046
1600	1757.57	1298.3	3.52364	791.2	5.804
1620	1782	1316.96	3.53879	834.1	5.574
1640	1806.46	1335.72	3.55381	878.9	5.355
1660	1830.96	1354.48	3.56867	925.6	5.147
1680	1855.5	1373.24	3.58335	974.2	4.949
1700	1880.1	1392.7	3.5979	1025	4.761
1750	1941.6	1439.8	3.6336	1161	4.328
1800	2003.3	1487.2	3.6684	1310	3.994
1850	2065.3	1534.9	3.7023	1475	3.601
1900	2127.4	1582.6	3.7354	1655	3.295
1950	2189.7	1630.6	3.7677	1852	3.022
2000	2252.1	1678.7	3.7994	2068	2.776
2050	2314.6	1726.8	3.8303	2303	2.555
2100	2377.7	1775.3	3.8605	2559	2.356
2150	2440.3	1823.8	3.8901	2837	2.175
2200	2503.2	1872.4	3.9191	3138	2.012
2250	2566.4	1921.3	3.9474	3464	1.864

Table L based on J. H. Keenan and J. Kaye,
Gas Tables, Wiley, New York, 1945.
Reproduced with permission.

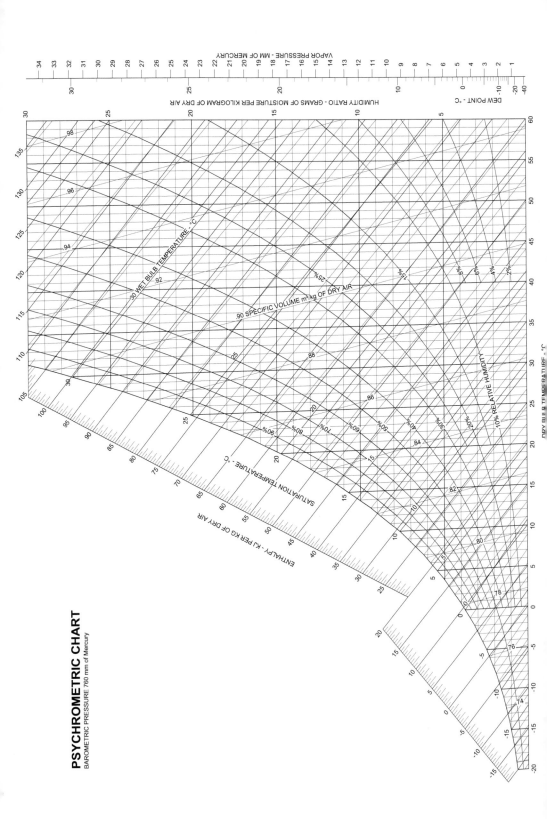

PSYCHROMETRIC CHART
BAROMETRIC PRESSURE 760 mm of Mercury

APPENDIX A

USE OF AIR TABLES

It may be recalled that a calorically perfect gas model stipulates that the ideal gas equation of state, $Pv = RT$ is obeyed and C_v is constant. When the range of temperature encountered in a particular application is large, then the latter assumption is unrealistic. In reality, C_v increases with temperature and consequently, addition of a given amount of heat will result in a lesser increase in temperature. A better model is one that still requires the gas to obey $Pv = RT$ and in addition, the gas to be thermally perfect i.e., u is a function of temperature alone.

Table L lists values for h and u for air at different temperatures. In addition, values for the specific entropy at the reference pressure (1 bar), denoted s°, are also provided. The value for specific entropy at any other pressure (and same temperature) may be evaluated using

$$s(T, P) = s^\circ(T) - R \ln \frac{P}{P_{ref}} \qquad (A.1)$$

This is is discussed in more detail in section 13.3.1. Note that $R = (h - u) \div T$, is a constant.

$$C_p \frac{dT}{T} - R \frac{dP}{P} = 0$$

Since C_p is a function of T, the above expression cannot be integrated to give the familiar $T^\gamma P^{1-\gamma} = $ constant.

Equation A.1 may be applied for any process between states 1 and 2 to give

$$s_2 - s_1 = s^\circ(T_2) - s^\circ(T_1) - R \ln \frac{P_2}{P_1}$$

For an isentropic process, we have

$$s^\circ(T_2) - s^\circ(T_1) - R \ln \frac{P_2}{P_1} = 0 \tag{A.2}$$

The following possibilities have to be considered now:

P_1, T_1 **and** P_2 **known :** In this case $s^\circ(T_2)$ can be evaluated from Eqn. A.2 and T_2 can then be retrieved from Table L.

P_1, T_1 **and** T_2 **known :** In this case, Eqn. A.2 may be written as

$$\frac{P_2}{P_1} = e^{\frac{s^\circ(T_2) - s^\circ(T_1)}{R}} = \frac{e^{s^\circ(T_2) \div R}}{e^{s^\circ(T_1) \div R}}$$

$$= \frac{P_{r,2}}{P_{r,1}} \tag{A.3}$$

Note that P_r is a function of temperature alone and is listed in Table L (after appropriate scaling). Values of P_r corresponding to T_1 and T_2 can be retrieved from the table and P_2 can then be evaluated.

P_1, T_1 **and** v_2 **known :** Since $Pv = RT$,

$$\frac{v_2}{v_1} = \frac{RT_2}{RT_1} \frac{P_1}{P_2}$$

$$= \frac{RT_2}{RT_1} \frac{P_{r,1}}{P_{r,2}} = \frac{\left(\frac{RT_2}{P_{r,2}}\right)}{\left(\frac{RT_1}{P_{r,1}}\right)}$$

$$= \frac{v_{r,2}}{v_{r,1}} \tag{A.4}$$

where we have used Eqn. A.3. As before with P_r, it may be seen that v_r is also a function of temperature alone. Values for v_r are also listed in Table L (again, after appropriate scaling). Since v_2, v_1 and $v_{r,1}$ are known, $v_{r,2}$ may be evaluated and the corresponding value for T_2 can then be retrieved from Table L.

temperature.

Solution : From Table L, we can retrieve $h_1 = 300.19$ kJ/kg and $P_{r,1} = 1.386$. Since $P_2 \div P_1 = 10$, we have, from Eqn. A.3, $P_{r,2s} = 10 \times P_{r,1} = 13.86$. From Table L, we can retrieve $h_{2s} = 579.865$ kJ/kg. From the definition of the isentropic efficiency,

$$\eta_{comp} = \frac{h_{2s} - h_1}{h_2 - h_1}$$

we can get $h_2 = 604.185$ kJ/kg. Therefore,

$$\frac{\dot{W}_{x,comp}}{\dot{m}} = h_2 - h_1 = 304 \, \text{kJ/kg}$$

From Table L, corresponding to $h = 604.185$ kJ/kg, we can retrieve $T_2 = 597$ K.

APPENDIX B

SOURCE FILES FOR SOLVING EXAM-
PLE 14.7 USING CANTERA

```
# -*- coding: utf-8 -*-
"""
Created on Sun Jun 23 12:56:09 2019
@author: aravi
"""
import cantera as ct
ct.suppress_thermo_warnings()
T_reactants = 298 # K
P_reaction = 1.013e5 # Pa
FuelAir = ct.Solution('HydCombustion.cti')
FuelAir.TPX=T_reactants, P_reaction,'H2:1, O2:0.75, N2:2.82'
FuelAir.equilibrate('HP')
EqbmFlameTemp = FuelAir.T }
```

File: HydCombustion.cti

```
units(length='cm', time='s', quantity='mol', act_energy='cal/mol')

ideal_gas(name='gas',
        elements="H C O N Ar",
        species="""h2                o2                n2
                   no                h2o
                   ar""",
        reactions='all',
        initial_state=state(temperature=300.0, pressure=OneAtm))

#------------------------------------------------------------------------------
# Species data
#------------------------------------------------------------------------------

species(name='no',
        atoms='O:1 N:1',
        thermo=(NASA([300.00, 1398.00],
                    [ 3.20971914E+00,  1.19199975E-03, -4.00785061E-07,
                      5.80069229E-11, -1.77636894E-15,  9.85406270E+03,
                      6.71393297E+00]),
                NASA([1398.00, 5000.00],
                    [ 3.30745011E+00,  1.02991918E-03, -3.14459911E-07,
                      4.42658687E-11, -2.38080924E-15,  9.81096664E+03,
                      6.16141074E+00])))

species(name='o2',
        atoms='O:2',
        thermo=(NASA([200.00, 1000.00],
                    [ 3.78245636E+00, -2.99673416E-03,  9.84730201E-06,
                     -9.68129509E-09,  3.24372837E-12, -1.06394356E+03,
                      3.65767573E+00]),
                NASA([1000.00, 6000.00],
                    [ 3.66096065E+00,  6.56365811E-04, -1.41149627E-07,
                      2.05797935E-11, -1.29913436E-15, -1.21597718E+03,
                      3.41536279E+00])),
        note='rus-89')

species(name='n2',
        atoms='N:2',
        thermo=(NASA([200.00, 1000.00],
                    [ 3.53100528E+00, -1.23660988E-04, -5.02999433E-07,
                      2.43530612E-09, -1.40881235E-12, -1.04697628E+03,
                      2.96747038E+00]),
```

```
            NASA([1000.00, 6000.00],
                  [ 2.95257637E+00,  1.39690040E-03, -4.92631603E-07,
                    7.86010195E-11, -4.60755204E-15, -9.23948688E+02,
                    5.87188762E+00])),
        note='g-8-02')

species(name='h2o',
        atoms='H:2 O:1',
        thermo=(NASA([200.00, 1000.00],
                  [ 4.19863520E+00, -2.03640170E-03,  6.52034160E-06,
                   -5.48792690E-09,  1.77196800E-12, -3.02937260E+04,
                   -8.49009010E-01]),
                NASA([1000.00, 6000.00],
                  [ 2.67703890E+00,  2.97318160E-03, -7.73768890E-07,
                    9.44335140E-11, -4.26899910E-15, -2.98858940E+04,
                    6.88255000E+00])),
        note='l-5-89')

species(name='h2',
        atoms='H:2',
        thermo=(NASA([200.00, 1000.00],
                  [ 2.34433112E+00,  7.98052075E-03, -1.94781510E-05,
                    2.01572094E-08, -7.37611761E-12, -9.17935173E+02,
                    6.83010238E-01]),
                NASA([1000.00, 6000.00],
                  [ 2.93286575E+00,  8.26608026E-04, -1.46402364E-07,
                    1.54100414E-11, -6.88804800E-16, -8.13065581E+02,
                   -1.02432865E+00])),
        note='tpis78')

species(name='ar',
        atoms='Ar:1',
        thermo=(NASA([300.00, 1000.00],
                  [ 2.50000000E+00,  0.00000000E+00,  0.00000000E+00,
                    0.00000000E+00,  0.00000000E+00, -7.45375000E+02,
                    4.36600100E+00]),
                NASA([1000.00, 5000.00],
                  [ 2.50000000E+00,  0.00000000E+00,  0.00000000E+00,
                    0.00000000E+00,  0.00000000E+00, -7.45375000E+02,
                    4.36600100E+00])),
        note='120186')
```

Index

Printed and bound by CPI Group (UK) Ltd, Croydon, CR0 4YY

22/10/2024

01777785-0001